T0350652

The Changing Frontier

**A National Bureau
of Economic Research
Conference Report**

The Changing Frontier
Rethinking Science and Innovation Policy

Edited by   **Adam B. Jaffe and Benjamin F. Jones**

The University of Chicago Press

Chicago and London

ADAM B. JAFFE is director and a senior fellow of Motu Economic and Public Policy Research, the Sir Douglas Myers Visiting Professor at Auckland University Business School, and a research associate of the National Bureau of Economic Research. BENJAMIN F. JONES is professor of management and strategy at the Northwestern University Kellogg School of Management and faculty affiliate at the Center for International Economics and Development and the Center for International and Comparative Studies at Northwestern University, where he also holds a courtesy appointment in the Department of Political Science. He is also a research associate of the National Bureau of Economic Research.

The University of Chicago Press, Chicago 60637
The University of Chicago Press, Ltd., London
© 2015 by the National Bureau of Economic Research
All rights reserved. Published 2015.
Printed in the United States of America

24 23 22 21 20 19 18 17 16 15   1 2 3 4 5

ISBN-13: 978-0-226-28672-3 (cloth)
ISBN-13: 978-0-226-28686-0 (e-book)
DOI: 10.7208/chicago/9780226286860.001.0001

Library of Congress Cataloging-in-Publication Data

The changing frontier : rethinking science and innovation policy /
   edited by Adam B. Jaffe and Benjamin F. Jones.
      pages cm — (National Bureau of Economic Research conference
  report)
   Includes bibliographical references and index.
    ISBN 978-0-226-28672-3 (cloth : alk. paper) — ISBN 978-0-226-
  28686-0 (e-book) 1. Science and state. 2. Technological innovations.
  I. Jaffe, Adam B., editor. II. Jones, Benjamin F., editor. III. Series:
  National Bureau of Economic Research conference report.
   Q125.C435 2015
   338.9'26—dc23
                                2014041834

♾ This paper meets the requirements of ANSI/NISO Z39.48-1992
(Permanence of Paper).

Relation of the Directors to the Work and Publications of the National Bureau of Economic Research

1. The object of the NBER is to ascertain and present to the economics profession, and to the public more generally, important economic facts and their interpretation in a scientific manner without policy recommendations. The Board of Directors is charged with the responsibility of ensuring that the work of the NBER is carried on in strict conformity with this object.

2. The President shall establish an internal review process to ensure that book manuscripts proposed for publication DO NOT contain policy recommendations. This shall apply both to the proceedings of conferences and to manuscripts by a single author or by one or more co-authors but shall not apply to authors of comments at NBER conferences who are not NBER affiliates.

3. No book manuscript reporting research shall be published by the NBER until the President has sent to each member of the Board a notice that a manuscript is recommended for publication and that in the President's opinion it is suitable for publication in accordance with the above principles of the NBER. Such notification will include a table of contents and an abstract or summary of the manuscript's content, a list of contributors if applicable, and a response form for use by Directors who desire a copy of the manuscript for review. Each manuscript shall contain a summary drawing attention to the nature and treatment of the problem studied and the main conclusions reached.

4. No volume shall be published until forty-five days have elapsed from the above notification of intention to publish it. During this period a copy shall be sent to any Director requesting it, and if any Director objects to publication on the grounds that the manuscript contains policy recommendations, the objection will be presented to the author(s) or editor(s). In case of dispute, all members of the Board shall be notified, and the President shall appoint an ad hoc committee of the Board to decide the matter; thirty days additional shall be granted for this purpose.

5. The President shall present annually to the Board a report describing the internal manuscript review process, any objections made by Directors before publication or by anyone after publication, any disputes about such matters, and how they were handled.

6. Publications of the NBER issued for informational purposes concerning the work of the Bureau, or issued to inform the public of the activities at the Bureau, including but not limited to the NBER Digest and Reporter, shall be consistent with the object stated in paragraph 1. They shall contain a specific disclaimer noting that they have not passed through the review procedures required in this resolution. The Executive Committee of the Board is charged with the review of all such publications from time to time.

7. NBER working papers and manuscripts distributed on the Bureau's web site are not deemed to be publications for the purpose of this resolution, but they shall be consistent with the object stated in paragraph 1. Working papers shall contain a specific disclaimer noting that they have not passed through the review procedures required in this resolution. The NBER's web site shall contain a similar disclaimer. The President shall establish an internal review process to ensure that the working papers and the web site do not contain policy recommendations, and shall report annually to the Board on this process and any concerns raised in connection with it.

8. Unless otherwise determined by the Board or exempted by the terms of paragraphs 6 and 7, a copy of this resolution shall be printed in each NBER publication as described in paragraph 2 above.

Contents

Preface   ix

Introduction   1
Adam B. Jaffe and Benjamin F. Jones

I. The Organization of Scientific Research

1. **Why and Wherefore of Increased Scientific
   Collaboration**   17
   Richard B. Freeman, Ina Ganguli, and Raviv
   Murciano-Goroff

2. **The (Changing) Knowledge Production Function:
   Evidence from the MIT Department of Biology
   for 1970–2000**   49
   Annamaria Conti and Christopher C. Liu

3. **Collaboration, Stars, and the Changing
   Organization of Science: Evidence from
   Evolutionary Biology**   75
   Ajay Agrawal, John McHale, and Alexander Oettl
   Comment: Julia Lane

4. **Credit History: The Changing Nature of
   Scientific Credit**   107
   Joshua S. Gans and Fiona Murray

II. The Geography of Innovation

5. The Rise of International Coinvention   135
   Lee Branstetter, Guangwei Li, and
   Francisco Veloso

6. Information Technology and the Distribution
   of Inventive Activity   169
   Chris Forman, Avi Goldfarb, and
   Shane Greenstein

III. Entrepreneurship and Market-Based Innovation

7. Innovation and Entrepreneurship in
   Renewable Energy   199
   Ramana Nanda, Ken Younge, and Lee Fleming

8. Economic Value Creation in
   Mobile Applications   233
   Timothy F. Bresnahan, Jason P. Davis, and
   Pai-Ling Yin

9. State Science Policy Experiments   287
   Maryann Feldman and Lauren Lanahan

IV. Historical Perspectives on Science Institutions and Paradigms

10. The Endless Frontier: Reaping What
    Bush Sowed?   321
    Paula Stephan
    Comment: Bruce A. Weinberg

11. Algorithms and the Changing Frontier   371
    Hezekiah Agwara, Philip Auerswald, and
    Brian Higginbotham
    Comment: Timothy Simcoe

    Contributors   415
    Author Index   419
    Subject Index   425

Preface

The idea for this volume was born in a conversation about science and innovation policy with Lawrence H. Summers, who wondered whether the classic, postwar perspective laid out in Vannevar Bush's classic, *Science: The Endless Frontier*, needed any substantial updates for the twenty-first century. To help answer this question we issued a call for papers and held two conferences, resulting in the eleven chapters and associated comments collected in this volume. While the subject of how science is changing is a vast one—perhaps endless itself—the chapters in this volume demonstrate numerous, essential changes in the scientific enterprise with potentially substantial policy implications. We hope that the perspective of the "changing frontier" will continue to spark new research on shifts in the systems of scientific and technological progress and the effectiveness of their support.

Funds for this project were provided by the National Bureau of Economic Research and the Erwin Marion Kauffman Foundation. We are indebted to these organizations for their support. We thank all of the authors and discussants for their contributions to this work, and thank Josh Lerner and Scott Stern for suggesting that we undertake this project. Special thanks are due to Rob Shannon at the NBER for expertly managing the conferences in Cambridge and Chicago and to the 1871 entrepreneurship incubator in Chicago for letting us use their space. We also thank our editor, Joe Jackson, at the University of Chicago Press, and Helena Fitz-Patrick at the NBER for her great help in managing the publication process.

Introduction

Adam B. Jaffe and Benjamin F. Jones

With the 1945 publication of *Science: The Endless Frontier*, Vannevar Bush established an intellectual architecture that helped define a set of public science institutions that were dramatically different from those that came before. Yet what was radical in 1945 remains largely in place today. At the start of the twenty-first century, many aspects of the science and innovation system—from its organization and scale to the role of geography and the nature of entrepreneurship—have witnessed important changes, with potentially substantial implications for the design of science policy and institutions both today and in the decades ahead.

This volume explores two overarching questions: What are critical dimensions of change in science and innovation systems? and What are the implications of these changes for policies and institutions in the twenty-first century? In this introduction, we present an overview of eleven new contributions that explore important dimensions of these questions.

Part I of the volume investigates the organization of scientific research, especially new norms around collaboration, which appears to be a central force reshaping the production of knowledge. These studies also lay some important foundations for part II, which considers shifts in the geography of scientific research and connects to a broader literature suggesting that geographic agglomeration remains an enduring and, in some ways, strength-

Adam Jaffe is director and senior fellow of Motu Economic and Public Policy Research, the Sir Douglas Myers Visiting Professor at Auckland University Business School, and a research associate of the National Bureau of Economic Research. Benjamin Jones is associate professor of management and strategy at the Kellogg School of Management, Northwestern University, and a research associate of the National Bureau of Economic Research.

For acknowledgments, sources of research support, and disclosure of the authors' material financial relationships, if any, please see http://www.nber.org/chapters/c13027.ack.

ening feature of innovative activity. Part III considers modern modes of entrepreneurship and market-based innovation, with chapters studying mobile applications, clean energy, and state-level entrepreneurship policies. Finally, in part IV, our contributors investigate changes in science institutions and science-innovation linkages within broader historical visions, including from the perspective of *Science: The Endless Frontier* itself.

The following sections discuss each of the volume's chapters, with the purpose of presenting key findings while drawing out common themes and potential policy implications. In a concluding section, we summarize the broad, fundamental changes these contributions inform and point to additional aspects of the science and innovation system that may be undergoing substantial shifts but remain for future study.

The Organization of Scientific Research

A primary theme, featured in four different contributions to this volume, considers the evolving role of collaboration in science—within institutions, across institutions, and through the scientific community as a whole. These contributions build primarily on two theories for increased collaboration in the sciences, both of which increase the return to collaboration. One theory emphasizes the benefits of increased collaboration as individual researchers become increasingly specialized. This tendency can be seen as a necessary response to the rising "burden of knowledge" as the stock of knowledge accumulates and the individual knows an increasingly narrow fraction of it (Jones 2009). The second theory emphasizes the declining costs of collaboration through the advance of information and communications technologies (Agrawal and Goldfarb 2008). An observation that persists across the contributions of this volume and elsewhere (Kim, Morse, and Zingales 2009; Agrawal, Goldfarb, and Teodoridis 2013) is that both forces appear to be operating. The following contributions add substantial and novel evidence to these dimensions, while also extending conceptions of collaboration in the organization of scientific research.

In "Why and Wherefore of Increased Scientific Collaboration," Richard B. Freeman, Ina Ganguli, and Raviv Murciano-Goroff establish several new facts about scientific collaborations, comparing colocated coauthors, geographically distant coauthors within the United States, and coauthors across countries. Freeman et al. study nanotechnology, subfields of biomedicine, and subfields of physics. An important innovation of this chapter is to conduct in-depth surveys of the authors, rather than relying purely on bibliometric databases; the surveys produce first-order, novel insights about the various collaborations.

One striking finding is that nearly all geographically distant coauthors were once colocated. Typically these distant coauthors were previously colocated either as colleagues, as visitors, or in an advisor-student relationship. These findings extend a body of work establishing that face-to-face interaction appears valuable even as communication technologies advance (e.g., Olson and Olson 2003; Olson, Zimmerman, and Bos 2008). A second finding is that the most common reason for collaboration, whether domestic or international, is access to specialized human capital, which is consistent with the burden of knowledge view of the demand for collaboration. Collaborations motivated by access to physical equipment or grant funding are, by comparison, less common.

In "The (Changing) Knowledge Production Function: Evidence from the MIT Department of Biology for 1970–2000," Annamaria Conti and Christopher C. Liu provide a rich and textured analysis of changes in scientific production by focusing on a leading biology department. The authors establish that later cohorts of students experience longer training periods, longer periods until the publication of their first paper, fewer first-author publications, and, consistent with much other literature, more coauthors per paper. The life cycle effects are consistent with the extended training phases associated with a rising burden of knowledge (Jones 2010), while the extended training period is also consistent with a declining number of future positions per student in biomedicine (Stephan 2012). Regardless, as the authors discuss, the incentive for entering biomedical careers may be decreasing; a striking fact in their data is that the length of training, including graduate and postdoctoral work, now exceeds ten years—a long road that may dissuade entry into these scientific careers.

Ajay Agrawal, John McHale, and Alexander Oettl, in "Collaboration, Stars, and the Changing Organization of Science: Evidence from Evolutionary Biology," examine how the locus of top research in evolutionary biology has changed with time. The chapter presents two intriguing and seemingly contradictory facts: the concentration of quality-weighted research produced by the top 20 percent of university departments is decreasing with time, yet the concentration of quality-weighted research produced by the top 20 percent of individual scientists is increasing with time. To reconcile these contrasting trends, the authors suggest that rising collaboration is a natural mechanism. In particular, the decline in the costs of distant collaboration, via advances in information and computing technology, may better connect lower-tier research departments to top researchers. A more specific mechanism may be the increasing capacity of star researchers to maintain collaborative relationships with their students once their students move away. More generally, the theme where information technology can link geographically distant players to centers of research excellence (here, stars) is repeated in various forms below—see the contributions of Branstetter, Li, and Veloso (chapter 5) and Forman, Goldfarb, and Greenstein (chapter 6).

In "Credit History: The Changing Nature of Scientific Credit," Joshua S. Gans and Fiona Murray explore collaboration in a broader frame, emphasizing that collaborations also occur across papers in the community of scholars pushing forward a scientific field. This notion, which is strongly grounded in the cumulative nature of innovation, emphasizes that scientific collaboration often proceeds through mechanisms other than the coauthor-based organizational form of a single paper. Taking classic Mertonian conceptions of scientific norms, this chapter then argues that the organizational form of collaboration that scientists take naturally hinges on credit considerations. On one dimension, credit considerations may influence coauthorship choices—both whether and with whom to coauthor. Moreover, the decision of when to call research "complete" and publish it, rather than continuing on one's own in private, may also naturally hinge on how credit is given when others build on the initial work. Thus both the unit of common analysis—the paper itself—and its coauthorship arrangement may be endogenous to credit considerations, and in important ways.

This chapter reviews collaborative choices under this broader frame, animates these choices with compelling examples that illuminate the diversity of organizational forms and concerns over credit, and provides a formal model to synthesize the analysis. The model develops conditions under which an author may "integrate" (keep their initial research results private in pursuit of gaining credit for a larger cumulative contribution), "collaborate" (draw in coauthors to improve the research potential), or "publish" (disclose the early results and gain credit as others build on the findings). The model thus links knowledge accumulation, collaboration, and credit sharing to inform many credit-related issues. Applications include the "salami slicing" of results into small, publishable pieces and the potential divergence between equilibrium organizational forms and the social optimum; for example, if peers assign excessive joint credit to coauthored research, then credit considerations will lead scientists to coauthor too much. More generally, this chapter nicely integrates credit considerations into research on collaboration and outlines a compelling and rich agenda for further work.

The Geography of Innovation

The geography of innovation has also undergone substantial changes. Three large forces appear to be at work. First, economic development has led many countries to catch up to the world technology frontier, introducing new players onto the global science and innovation landscape. Second, the advance of information and communication technologies has allowed people at geographically distant points to interact more easily in the production and consumption of new ideas. This force has led some observers to declare a "death of distance" (Cairncross 1997) or that the "world is flat" (Friedman 2005), with possible fundamental implications for economic geography. Third, and in contrast to the last forces, increased specialization of human capital or other inputs may encourage further geographic agglomeration. This force, which can link burden of knowledge reasoning (driving increasing specialization) with a classic Marshallian analysis of geographic agglomeration, suggests that the primacy of place (e.g., in Silicon Valley or other clusters) may increase with time rather than dissolve.

The policy implications of these forces are substantial. Should regions increasingly pluck the fruit of research insights produced elsewhere, local taxation to support such public goods may be more difficult to sustain politically. Meanwhile, local investments to promote innovative clusters, often attempted by polities seeking to replicate other region's successes, may be either more or less well motivated or sustainable depending on the balance of the above forces.

This section considers two valuable contributions that speak to these issues. In "The Rise of International Coinvention," Lee Branstetter, Guangwei Li, and Francisco Veloso examine the explosion of patenting from inventors in China and India. They start by noting a puzzle: both countries appear to have remarkably high patenting rates despite low per-capita income, which appears to contradict a basic idea of economic development where developing countries grow primarily through capital accumulation and the adoption of existing technologies, rather than through the innovation of new technologies. In Branstetter and colleague's contribution, the puzzle is resolved through two kinds of empirical analysis. First, studying patents issued in the United States by Chinese and Indian inventors, they find the vast majority of patents coming from these developing countries occur through multinational corporations. Moreover, these patents typically involved collaborations between inventors located in China or India and inventors located in advanced economies. One implication is that the rise in patenting by China and India may not be undermining the technological leadership of advanced economies and their multinational corporations, but rather assisting it.

While these results are based on patents issued in the United States (which are presumably the inventions with more substantial global value), this chapter also provides a detailed assessment of patents issued domestically in China by the State Intellectual Property Office (SIPO). China's domestic patent rates have recently soared, which has suggested to some observers that domestic Chinese firms have become highly innovative. However, Branstetter, Li, and Veloso find that only 20 percent of these SIPO patents qualify as patents in the usual sense (being new and useful, and evaluated as such). Moreover, half of these patents are already filed in foreign jurisdictions and are simply seeking protection in China. Among the remainder, many come from multinational subsidiaries.

This chapter thus takes an especially deep look at the first force for geographic change noted above by studying the entry of newly developing countries onto the innovation landscape. The chapter finds that China and India neither overturn conventional wisdom about the development process nor suggest much innovation independent from multinational enterprises. At the same time, these countries are increasingly connected through collaboration into multinational research and development (R&D) efforts, suggesting a dimension on which the world has become flatter, but in dependence with global collaboration.

In "Information Technology and the Distribution of Inventive Activity," Chris Forman, Avi Goldfarb, and Shane Greenstein turn the geographic lens to the concentration of patenting within the United States and explore linkages between geography and information technology. Studying patenting at the county level, they find that counties saw larger patenting growth rates when they were both patenting laggards in 1990 but Internet adoption leaders in 2000. Echoing the prior study, the authors also find some evidence that it is distant collaboration in the context of multiestablishment firms, rather than purely local innovation, that information technologies appear to assist.

Nonetheless, despite this evidence, a primary finding of Forman and colleagues is that the overall geographic concentration of patenting activity has substantially increased with time. While the rate of patenting increased 27 percent over their study period (1990–2005), it increased 50 percent among the initial top quartile of patenting counties. In initially below-median counties, patent rates did not grow.

This chapter comes close to an explicit analysis of the contest between second and third forces noted above, with emphasis on measuring overall concentration trends while explicitly accounting for variation in access to information and communication technologies. Increasing concentration appears to win out, suggesting the dominance of some version of the third force, while information technologies somewhat soften the concentration tendency. Overall, these chapters paint a picture where concentration appears to be increasing, and any tendency for a death of distance occurs primarily through collaboration with the agglomerated regions. From a policy perspective, these findings suggest that the presumption of substantially "local" gains may be a surprisingly durable basis for public R&D support, both in the robustness of clusters and the dominance of advanced economies, or at least their multinationals, in the invention process.

Entrepreneurship and Market-Based Innovation

The words "entrepreneur" and "entrepreneurship" do not appear in *The Endless Frontier*. Today, many analysts of the science/innovation system see them as crucial to reaping the potential social and economic rewards of the public investment in science. While other National Bureau of Economic Research (NBER) volumes have been devoted to the role of entrepreneurship in this system, in this volume we have two chapters that focus on entrepreneurship in specific emerging sectors (renewable energy and mobile applications software), and one that looks at the history of the "policy innovation" of state-level programs designed to foster local/regional innovation and entrepreneurship.

Ramana Nanda, Ken Younge, and Lee Fleming explore the nature of the patents of venture capital-backed firms in the renewable energy sector in "Innovation and Entrepreneurship in Renewable Energy." Given climate change challenges and the role that venture capital-backed firms have played in biotechnology and information technology, this chapter examines VC's role in the renewable energy sector. Using a new data set of the renewable energy patents of both VC-backed and incumbent firms, the authors find that most such patents still come from incumbent firms. However, patents from VC-backed firms are more novel (defined by a measure derived from textual analysis of patent claims) and have greater technological impact (based on the number of later citations to the patent from subsequent patents) than those of incumbent firms. The authors also show that a surge of VC funding in this sector early in the first decade of the twenty-first century was associated with an increase in patenting by start-ups. Finally, the chapter discusses structural aspects of this sector that may limit the ability of venture capital to provide the support needed if rapid technological improvement is a policy goal.

In "Economic Value Creation in Mobile Applications," Timothy F. Bresnahan, Jason P. Davis, and Pai-Ling Yin characterize the state of innovation and entrepreneurship in a new sector: mobile software applications. The authors note that in just a few years the installed base of mobile devices already vastly exceeds that of any other programmable device; this large base combined with the ease of entry into the two mobile programming platforms (iOS and Android) has allowed three-quarters of a million programming innovations (apps) to be created. The chapter proceeds to analyze the ways in which this innovation wave resembles and differs from previous waves. The authors note the tremendous importance of the last step in the chain from technical discovery to creation of economic value, whereby creating new markets may itself require innovations that are distinct from the technological ones. The scale of the mobile sector is qualitatively greater than we have seen before, with market-dominant personal computer (PC) applications such as the spreadsheet having emerged when the quantity of software created for that platform numbered in the hundreds rather than the hundreds of thousands already in existence for mobile. The authors argue that this vastly greater scale creates a bottleneck whereby a new app and the subset of potential customers who might use it have trouble finding each other. Currently, existing firms (e.g., Starbucks or airline companies) have been most successful at solving this problem in mobile apps because they start with an existing customer base, but it remains to be seen what market mechanisms will evolve in the future and what firms will be most successful with those mechanisms.

If *The Endless Frontier* launched science and innovation as a central concern of the federal government, it was several decades before states began to consider their own policy choices. Maryann Feldman and Lauren Lanahan describe the emergence and evolution of state-level interventions in "State Science Policy Experiments." On one level, states invest in science and innovation for the same reason as the federal government, to create public goods and derive the spillover benefits therefrom. But this raises the obvious question of why states would not just leave this to the federal government and enjoy the benefits within their borders without having to invest their own resources. The answer, of course, is that the spillovers may be partially localized, so that states invest to increase local innovation and local economic growth. This chapter looks at the factors affecting states' adoption of the three main categories of state programs: "eminent scholars," designed to attract scientific talent to the state; "centers of excellence," designed to build research expertise that involves industry; and "university research grants," which provide funding for specific research projects. The results indicate that eminent scholar and university research grant programs seem to build on existing strengths in research, while the centers of excellence seem motivated by more generic economic growth concerns.

Given the apparent durability of geographic agglomeration in anchoring innovation (see the above section on geography and innovation), state-level policies may arguably be quite fruitful in bringing local benefits if these policies are well designed. The Feldman and Lanahan analysis thus appears to push forward an important research agenda. State policy to encourage innovation is widespread and expanding, thus calling for a detailed assessment of its effectiveness, especially given the variety of policy approaches states can undertake.

These chapters speak to entrepreneurship but more generally speak to market-based innovation and its potential interfaces with policy. If Bush's vision in *The Endless Frontier* centered on a robust public commitment to R&D, and the postwar period initially saw enormous growth in public R&D expenditure, the story since the early 1960s has been quite different, where private sector R&D funding has grown much faster than public funding.[1] The above chapters suggest specific mechanisms—including the roles of venture capital and platform formation—that go beyond the vision of Bush and appear to be central features of the modern innovation system. The role of standard setting, which can be assisted by public institutions, and market-based innovation policies such as the R&E tax credit and the tax treatment of early stage finance may then be increasingly important policy levers to encourage innovation, suggesting a broader and retuned vision from the emphasis on basic science that Bush articulated. We further take up these themes below, where the next two contributions consider the reali-

---

1. For example, in 1960 US federal government R&D funding and US private sector R&D funding were nearly 2 percent and 1 percent of GDP, respectively. By 2000, these shares had reversed. (See the National Science Foundation Science and Engineering Indicators 2012 at http://www.nsf.gov/statistics/seind12/).

zation and limits to Bush's vision and the shifting technology paradigms that may help define where innovation occurs.

Historical Perspectives on Science Institutions and Paradigms

The changes in science over the last half century encompass institutional evolutions as Bush's vision came to be implemented and also evolutions in the science-innovation paradigm itself as the types of technologies driving economic progress have evolved. Two chapters in this volume confront these central historical developments in the science and innovation system and offer rich, novel, and intriguing assessments of such changes. This volume closes with these broader historical analyses.

In "The Endless Frontier: Reaping What Bush Sowed?" Paula Stephan compares the current state of the basic research system with the vision that Bush originally articulated. On the surface we got what Bush wanted: a large basic research enterprise, centered in the university system, and funded by the federal government. But the system differs in some important ways from that envisioned by Bush, and Stephan argues that these differences are connected to a number of problems or issues in the existing system. First, the dependence on federal research funds for academic year salaries and the investments in buildings and equipment universities have made in order to compete for federal research funds have made universities dependent on a perpetually growing funding pie that no longer seems likely to grow at the same rate. Second, Bush envisioned research funded by research grants, while the building of human capital would be funded by fellowship programs. But today the salaries of PhD students and postdoctoral scholars are paid largely out of research grants. The result is that the size of education and training programs is determined not by the number of positions available for graduates, but by the needs of existing research labs for research staff. Such a system can operate in balance if the total research funding grows continuously, but creates another source of system instability as research funding remains flat. Third, perhaps as a result of the funding pressure created when a system built for growth confronts static funding levels, the need for public funding to facilitate high-risk breakthrough research seems to be giving way to a demand for incremental projects with a higher likelihood of success. Finally, while Bush envisioned a public investment in research that would be something like one-third medical and biosciences and two-thirds physical sciences, the political process that determines funding allocation has instead consistently devoted more than half of the federal research resources to biomedical sciences.

A second chapter providing a broad historical analysis argues that the scientific frontier discussed by Bush was not, in fact, endless, but was rather one in a succession of frontiers that sometime around the millennium was replaced by the "algorithmic frontier." "Algorithms and the Changing Frontier," by Hezekiah Agwara, Philip Auerswald, and Brian Higginbotham, argues that while the defining attribute of the world technological frontier in the mid-twentieth century was the application of science to product and process innovation, the current defining feature of the technological frontier is the ever-improving connections and interoperability among firm-level production algorithms, which are in turn made possible by the adoption of standards. Just as the transition from the industrial frontier of the nineteenth century to the scientific frontier of the late twentieth century meant that economists needed new analytical tools such as the knowledge production function and endogenous growth models, economists are now embarking on the development of new tools to understand algorithm-based innovation and growth.

An implication of this chapter is to emphasize that standard-setting institutions, in addition to basic science institutions, may be crucial for encouraging technological progress both today and in the decades ahead. Standard setting, like research and development, happens through both private sector and public sector mechanisms. To the extent that Agwara, Auerswald, and Higginbotham's analysis is accurate, research to improve standard-setting mechanisms becomes an increasingly impactful area of study. One may look no further than the recent development of mobile operating standards like iOS and Android to see an example of standards knitting together downstream demand and encouraging massive innovation and entrepreneurship in software applications—as detailed in Bresnahan, Davis, and Yin (chapter 8).

Concluding Comments

In July 1945, when Vannevar Bush wrote *Science: The Endless Frontier*, the world's scientific enterprise was a tiny fraction of its current scale. By articulating a compelling case for the impact of science and the need of public support (the first two sections of his introduction are entitled "Scientific Progress is Essential" and "Science is a Proper Concern of Government"), he helped set the United States, and ultimately many other countries, on a path toward strong and well-funded institutions of science, centered on universities and government labs, which can provide basic research insights and/or develop scientific human capital. Both of these outputs—ideas and people—Bush saw as the primary and essential way in which government can support industrial R&D.

Now we approach the seventieth anniversary of his seminal work and Bush's ensuing efforts within the government to create the modern science architecture. It is clear, based on the analysis in this volume, that major changes in the nature of science and innovation have occurred. One fundamental shift has occurred in the organization of scientific research. At a microlevel the shift toward collaboration, and the increasingly long period of PhD and postdoctoral study before researchers establish their own labs, impacts the scientific workforce considerations that center in the Bush vision. As articulated by both Paula Stephan and Conti and Liu, the system of human capital formation appears increasingly arduous, with a funding system that may redirect students from efficient skill building to faculty research needs. If the burden of knowledge is raising human capital demands on scientists, efficient training may be increasingly important; yet, as Stephan argues, our training systems may be pushing the other way. The shift toward collaboration also suggests a shift in the character of training, where learning collaborative and management skills may become an increasingly high-return investment, ultimately in furtherance of the individual's career and the overall science enterprise.

The shifts in organization, especially in collaboration, also link to shifts in the geography of innovation. Vannevar Bush wrote at a time when the United States sat uniquely as the only advanced economy left largely undamaged by war. It is not surprising that issues of the geography of innovation did not feature in *Endless Frontier*, while it is also not surprising that in today's globalized economy they are central to science and innovation policy debates. As discussed above, the chapters in this volume add to other recent empirical evidence (e.g., Glaeser and Kerr 2009; Puga 2010; Glaeser 2010) that suggests agglomeration economies remain a profoundly important aspect of innovation geography. The world may be getting flatter with respect to tasks that depend on codified knowledge and that can therefore be made routine, but fundamentally creative processes such as innovation appear to remain dependent on complex interactions among people that are facilitated by geographic concentration. While important aspects of geography—where distant researchers are increasingly connected, especially those who were once colocated—flattens the world in some respects, it appears that agglomerative tendencies continue to be strong, suggesting that local spillovers may remain a potentially credible basis for motivating a polity to bear costs in pursuit of science and innovation's public goods.

It seems plausible to imagine that a major force compelling Bush's vision of the long-run benefits of public science was the contributions that technologies such as radar, aircraft, and the atomic bomb had made to the war effort. These are examples of science harnessed for social goals essentially outside of the market system. But today our innovation goals—even those greatly enmeshed in public policy such as environment and health—are typically met by bringing products and processes to the marketplace successfully. Moreover, the private sector is the increasingly dominant source of R&D funding in the United States. This means that issues of market behavior and institutions, such as entrepreneurship and standard setting, play a significant role in the success of the overall system in delivering ultimate social and economic benefits from scientific research. From a policy perspective, these issues raise many possibilities for market failure. The chapters in this volume on innovation and entrepreneurship in clean energy and mobile applications, and on state science/innovation programs, illuminate important aspects of these issues.

Other issues, not studied here, suggest further substantial changes in the science and innovation system. The university-market interface has evolved, especially with the Bayh-Dole Act, the rise of technology transfer offices, and the interest of nonprofit research institutions in both creating and tapping royalty streams. Intellectual property regimes including patenting, copyright, and even noncompete agreements, have experienced changes in their strength, scale, and strategic use through evolutions of law, court interpretation, and with the rise of new types of codified knowledge, like software and gene sequences, that challenge standing intellectual property systems. Constraints imposed on the basis of social ethics, too, have evolved, with more oversight and restrictions upon human experimentation, especially through institutional review boards, even as ever-expanding consumer data resources are unleashing new innovative opportunities in the private sector, often at the expense of consumer privacy. These subjects and others are also worthy of substantial consideration in any holistic assessment of the "changing frontier." What is clear is that science and innovation landscape has undergone profound transformations since Vannevar Bush shaped the US science institutions based on the landscape he observed.

References

Agrawal, Ajay, and Avi Goldfarb. 2008. "Restructuring Research: Communication Costs and the Democratization of University Innovation." *American Economic Review* 98 (4): 1578–90.

Agrawal, Ajay, Avi Goldfarb, and Florenta Teodoridis. 2013. "Does Knowledge Accumulation Increase the Returns to Collaboration? Evidence from the Collapse of the Soviet Union." NBER Working Paper no. 19694, Cambridge, MA.

Cairncross, Frances. 1997. *The Death of Distance*. Cambridge, MA: Harvard University Press.

Friedman, Thomas L. 2005. *The World is Flat: A Brief History of the Twenty-First Century*. New York: Farrar, Straus, and Giroux.

Glaeser, Edward L. 2010. *Triumph of the City: How Our Greatest Invention Makes Us Richer, Smarter, Healthier, and Happier*. New York: Penguin Press.

Glaeser, Edward L., and William R. Kerr. 2009. "Local Industrial Conditions and Entrepreneurship: How Much of the Spatial Distribution Can We Explain?" *Journal of Economics & Management Strategy* 18 (3): 623–63.

Jones, Benjamin F. 2009. "The Burden of Knowledge and the 'Death of the Renaissance Man': Is Innovation Getting Harder?" *Review of Economic Studies* 76 (1): 283–317.

———. 2010. "Age and Great Invention." *Review of Economics and Statistics* 92 (1): 1–14.

Kim, E. Han, Adair Morse, and Luigi Zingales. 2009. "Are Elite Universities Losing Their Competitive Edge?" *Journal of Financial Economics* 93 (3): 353–81.

Olson, Gary, and Judith Olson. 2003. "Mitigating the Effect of Distance on Collaborative Work." *Economics of Innovation and New Technology* 12 (1): 27–42.

Olson, Gary, Ann Zimmerman, and Nathan Bos, eds. 2008. *Science Collaboration on the Internet*. Cambridge, MA: MIT Press.

Puga, Diego. 2010. "The Magnitude and Causes of Agglomeration Economies." *Journal of Regional Science* 50 (1): 203–19.

Stephan, Paula. 2012. *How Economics Shapes Science*. Cambridge, MA: Harvard University Press.

I

The Organization of Scientific Research

1

Why and Wherefore of Increased Scientific Collaboration

Richard B. Freeman, Ina Ganguli, and
Raviv Murciano-Goroff

Scientists increasingly collaborate on research with other scientists, producing an upward trend in the numbers of authors on a paper (Jones, Wuchty, and Uzzi 2008; Wuchty, Jones, and Uzzi 2007; Adams et al. 2005). Papers with larger numbers of authors garner more citations and are more likely to be published in journals with high impact factors than papers with fewer authors (Lawani 1986; Katz and Hicks 1997; deB. Beaver 2004; Wuchty, Jones, and Uzzi 2007; Freeman and Huang 2014), which seems to justify increased collaborations in terms of scientific productivity. The trend in coauthorship extends across country lines, with a larger proportion of papers coauthored by scientists from different countries (National Science Board 2012; Adams 2013). In the United States and other advanced economies, the proportion of papers with international coauthors increased from the 1990s through the first decade of the twenty-first century, while the proportion of papers with domestic coauthors stabilized. In emerging economies, where collaboration has not yet reached the proportions in the United States and

Richard B. Freeman holds the Herbert Ascherman chair in economics at Harvard University and is a research associate of the National Bureau of Economic Research. Ina Ganguli is assistant professor at the University of Massachusetts Amherst, a research affiliate at the Stockholm Institute of Transition Economics (SITE) at the Stockholm School of Economics, and a postdoctoral affiliate at the Center for International Development, Kennedy School of Government, Harvard University. Raviv Murciano-Goroff is a PhD candidate in economics at Stanford University.

We appreciate assistance with the survey from John Trumpbour and input from Jennifer Amadeo-Holl, Paula Stephan, and Andrew Wang. We received helpful comments from Adam Jaffe, Ben Jones, Manuel Trajtenberg, and participants at the NBER "The Changing Frontier: Rethinking Science and Innovation Policy" conferences. This research was supported in part by the National Science Foundation's National Nanotechnology Initiative, award 0531146. For acknowledgments, sources of research support, and disclosure of the authors' material financial relationships, if any, please see http://www.nber.org/chapters/c13040.ack.

other advanced countries, the share of papers with domestic collaborations and the share with international collaborations have both increased.

The spread of scientific workers and research and development activity around the world (Freeman 2010) has facilitated the increase in international collaborations. The growing number of science and engineering PhDs in developing countries, some of whom are international students and postdocs returning to their country of origins (Scellato, Franzoni, and Stephan 2012) has expanded the supply of potential collaborators outside the North American and Western European research centers. A rising trend in government and industry research and development (R&D) spending in developing countries and grant policies by the European Union and other countries favor international cooperation. At the same time, the lower cost of travel and communication has reduced the cost of collaborating with persons across geographic locales (Agrawal and Goldfarb 2008; Catalini, Fons-Rosen, and Gaulé 2014). The increased presence of China in scientific research, exemplified by China's move from a modest producer of scientific papers to number two in scientific publications after the United States, has been associated with huge increases in collaborations between Chinese scientists and those in other countries.[1]

Finally, the location of scientific equipment and materials, such as the European Organization for Nuclear Research (CERN)'s Large Hadron Collider, huge telescopes in particular areas, or geological or climatological data available only in special localities, have also increased collaborations. The United States was not a prime funder for CERN, but Americans are the largest group of scientists and engineers working at CERN. China eschewed joining the CERN initiative as an associate member state, but many China-born scientists and engineers work at CERN as members of research teams from other countries.

How successful are collaborations across country lines and across locations in the same country? How do collaborators meet and develop successful research projects? What are the main advantages and challenges in collaborative research?

To answer these questions, we combine data from a 2012 survey that we conducted of corresponding authors on collaborations *with at least one US coauthor* with bibliometric data from Web of Science (WoS) (Thomson Reuters 2012) in three growing fields—particle and field physics, nanoscience and nanotechnology, and biotechnology and applied microbiology. The survey data allow us to investigate the connections among coauthors in collaborations and the views of corresponding authors about collaborations. The WoS data allows us to examine patterns of collaborations over

1. Science and Engineering Indicators 2013, appendix table 5–27, gives scientific papers for the top five countries in 2009: United States, 208,601; China, 74,019; Japan, 49,627; United Kingdom, 45,649; and Germany, 45,002.

time and to compare patterns found in our fields to those found in scientific publications broadly. To determine whether borders or space are the primary factors that affects the nature and impact of collaborations, we contrast collaborations across locations in the United States, collaborations in the same city in the United States, and collaborations with international researchers.

We find that US collaborations increased across US cities as well as internationally and that scientists involved in these collaborations and those who collaborate in the same locale report broad similarities in their experiences. Most collaborators first met while working in the same institution. Most say that face-to-face meetings are important in communicating with coauthors across distances. And most say that specialized knowledge and skills of coauthors drive their collaborations. We find that international collaborations have a statistically significant higher citation rate than domestic collaborations only in biotech, a modestly higher citation rate in particle physics, but a lower rate in nanotech. Because international collaborations have a greater number of authors than other collaborations, once we account for the number of coauthors on papers, the higher citation rate for biotech and particle physics international collaborations also disappear. Our results suggest that the benefits to international collaboration in terms of citations depend on the scientific field in question, rather than from any "international magic" operating on collaborations with the same number of researchers. By limiting our sample to papers with at least one US-based author, however, we exclude the possibility that international collaborations greatly benefit researchers in countries with smaller research communities by linking them to experts outside their country, the United States aside.

1.1 The Growing Trend of International Collaboration

We analyze data from corresponding authors and articles in which researchers collaborate in particle and field physics, nanoscience and nanotechnology, and biotechnology and applied microbiology. These three fields cover a wide span of scientific activity, with different research tools and methodologies.

Particle physics has a theoretical part and an empirical part. Leading edge empirical research requires massive investments in accelerators and colliders, of which the Large Hadron Collider is the most striking. Europe's decision to fund the Hadron Collider while the United States' rejection to build a large collider in Texas shifted the geographic locus of empirical research from the United States to Europe and arguably spurred the greater growth of string theory (which does not need direct access to the Collider) in the United States than in Europe. Particle physics is the most mathematically and theoretically sophisticated of the sciences we study, where pathbreaking mathematical analysis guides empirical work, and where the massive equipment exemplifies big science.

Nanotechnology is a general interdisciplinary applied technology, where engineers often collaborate with material scientists. The electron microscope is a pivotal research tool. The United States made sizable investments in nanotechnology beginning at the turn of the twenty-first century, when President Clinton called for greater investment in nano-related science and technology. This led to the 21st Century Nanotechnology Research and Development Act that President Bush signed in 2003. Other countries undertook similar initiatives in the same period.

Biotechnology is lab-based, in which the National Institutes of Health (NIH) dominates basic research funding, but where big pharmaceutical firms also fund considerable research. The most important change in biotech research technology has been the US-sponsored Human Genome Project and associated new methods of genetic analysis and engineering that allow labs around the world to modify the biological underpinnings of living creatures to advance medicine and improve biological products and processes.

To measure collaboration patterns in the three fields, we use publication data from the WoS. We identified all papers in the WoS database from 1990–2010, with at least one US coauthor in journal subject categories particle and field physics; nanoscience and nanotechnology; and biotechnology and applied microbiology. From these papers, we identify teams by the names of coauthors and locate the authors by author affiliations. This sample includes 125,808 papers.

Using the location of the authors on each paper, we define four types of collaborations:

US-only collaborations, divided into US colocated, in which all US authors are in the same city; US non-colocated, in which US coauthors are in at least two different cities; international collaborations, divided into international/US colocated, in which US coauthors are in the same city with at least one foreign coauthor; and international/US non-colocated, in which US coauthors are in two or more cities with at least one foreign coauthor.

Distinguishing between these forms of collaborations allows us to identify differences between papers with international collaborations and papers with collaborations in different locations, whether they are in the United States or overseas, as well as between papers with collaborations across locations within the United States. By focusing only on papers in which there is some US presence, our analysis may not generalize to papers written in which all authors are based outside the United States; by differentiating city location only for US coauthors, our findings do not address the potential effects of colocation or non-colocation of non-US-based researchers on paper outcomes.

Figure 1.1 displays the proportion of papers in our four categories and the proportion with single authors in the three fields taken together in each year. The solid top line gives the share of papers in which a US-based author collaborates solely with authors colocated in the same city. It shows a marked

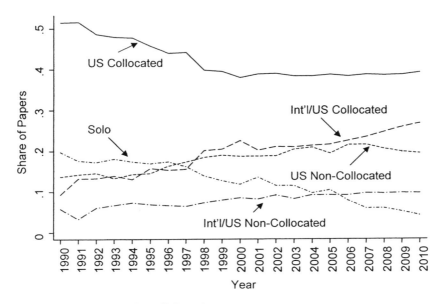

Fig. 1.1 Share of papers by collaboration type

Notes: Includes all papers in the Web of Science database with at least one US author, and with journal subject categories of particle and field physics, nanoscience and nanotechnology, and biotechnology and applied microbiology, published from 1990 to 2010.

decrease in collaborations between these authors from 1990 through 2000, which then stabilizes at about 40 percent of papers. The line labeled Solo shows the proportion of papers that are solo authored. It drops from 20 percent to about 5 percent from 1990 to 2010. The line for International/US Colocated papers gives the share of papers for which at least one of the authors is in another country while all US authors are in the same city. It increases by 18 percentage points from 1990 to 2010. The line for International/US non-colocated increases by about 5 percentage points from 1990 to 2010. Most of the increase in international collaborations was between US scientists based in one location and persons in another country. Overall, while papers with authors in different US cities increased less than international collaborations, the data shows that increased geographic scope of collaborations involved more than crossing national boundaries.

To see whether the trend in collaborations varied noticeably among fields, figures 1.2A, 1.2B, and 1.2C display the proportion of papers by collaboration type for the three fields separately. The data for particle physics in figure 1.2A show the highest level of international collaborations, due presumably to the importance of particle accelerators and other equipment that are available at only some sites. Figures 1.2B and 1.2C show that in nano and biotech, the most common form of collaborations are US-colocated teams, while

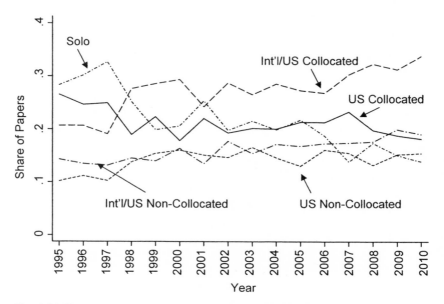

Fig. 1.2A Share of papers by collaboration type, particle physics

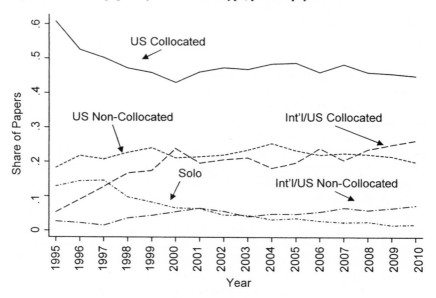

Fig. 1.2B Share of papers by collaboration type, nanotechnology

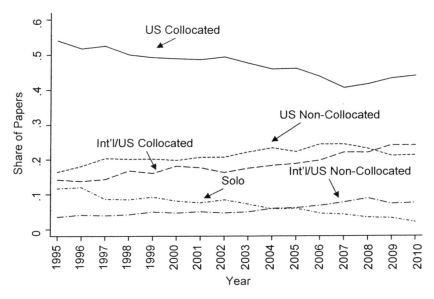

Fig. 1.2C Share of papers by collaboration type, biotechnology

international/US-colocated collaborations are the second most common and US non-colocated collaborations are third in frequency. International collaborations were roughly as common as US non-colocated collaborations in nano and biotech until late in the first decade of the twenty-first century, when international collaborations increased sharply. In all fields, the proportion of papers by sole researchers and by researchers collaborating in the same city falls.

The increase in international collaborations in our three fields resembles the patterns in National Science Board (2012) and in Adams (2013) for science more broadly. Similarly, the increased geographic dispersion of coauthorship in our fields reflects the pattern in science more broadly in the United States as well.

1.2 Survey of Corresponding Authors

To go beyond bibliometric data on collaborations, in August 2012 we conducted an online survey of the corresponding authors of papers published in 2004, 2007, and 2010 in the Web of Science nano, biotech, and particle physics subject categories with at least one US coauthor. We identified unique corresponding authors based on e-mail addresses in these categories and selected one paper for each author, randomly choosing the paper from authors who had more than one paper in the database. Using the e-mail address of the corresponding author, we sent a personalized e-mail in

English that invited them to complete the survey by clicking a link that connected them to the online survey instrument. If a paper had more than one corresponding author, we selected the one that appeared first. We sent two follow-up e-mail reminders in August and September 2012. We used Qualtrics Survey Software and respondents accessed it from the Qualtrics server.

We customized each survey to ask the respondent about the specific collaboration and individual team members. The survey had twenty-five questions and was designed so that respondents could complete it in ten to fifteen minutes. The questions sought to discover how the team formed, how it communicated and interacted during the collaboration, the contribution of each coauthor, types of research funding, and the advantages and disadvantages of working with the team. The survey also included an open-ended question for respondents to make comments. Several respondents sent e-mails with additional thoughts and information about the collaboration.

Between August 13, 2012, and August 20, 2012, we e-mailed a total of 19,836 individuals. Since some e-mail addresses had expired, changed, or some individuals were deceased, the number of individuals who received the e-mail is lower. We received 3,925 responses, which implies a response rate of 20 percent—a proportion that is in line with other surveys of scientists (Sauermann and Roach 2013). For individuals who published their papers in the most recent year of our survey (2010), the response rate was 26 percent. Taking account of the proportion of e-mails that likely did not reach respondents, we estimate approximately 29 percent of recipients of e-mails answered them.[2]

The survey asked the respondent which country each coauthor was "primarily based in during the research and writing" of the article. This gives us a more accurate measure of whether teams are international than in the WoS data, which are based on author affiliations at the time of publication, which can differ from those during the work either because affiliations change between the time of the research and the time of publication, or because some people have multicountry affiliations.

Table 1.1 compares the characteristics of collaborations in the papers we analyze to those in the full sample of WoS papers and those in the 2004, 2007, and 2010 WoS sample from which we drew the survey. Our final sample includes 3,452 respondents, which is lower in part than the returned responses due to the fact that some papers with US addresses on the publication did not meet our requirement that at least one author be primarily based in the United States at the time of the research. Our analysis uses the respondents' information to define US colocated, US non-colocated,

---

2. Of those who received the e-mail, 5,744 opened the survey, and 3,925 completed and submitted their answers. While we are unable to precisely count how many e-mails reached active mailboxes, based on the number of e-mails that "bounced" back from a sample of the messages sent, we estimate that approximately 32 percent of e-mails sent were undeliverable. Given this estimate, we approximate a response rate of 29 percent from the deliverable messages.

Table 1.1   **Distribution of papers by characteristics, Web of Science papers and survey respondents**

	Papers, 1990–2010 (1)	Papers in 2004, 2007, and 2010 (2)	Survey sample, papers in 2004, 2007, and 2010 (3)	Difference (3)–(2)
Collaboration type				
US collaboration only	66.29	63.65	62.25	−1.4
US colocated	44.81	41.56	46.84	5.28
US non-colocated	21.47	22.09	15.41	−6.68
Int'l collaboration	33.71	36.35	37.75	1.4
Int'l/US colocated	24.04	26.04	26.94	0.9
Int'l/US non-colocated	9.68	10.31	10.81	0.5
Int'l collaboration survey			34.01	
Year				
2004	6.08	25.38	18.42	−6.96
2007	8.05	33.61	29.46	−4.15
2010	9.83	41.01	52.11	11.1
Field				
Particle physics	25.19	21.75	19.55	−2.2
Nano	23.82	32.85	30.5	−2.35
Biotechnology	50.99	45.40	49.94	4.54
N	125,808	30,141	3,452	

Notes: Column (1) includes all papers in the Web of Science with more than one author, at least one US coauthor, and with journal subject categories of particle and field physics, nanoscience and nanotechnology, and biotechnology and applied microbiology, published from 1990 to 2010. Column (2) includes those papers in 2004, 2007, and 2010. Column (3) includes the respondents to our survey, which was a sample based on unique corresponding authors appearing in column (2) that had more than one author.

and international teams.[3] The column giving the difference between the distribution of our sample in column (3) and the distribution of the WoS sample in column (2) shows that our survey sample is overrepresented by US colocated teams, the more recent publication year (2010), and publications from biotechnology.

1.3 Collaborations over Distance

In what ways, if any, do papers with international collaborations differ from collaborations that occur solely in the United States?

---

3. Comparing the 34.01 percent in row "int'l collaboration survey," which is based on the respondent's answers regarding the location of coauthors, and the 36.35 percent in "int'l collaboration," table 1.1 shows that using only reported author affiliations from publications overestimates the number of international teams by 2.35 percentage points.

As others have found (e.g., Katz and Hicks 1997; Rigby 2009; Guerrero Bote, Olmeda-Gómez, and Moya-Anegón 2013; Lancho Barrantes et al. 2012; Adams 2013), international collaborations tend to produce more highly cited papers than collaborations of persons in a single country. Taking all of our fields together gives a similar pattern, where the United States is the single country to which we compare the international collaborations. We examined citations for papers published in 1990–2007—dates chosen to allow time for papers to gain substantial numbers of citations. In this group, US papers with foreign authors obtained 26.59 citations compared to 25.65 citations in which collaborations were solely with fellow residents of the United States. Since US-authored papers average more citations than papers worldwide, it would have been reasonable to expect the opposite: fewer citations for US-based scientists collaborating with persons outside the country than for US-based scientists collaborating with other US scientists.

Does this mean that international collaborations per se produce better science as reflected in numbers of citations?[4]

We answer this question by comparing citations for papers with international collaborations and citations for papers with collaborations across locales in the United States. If the observed international effect is due to something special about international collaborations, the average citations for international collaborations would exceed average citations for collaborations among non-colocated authors in the United States as well as exceed the average citations for colocated authors in the United States. Figure 1.3 shows the average number of citations for papers published between 1990 (with twenty-one years of potential citations) and 2007 (with three years of potential citations) for these three forms of collaboration. The number of citations varies over time, from approximately thirty for the older papers to three to four citations for the newer papers. In almost all years, papers with international collaborators and papers with non-colocated US collaborators have more citations than those published by collaborators in the same US city. But there is no clear pattern of differences in citations for papers coauthored by people in different US cities than for papers coauthored by people in the United States and in a foreign location. Among papers published between 1998 and 2007, US non-colocated collaborations obtain more citations than international papers, but among papers published between 1990 and 1997, there is no clear difference. That cites per year between US non-colocated papers and international collaborations are reasonably similar and that both are notably larger than cites to US-colocated papers suggests that the greater cites of international collaborations reflect multiple locations more than having authors across national borders.

4. Citations measure the attention given to a paper, which is an imperfect measure of its scientific contribution since citation behavior can be driven by factors besides its contribution to knowledge (see, e.g., Simkin and Roychowdhury 2003). But it is still a sufficiently valuable indicator of the impact of a paper and is the most widely used measure in the science of science.

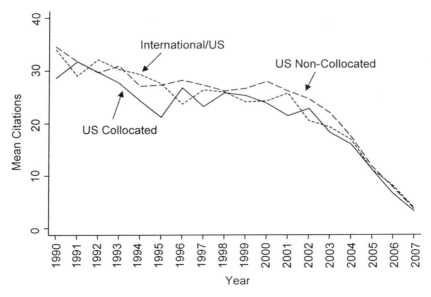

Fig. 1.3 Citations by the nature of collaboration, all fields by year of publication

Notes: Figure shows forward citations of all papers in the Web of Science database with at least one US author, and with journal subject categories of particle and field physics, nanoscience and nanotechnology, and biotechnology and applied microbiology, published from 1990 to 2007. Year indicates the year of publication of the cited paper.

We pursue the comparison of citations across types of collaborations for each of our three fields separately. The style of research in the fields differs greatly between particle physics, where empirical work often involves huge collaborations around particular pieces of equipment, and the smaller collaborations of nanotechnology and biotechnology research. This difference shows itself in the much higher average number of authors per paper in physics than in the other two fields (see appendix table 1A.1). The difference is concentrated in the upper tail of the distribution of authors per paper. In particle physics, the upper 95th percentile of the number of authors per paper have 100 authors, while those in the 99th percentile have 523 authors—which far exceed the upper percentile numbers for authors in nano and biotech.

Reflecting the "big science" nature of some of the physics projects, the corresponding author on a physics paper with over 450 coauthors noted in our survey:

This research was carried out as part of a very large collaboration in which every member gets authorship and this is listed in alphabetical order on our papers. The collaboration consists of scientists and engineers with a wide range of expertise—many primarily involved in designing, building, and running instrumentation, and many analyzing data for various kinds of signal. This particular research was primarily carried out by myself, and the majority of the listed coauthors (including three of the selected authors in this survey) had no direct involvement in its preparation other than through collaboration membership.

We next use regression analysis to examine the relation between the modes of collaboration and the number of researchers listed as authors and the number of references in a paper, both of which tend to be positively related to citations. To the extent that the references influence the paper by providing information and ideas from other scientists, they can be viewed as indicators of "invisible coauthors," self-citations aside.

Table 1.2 records the regression coefficients and standard errors for regressions of numbers of coauthors and references on the type of collaboration and a year trend for each field. While there is a broad similarity in the estimated effect of the collaborations on the number of coauthors and references across the fields, there are also differences that presumably reflect differences in their research technologies. In all of the fields, the regression of number of coauthors on the dummy variable for whether or not the paper had an international coauthor gives a positive coefficient on the dummy variable. But the magnitudes of the coefficients differ greatly. The estimated coefficient on international collaborations in particle physics in column (1) (43.8) shows that the number of authors on papers is much higher for those than for the US collaboration reference group, whereas the estimated coefficients for the relation between international collaborations and coauthors in nanotech and biotech are magnitudes smaller: 1.3 more authors on international papers than papers written by authors solely in the United States for nano (column [2]) and 2.2 more authors on international papers than US-only papers for biotech (column [3]). The more detailed measures of collaborations in columns (4), (5), and (6) show that this difference is largely driven by international collaborations in which the US scientists doing particle physics are from many locations as well. This reflects the big science nature of empirical particle physics, where huge numbers of collaborators work together with massive instruments and machines compared to the smaller lab science of nano and biotech.

The regressions in columns (7)–(9) show greater differences among the fields in the number of references on international papers relative to US non-colocated papers, and differences among the fields in the relation between numbers of coauthors and numbers of references. In particle physics, numbers of references for papers with international collaborations exceed those for US non-colocated papers, which in turn exceed both those for US colocated international papers and those US colocated (column [7]). In biotech, numbers of references for papers with international collaborations and non-colocated US collaborations exceed those for US non-colocated papers, which in turn exceed those for US colocated papers (column [9]). A potential explanation is that persons in a given location are more likely to cite papers written in their location, so that the greater the number of locations, the greater the number of references. But the regression for number of references in the nano papers

Table 1.2   Estimated relation between number of coauthors and number of references on papers by nature of collaboration, by field

	Coauthors						References		
	Particle physics	Nano	Biotech	Particle physics	Nano	Biotech	Particle physics	Nano	Biotech
	(1)	(2)	(3)	(4)	(5)	(6)	(7)	(8)	(9)
US collaboration Only									
US colocated									
US non-colocated				2.654**	1.450**	1.688**	3.453**	−0.879**	0.727**
				(0.150)	(0.033)	(0.029)	(0.377)	(0.232)	(0.179)
Int'l collaboration	43.776**	1.331**	2.168**						
	(0.924)	(0.032)	(0.040)						
Int'l/US colocated				12.017**	1.458**	1.973**	4.737**	−0.963**	0.275
				(0.641)	(0.033)	(0.032)	(0.313)	(0.272)	(0.189)
Int'l/US non-colocated				99.983**	3.075**	5.015**	4.590**	0.168	3.131**
				(2.091)	(0.073)	(0.126)	(0.400)	(0.400)	(0.359)
No. coauthors							0.001	−0.060	0.435**
							(0.001)	(0.042)	(0.031)
Year trend	−0.214*	0.039**	0.078**	−0.183*	0.038**	0.064**	0.796**	1.491**	0.535**
	(0.094)	(0.003)	(0.002)	(0.090)	(0.003)	(0.002)	(0.024)	(0.024)	(0.013)
Constant	433.018*	−73.670**	−151.290**	368.918*	−71.578**	−125.213**	−1.6e+03**	−3.0e+03**	−1.0e+03**
	(188.702)	(6.828)	(4.713)	(179.207)	(6.581)	(4.484)	(47.520)	(48.749)	(26.100)
R^2	0.055	0.068	0.091	0.170	0.144	0.159	0.046	0.116	0.044
No. of obs.	31,690	30,761	64,153	31,690	30,761	64,153	31,690	30,761	64,153

Notes: Includes all papers in the Web of Science with more than one author, at least one US coauthor, and with journal subject categories of particle and field physics, nanoscience and nanotechnology, and biotechnology and applied microbiology, published from 1990 to 2010.

**Significant at the 1 percent level, OLS estimation.

*Significant at the 5 percent level, OLS estimation.

+ Significant at the 10 percent level, OLS estimation.

shows a different relation between references and collaborations (column [8]). Finally, the estimated coefficients on the number of authors also shows no consistent pattern among the fields: negligible effects for particle physics (potentially because the number of authors can be extremely high), slight negative effects in nano, but substantial positive effects in biotech (column [9]).

All told, table 1.2 shows that simple comparisons of papers with international and national collaborations can present a misleading picture about the science involved in various types of collaborations. The collaborations can involve huge differences in the numbers of coauthors and differing relations to the numbers of references. Given these results, we examine the relation between the citations to a paper and the form of collaboration separately for each filed using a regression that includes the number of coauthors and the number of references in the paper. To deal with the life cycle of citations in which the number of citations increases sharply in the first five to seven years after publication and then grows more slowly, we include dummy variables for the year the paper was published as well.

Tables 1.3A, 1.3B, and 1.3C give the results of this analysis. Column (1) of each of the tables estimates the difference in citations between international papers and US-only collaborations. The estimates show a disparate pattern across the fields: an insignificant positive relation between international collaborations and citations for particle physics, a negative relation in nano, and a positive relation in biotech. Column (2) of each table adds the number of coauthors to the regression. In each of the fields, the addition of numbers of authors reduces the coefficient on international collaborations. In biotech it turns the coefficient from positive to negative.[5] With the addition of numbers of references in column (3), the estimated relation of international collaborations to citations is significantly negative in all three fields. The disaggregation of types of collaborations in columns (4) in tables 1.3A–1.3C show sufficiently weak and different patterns across the fields to suggest that there is nothing universal in the link between international collaborations and ensuing citations to papers.

All told, the regression analysis in tables 1.2 and 1.3A, 1.3B, and 1.3C document the changing patterns of cooperation across locations in the three fields and their disparate relation with citations. While invaluable as descriptions about collaborations, such bibliometric analysis cannot, however, provide insight into the ways collaborating scientists work together to conduct the research that leads to published papers. To gain insight into what goes on in collaborations, we turn to the survey of corresponding authors described in section 1.1.

---

5. To see if this is a more general pattern, we ran similar regressions for other scientific fields in the WoS and find variation across fields in the difference between citation rates for international collaborations and domestic collaborations; the addition of the number of coauthors to citation regressions reduces the coefficient on international collaborations in almost all fields.

Table 1.3A   The estimated relation between number of citations to a paper and the type of collaboration that produced the paper, particle physics

	(1)	(2)	(3)	(4)
US collaboration only				
US colocated				
US non-colocated				1.664*
				(0.691)
Int'l collaboration	0.718	0.096	−1.212**	
	(0.469)	(0.452)	(0.464)	
Int'l/US colocated				−1.418**
				(0.532)
Int'l/US non-colocated				1.402
				(0.856)
No. coauthors		0.014**	0.014**	0.010**
		(0.003)	(0.002)	(0.002)
No. references			0.398**	0.396**
			(0.017)	(0.017)
Constant	24.030**	24.031**	15.404**	14.817**
	(1.953)	(1.945)	(1.894)	(1.901)
Year FE	Yes	Yes	Yes	Yes
R^2	0.030	0.031	0.072	0.073
No. of obs.	31,690	31,690	31,690	31,690

Notes: Sample is all papers in the Web of Science with more than one author, at least one US coauthor, and with a journal subject category of particle and field physics, published from 1990 to 2010.
**Significant at the 1 percent level, OLS estimation.
*Significant at the 5 percent level, OLS estimation.
+ Significant at the 10 percent level, OLS estimation.

1.4 Survey Evidence

I think the best example of collaboration I have done is . . . where all the authors are from different countries and we met at the Bellagio Conference Center of the Rockefeller Foundation.

I think that it is absolutely indispensable to meet people in person to have effective collaborations.

Skype was not available . . . at the time we completed this work. We now use Skype or ITV connection to meet and discuss data with collaborators on a weekly basis.

The international collaboration worked so well because of my frequent trips to Brazil during the project.[6]

6. The four quotes are based on comments from the open-ended section of our survey.

Table 1.3B   The estimated relation between number of citations to a paper and the type of collaboration that produced the paper, nanotechnology

	(1)	(2)	(3)	(4)
US collaboration only				
US colocated				
US non-colocated				-3.971**
				(0.423)
Int'l collaboration	-2.300**	-3.732**	-3.637**	
	(0.358)	(0.388)	(0.387)	
Int'l/US colocated				-4.849**
				(0.470)
Int'l/US non-colocated				-6.305**
				(0.621)
No. coauthors		1.074**	1.110**	1.294**
		(0.083)	(0.080)	(0.085)
No. references			0.295**	0.293**
			(0.068)	(0.068)
Constant	26.252**	21.712**	14.660**	14.747**
	(4.749)	(4.683)	(4.805)	(4.819)
Year FE	Yes	Yes	Yes	Yes
R^2	0.039	0.045	0.068	0.070
No. of obs.	30,761	30,761	30,761	30,761

Notes: Sample is all papers in the Web of Science with more than one author, at least one US coauthor, and with a journal subject category of nanoscience and nanotechnology, published from 1990 to 2010.

**Significant at the 1 percent level, OLS estimation.

*Significant at the 5 percent level, OLS estimation.

+ Significant at the 10 percent level, OLS estimation.

For scientists to collaborate, they must meet and decide to work together, communicate during the collaboration, and combine their knowledge and skills to create sufficient new knowledge to generate a publishable paper.

1.4.1 Meeting and Communicating

We asked corresponding authors to answer the following question about their coauthors: "How did you FIRST come in contact with each of these coauthors?" For papers with up to six authors, we asked about each coauthor. For papers with more than six we asked about the first and the last authors, if they were not the corresponding author, and about randomly selected authors from the list of coauthors to obtain information on a maximum of six collaborators.

Figure 1.4 displays the proportion of persons of each collaboration type who the corresponding author first met as advisor-student/postdoc; colleagues in the same department/institution; through contact without an introduction; at a conference, seminar, or other meeting; or by visiting the

Table 1.3C	The estimated relation between number of citations to a paper and the type of collaboration that produced the paper, biotechnology			
	(1)	(2)	(3)	(4)
US collaboration only				
US colocated				
US non-colocated				1.109*
				(0.531)
Int'l collaboration	1.800**	−1.466*	−1.583*	
	(0.597)	(0.680)	(0.677)	
Int'l/US colocated				−2.138**
				(0.647)
Int'l/US non-colocated				2.394
				(1.891)
No. coauthors		1.506**	1.412**	1.333**
		(0.103)	(0.101)	(0.110)
No. references			0.193**	0.191**
			(0.015)	(0.015)
Constant	34.629**	29.522**	24.805**	24.917**
	(2.047)	(2.054)	(2.075)	(2.067)
Year FE	Yes	Yes	Yes	Yes
R^2	0.025	0.032	0.036	0.036
No. of obs.	64,153	64,153	64,153	64,153

Notes: Sample is all papers in the Web of Science with more than one author, at least one US coauthor, and with a journal subject category of biotechnology and applied microbiology, published from 1990 to 2010.
**Significant at the 1 percent level, OLS estimation.
*Significant at the 5 percent level, OLS estimation.
+ Significant at the 10 percent level, OLS estimation.

department/institution. The figure shows that for all forms of collaboration, most first meetings occurred when the corresponding author and the other person worked in the same institution. For papers written in the same location, the predominant contact was through advisor-student or postdoc relationships, but that over one-third of the meetings came about as colleagues. For papers with authors from other US locations or foreign locations, the corresponding authors met through working in the same place, primarily as a colleague, but with nearly 10 to 16 percent meeting the person as a visitor. Conferences also accounted for a substantial proportion of the first meetings between corresponding authors on papers written with persons in other US locations or in foreign locations.[7] Overall, figure 1.4 shows broad similarity in the mode of meeting between non-colocated US authors and

7. The time series data in appendix figures 1A.1 and 1A.2 show that conferences have become a less important way to meet future coauthors, while students/postdocs have become more important, possibly due to their increased importance in the scientific production process.

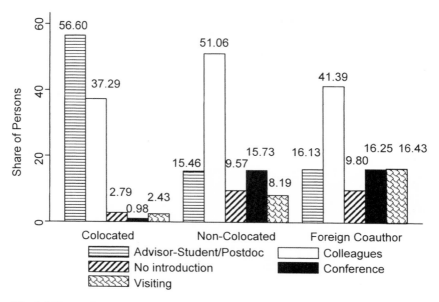

Fig. 1.4 Share of persons who first met in a given way by the nature of collaboration

Notes: Share of all coauthors on papers for a given collaboration type. Question was phrased as "How did you FIRST come in contact with each of these coauthors?"

in the mode of meeting between US and foreign-located authors compared to the mode of meeting for coauthors in US-colocated collaborations.

We asked corresponding authors the frequency with which they communicated with one or more of their coauthors from "every week" to "never." Because collaborations that include persons in the same locale and persons in other locales as the corresponding author allow the corresponding author to meet face-to-face easily with some coauthors but only infrequently with coauthors in other locations, the question does not pin down differences associated with distance. To overcome this problem, we show in figure 1.5 modes of communication between coauthors on two-authored papers, which differentiate properly communication between colocated, US non-colocated, and foreign coauthors.

The results show that the corresponding author relies extensively on face-to-face meetings when all authors are in the same location. But figure 1.5 also shows that while face-to-face meetings are much lower for authors across distances, such meetings are still frequent. Among the two-author papers, just over 50 percent of corresponding authors on international teams report meeting face-to-face at least a few times per year, while 64 percent of those on US non-colocated papers reported face-to-face meetings at least a few times a year. By contrast, the figure shows no noticeable differences in using e-mail by distance. Corresponding authors in all forms of col-

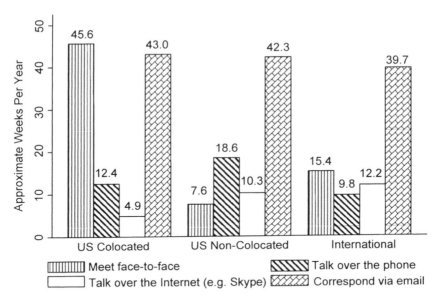

Fig. 1.5 Overcoming distance: Frequency of communication modes for two-author papers by the nature of collaboration (approx. weeks per year)
Notes: Question was phrased as "When carrying out the research and writing for this article, how frequently did you use the following forms of communication with one or more of your coauthors?" The possible choices were transformed into approximate number of weeks per year that each communication type was used: 6 = every week (52), 5 = almost every week (45), 4 = once or twice a month (15), 3 = a few times per year (5), 2 = less often than that (2), and 1 = never (0).

laborations use e-mail frequently to communicate with their collaborators, approximately forty weeks during the year. There are substantial differences in use of telephone versus Internet (e.g., Skype) between US-based teams and international teams that are readily explained by the differential in cost of international and within US telephone calls.

Our survey findings that face-to-face meetings are important in both the initiation of research collaborations and the working of distant collaborations are consistent with evidence on the role of colocation in the formation of research collaborations (such as Boudreau et al. 2014) and the need for periodic colocation to maintain the effectiveness of distant collaborations even with advances in long-distance communication technologies (see, e.g., Olson and Olson 2000, 2003; Cummings and Kiesler 2005).

1.4.2 What Coauthors Bring to Collaboration

To understand what factors helped produce the collaborations, we asked the corresponding author to specify the unique contribution of each team member. Our question was "Did any of the team members working on this

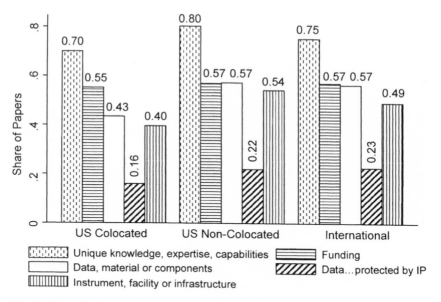

Fig. 1.6 Contribution of coauthors by the nature of collaboration

Notes: Share of US and foreign coauthors on two-author papers only, as reported by the corresponding author.

article (including yourself) have access to one of the following resources that the other team members did NOT have, which made it important for you to all work together on this topic?" The possible choices were: access to data, material, or components; data, material, or components protected by intellectual property; a critical instrument, facility, or infrastructure; funding; or unique knowledge, expertise, or capabilities.

Figure 1.6 shows that the major factor cited for all collaborations was "unique knowledge, expertise, or capabilities." That access to specialized human capital seems to drive collaborations, whether in the United States or international, implies that a theory of collaboration should focus on the complementarity of skills and knowledge of collaborators just as the theory of trade focuses on comparative advantage in creating trade among countries. But there are differences in the importance of other factors across forms of collaboration. Non-colocated and international teams were more likely to have a coauthor contributing data, material, or components than US colocated teams—a pattern that has increased over time (see appendix figure 1A.3).

While most corresponding authors reported the contribution and role of their coauthors, those on huge collaborations told a different story. As one respondent remarked, "Many of the questions are hard to translate to the field of experimental particle physics, where an international collaboration of hundreds of scientists work on the same project with funding from

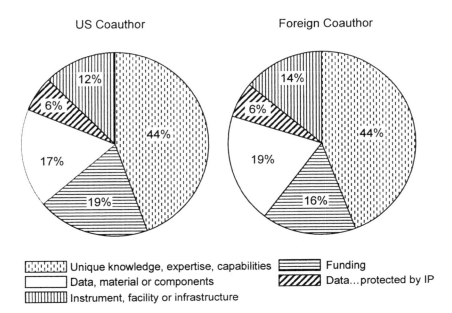

US Coauthor   Foreign Coauthor

Unique knowledge, expertise, capabilities   Funding
Data, material or components   Data...protected by IP
Instrument, facility or infrastructure

Fig. 1.7 Contribution of US and foreign coauthors for two-author papers

Notes: Share of papers for which the corresponding author reported at least one coauthor contributing the given resource. Question was phrased as "Did any of the team members working on this article (including yourself) have access to one of the following resources that the other team members did NOT have, which made it important for you to all work together on this topic?"

many countries. I can only guess where the funding from each of the ~300 coauthors comes from, many of whom I have not even met. The published research is primarily the work of a single person (myself), but would not have been possible without having access to custom software and data provided by the collaboration."

Finally, taking advantage of the unique identification of authors in two-authored papers, we compare the specific contributions of foreign-located coauthors and domestic coauthors on those papers (see figure 1.7). The US and foreign coauthors were equally likely to contribute "unique knowledge, expertise, or capabilities" and "data, material, or components protected by intellectual property." Foreign coauthors are slightly more likely to contribute access to "data, material, or components" or "a critical instrument, facility, or infrastructure," while the US coauthor was slightly more likely to contribute funding.

1.4.3 Advantages and Challenges

To assess the effects of the different forms of collaboration on the production and output of scientific activity, we use our survey, where we asked the corresponding authors their views of the advantages and challenges on their collaboration, and the bibliometric data, where we estimated a regression model linking the number of citations to a paper to the attributes of the collaboration reported on the survey.

Table 1.4 summarizes the responses of corresponding authors on the advantages and challenges of the collaborations. It records the average score on a five-point scale of agreement (5) or disagreement (1) with statements regarding the attributes of the collaboration. The corresponding authors agreed that their collaboration had substantial advantages in harnessing human capital to produce a scientific outcome. "Complementing our knowledge, expertise, and capabilities" and "learning from each other" are the only items with average scores greater than (4) in the table. The next highest score was that collaborations made the research experience more pleasant. There is little variation here in the responses between US non-colocated and international teams. Corresponding authors on the both of those collaborations gave modestly higher scores to the knowledge advantages than the colocated teams. Similarly, all three groups ranked highly "gaining access to data, material, or components," with the highest assessment coming from the corresponding authors of US non-colocated teams.

The corresponding authors of international teams gave higher scores to the advantage of "our research reached a wider audience" than did the corresponding authors of US non-colocated teams, who in turn gave higher scores than the corresponding authors of US colocated only teams. Viewing "wider audience" in terms of the geographic distribution of citations, this suggests that the wider the geographic distribution of authors, the wider is the distribution of citations, possibly even among papers with the same numbers of citations.

Regarding the challenges of collaborations, US non-colocated and international teams reported similarly that there was "insufficient time for communication," "problems coordinating with team members' schedules," and "insufficient time to use a critical instrument, facility, or infrastructure" than did US colocated teams. As in the bibliometric analysis in section 1.3, geographic location appears to be more than national boundaries in the way teams operated.

We also asked whether the corresponding authors viewed teams as having the optimal size. The responses, given in appendix table 1A.2, show that most corresponding authors viewed their team as having the right size. Presumably the principal investigator(s) would have modified the team if they did not think that was the case, but there are some differences by collaboration type. The US colocated teams were more likely to say that they needed additional collaborators (7.58 percent vs. 3.48 percent, and 3.38 percent for US non-colocated and international), whereas international teams were more likely to say that fewer team members were needed (6.67 percent vs. 3.37 percent for US colocated). Reflecting the role of government policies,

Table 1.4   Advantages and challenges to working with the team

	US colocated	US non-colocated	Int'l
Advantages			
Learning from each other	4.26	4.33	4.36
Complementing our knowledge, expertise, and capabilities	4.39	4.58	4.57
Gaining access to data, materials, or components	3.21	3.56	3.32
Gaining access to data, materials, or components protected by IP	2.14	2.30	2.29
Our research reached a wider audience	3.24	3.37	3.48
The research experience was more pleasant	3.96	3.92	4.02
Challenges			
Insufficient time for communication	1.82	2.13	2.11
Less flexibility in how the research was carried out	1.73	1.99	1.93
Unable to unequivocally portray my contribution	1.55	1.59	1.65
Problems coordinating with team members' schedules	1.96	2.18	2.11
Insufficient time to use a critical instrument, facility, or infrastructure	1.45	1.67	1.67
Observations	1,693	585	1,174

Note: Respondents were asked to indicate their level of agreement with these statements regarding the main advantages/disadvantages of "carrying out the research for this article with your team members," where (5) = agree and (1) = disagree.

24 percent of the international teams received funding aimed at supporting cross-country collaboration, with 6.65 percent receiving US government funding, 4.64 percent receiving EU funding, and the remainder from other government sources.

As our second way to assess how the attributes of collaborations affect outcomes we added the corresponding authors' descriptions of the collaboration to the table 1.3 regressions of the number of citations on attributes of papers. Because publication of the paper preceded the survey, some of the corresponding author views of the collaboration will presumably have been affected by the success of the paper, which would give a distorted view of the link from collaboration to outcome. To deal with this problem, we limit analysis to the survey responses that seem least prone to be affected by the outcome—relatively objective questions about the way corresponding authors met coauthors, what coauthors contributed, and funding support.

Table 1.5 gives the results of this analysis. Columns (1) and (2) replicate the regression estimates in table 1.3 for the dichotomous international collaboration variable. The results in table 1.5 show some differences in the regression coefficients from that found in the larger WoS sample. The positive coefficient on international collaborations in column (1) in table 1.5 is larger than the coefficient in the comparable regression using the larger WoS sample papers in our three fields. The coefficients on the number of coauthors and number of references variables are positive and significant in column (2) of table 1.5 but the coefficient on coauthors is larger than that of references, contrary to the result in the larger WoS sample. Subject to these differences, which suggest some modest differences between the papers of respondents to the survey and the population of papers, the estimated coefficients on the survey variables in columns (3), (4), and (5) tell a clear story. They show that papers in which at least one coauthor met at a conference had higher citations, that papers for which a coauthor contributed funding had lower citations, and that papers that got funding specifically for cross-country collaborations had lower citations.[8] The natural interpretation of these patterns is that collaborations based on ideas or relations developed at conferences produce more cited and potentially better science than collaborations based on funding.

1.5 Toward an Economics of Scientific Collaborations

Scientific collaborations have become increasingly important in scientific research, but the nature of collaborations, their determinants, effects on scientific outcomes, and the incentives that drive scientists to collaborate or

8. We also estimated the model including dummies for whether the corresponding author did not view the team size as optimal, and an average of the scores assessing the advantages and disadvantages to the collaboration, but found no effect of these measures on citations.

Table 1.5   The estimated relation between number of citations to a paper and the type and characteristics of collaboration, survey sample

	(1)	(2)	(3)	(4)	(5)
US collaboration only					
US colocated					
US non-colocated			−0.355	0.434	0.444
			(0.779)	(0.779)	(0.773)
Int'l collaboration	0.878+	0.192	−0.579	0.370	0.495
	(0.529)	(0.538)	(0.649)	(0.600)	(0.586)
No. coauthors		0.161*	0.157*	0.160*	0.161*
		(0.066)	(0.066)	(0.066)	(0.066)
No. references		0.099**	0.098**	0.099**	0.098**
		(0.015)	(0.015)	(0.015)	(0.015)
How they met					
Advisor-stu./postdoc			−0.734		
			(0.656)		
Colleagues			0.592		
			(0.547)		
Visiting			0.703		
			(0.877)		
Conference			2.939**		
			(0.993)		
No introduction			0.575		
			(0.890)		
Coauthor contributions					
Knowledge, etc.				0.498	
				(0.682)	
Funding				−1.327*	
				(0.553)	
Data, etc.				−0.305	
				(0.520)	
IP data, etc.				0.124	
				(0.630)	
Instrument, etc.				0.166	
				(0.567)	
Cross-country funding					−1.207*
					(0.610)
Constant	17.433**	14.655**	14.654**	14.925**	14.676**
	(2.497)	(2.513)	(2.593)	(2.607)	(2.548)
R^2	0.076	0.114	0.119	0.116	0.115
No. of obs.	3,452	3,452	3,452	3,452	3,452

Notes: All regressions include year, field, and year × field fixed effects. Sample is the survey sample described in section 1.2. "How they met" and "coauthor contribution" variables are dummies indicating whether any coauthor on the team met that way/contributed the resource.
**Significant at the 1 percent level, OLS estimation.
*Significant at the 5 percent level, OLS estimation.
+ Significant at the 10 percent level, OLS estimation.

not, and with whom to collaborate, is not well understood. From the perspective of economics, collaborations occur because they enhance scientific productivity, but collaborations have costs as well as benefits—the costs of communicating ideas and coordinating disagreements among collaborators and the expenses of getting them together in one place or linking them with unique data and equipment. But arguably the biggest problem collaborations must solve in order to succeed is to find ways to divide the credit for enhanced productivity among persons, so that each collaborator prefers working on the team rather than by themselves, where they gain full credit for research outcomes.

This study has linked a unique survey of corresponding authors in particle physics, nanotechnology, and biotechnology with bibliometric data from the same fields to identify some of the key empirical relations in the growth of collaborations from the 1990s to 2010. The bibliometric data shows that the share of papers with a US address written by collaborators in different locations increased in the period studied and that the largest increase occurred across country lines, followed by collaborations in different locations in the United States. Commensurately, the share of papers written by single scientists or by groups of scientists in single locations declined.

The survey of corresponding authors shows similarities in the way collaborators first meet and later communicate and work together among different types of collaborations, save for very large physics projects. The bibliometric data shows that the number of collaborators on a paper is positively associated with the numbers of citations to the paper, but the data is mixed on whether international collaborations are more productive in terms of citations than domestic collaborations. In biotech, international collaborations obtain more citations than domestic collaborations, in nanotech they obtain fewer citations, while particle physics shows no significant differences between international and domestic collaborations. In all three fields, papers with the same number of coauthors had lower citations if they were international collaborations, which suggests that a main advantage of international collaborations for US authors is that they allow researchers to increase the number of collaborators more easily than if the supply of potential coauthors was limited to US-based scientists.

The data are thus consistent with the notion that collaborative work has greater productivity or impact as reflected in citations, but do not give a clear message about whether the fastest growing form of collaborations—that across country lines—has any productivity edge over collaborations across space within the United States. The most likely reason for the rapid growth of international collaborations is the more rapid growth of science and engineering PhDs and researchers in other countries than in the United States, which creates a large supply of potential collaborators overseas.

Wherein does the productivity advantage of collaborations lie? Viewing science as an aggregate process for producing new knowledge, the most plausible answer is that the knowledge base has become increasingly complex and specialized (Jones 2011), and thus scientific advances require increased numbers of researchers combining their skills and expertise. Consistent with this, our survey of corresponding authors shows that access to specialized human capital is the main driver of collaborations. The growing number of references within papers suggests that each forward step in science builds on a large base of previous knowledge. And the positive link between numbers of references and citations suggests that the greater the knowledge that goes into a paper, the greater the scientific contribution of the paper—at least to the extent that these measures are a valid "paper trail" of flows of knowledge. All of which is consistent with the view that the productivity advantage from collaborations depends on the combination of ideas/knowledge from persons with different expertise,[9] though it does not prove the validity of this interpretation.

But, as noted, collaborations have costs as well as benefits, and decisions to collaborate involve balancing the benefits against the costs. On one side are problems of coordinating the ideas of persons with different expertise or viewpoints or who are in different locations, and the expenses and difficulty of getting collaborators together or linking them with data and key pieces of machinery. Our survey finding that researchers meet most collaborators through personal connections made at their institution or, to a lesser extent, at conferences, suggests that there is some role of chance in creating collaborations. Most important, the survey finding that corresponding authors view face-to-face meetings as critical in their collaborations, and hold them relatively often, suggests that the improvements in communication and reductions in their cost do not fully substitute for human interactions in collaborations.

To an individual researcher, the biggest issue in collaboration is getting credit for a joint production. In a one-author paper, the one takes credit or blame. In a two-author paper, many fields adhere to the convention that the senior person's name comes last and the junior person comes first, which potentially gives substantial credit to each. Freeman and Huang (2014) find that the impact factor of the placement of the paper and citations received depend more on the characteristics of the last named, typically senior, author rather than of the first named author. The senior person thus appears to play a greater role in gaining attention to the research, which can help the junior author as long as other scientists view the first author as more than a pair of hands in the lab. In papers with more than two authors, the decision on who is the first author and the placement of the non-first or non-last authors can create disputes. On papers having huge numbers of names where tasks are highly specialized, the credit to a given author is presumably related to their specialty, much like the credits that give the names of specialists in a movie

9. Weitzman's (1998) model of the growth of useful knowledge from combining the growing supply of past ideas and knowledge in new ways offers a way to structure such a model.

production. Only people who know how the research (movie production) proceeded and what the particular person's function was would understand how to evaluate their contribution in the author (credit) list.

From the perspective of economic rationality, the decision of scientists to collaborate depends on both the productivity of the collaboration and the distribution of credit. To get some notion of the interplay of the factors, consider the situation in which a scientist compares the value of collaborating on a paper with one or more other scientists to writing a solo paper. On the productivity/citation side, assume that a paper with N collaborators gains proportionately more citations (C) than a solo-authored paper according to a linear productivity parameter $p > 1$ that links citations to numbers of authors by $C = pN$. If a single-authored paper gets p citations, then this function gives two-authored papers $2p$ citations, a paper with three authors $3p$ citations, and so on.

But whereas each author gets full credit for an individual paper, they get only partial credit for joint work. Assume that the science community allocates credit for joint work with a citation crediting function $\gamma(n)$ that has the value 1 for $\gamma(1)$ and in which the derivative $\gamma'(n) < 0$ so that $\gamma(1) = 1 > \gamma(2) > \gamma(3)$ and so on. The only restriction on the citation crediting function is that each author gets less credit the larger the collaboration. Someone seeking to maximize the number of citations credited to them would collaborate only if $p\gamma \geq 1$—that is, if the gain in productivity from the collaboration exceeds the loss of credit associated with γ. If the crediting function was based on simple fractionalization of credit, so that each author in a two-authored paper would be credited with one-half of the paper and thus one-half of the cites, p would have to exceed 2 for the two-authored paper to be worthwhile. Similarly, p would have to exceed n for an n-sized collaboration to be attractive. But estimates of the extent to which citations increase with number of authors falls far short of such proportionality. Depending on field and specification, our table 1.3 estimates showed that additional authors raises citations by at most one to two citations per additional author. In this case, if a solo-authored paper gained ten citations, there would be little incentive to write a joint paper that gained twelve citations for which each author obtained credit for just six citations.

Why then have collaborations increased so much in the sciences?

One possibility is that scientists who collaborate with others are able to write so many more papers through division of labor than by themselves that the increased number of papers offsets the lower credit set by the crediting function. For most scientists, this seems unlikely. The average number of collaborators in science articles has roughly doubled in the past four to five decades, while the number of papers written per researcher has not shown any such doubling. Most of the increase in papers over time has been associated with increased numbers of researchers rather than increased papers per researcher.

The other possibility, which we view as the likely solution to the question, is that the crediting function diverges greatly from fractionalization. Fractionalization imposes the constraint that the sum of credits to all authors is one, but there is no law that rules out allocating credit differently. The first author of a highly successful paper gains lots of credit for their contribution. The last author can also gain lots of credit for the different contribution they made. The authors in the middle of the author list presumably gain some credit, but less than the other two authors. The purpose of the crediting is to propel the careers of persons who are part of a successful team activity. Thus, we would expect first and last authors to benefit most from a successful collaboration in their future careers, and for intermediate authors to benefit proportionate to their role on the research, subject to the imperfections in markets and market information. While we cannot test this interpretation with our data, it is testable with information on the future careers of persons who work on papers with different numbers of collaborators. Our prediction would be that two coauthors of a well-cited paper would gain more in their careers from the joint paper than if each had written a solo paper with half as many citations.

Finally, to the extent that the interplay between the productivity of working with other scientists and the distribution of credit affect collaboration decisions as hypothesized above, we would expect to find at most modest differences between the nature and effects of collaborations across national borders as within the United States, as our survey and WoS data analysis seem to show.

Appendix

Table 1A.1   **Team size summary statistics**

	Particle physics	Nano	Biotech
Mean	21.86	4.56	4.74
Standard deviation	82.49	2.49	3.61
Maximum	1,062	32	202
Percentiles			
10th	1	2	2
50th	3	4	4
75th	4	6	6
95th	100	9	11
99th	523	13	16
N	40,474	31,934	68,731

Notes: Measures of number of authors on papers in the Web of Science published 1990–2010, with a US author (including solo author papers), and with journal subject categories of particle and field physics, nanoscience and nanotechnology, and biotechnology and applied microbiology.

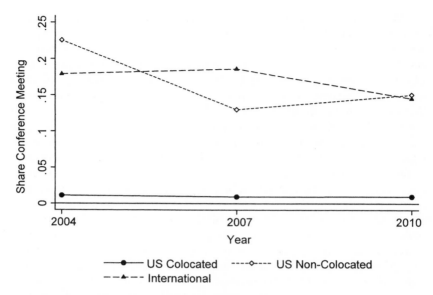

Fig. 1A.1 Share of coauthors who first met at a conference

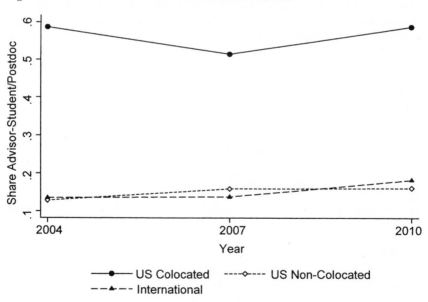

Fig. 1A.2 Share of coauthors who first met as advisor-student/postdoc

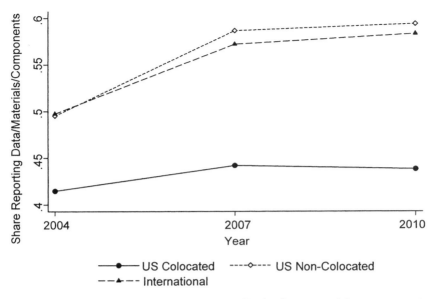

Fig. 1A.3 Share of papers with a coauthor contributing data, material, or components

Table 1A.2   Optimal team size by nature of collaboration

	US colocated	US non-colocated	All int'l	Int'l with cross-country funding
Yes	89.06	91.11	89.95	92.50
No, additional	7.58	3.48	3.38	2.86
No, fewer	3.37	5.40	6.67	4.64
N	1,663	574	1,154	280

Notes: Question phrased as "Do you think that the size of your team was optimal?" The cross-country funding question was phrased as "In carrying out the research for this article, did any of the coauthors receive funding that was specifically aimed at supporting cross-country scientific collaboration?"

References

Adams, J. 2013. "Collaborations: The Fourth Age of Research." *Nature* 497 (7451): 557–60.

Adams, J. D., G. C. Black, J. R. Clemmons, and P. E. Stephan. 2005. "Scientific Teams and Institutional Collaborations: Evidence from US Universities, 1981–1999." *Research Policy* 34 (3): 259–85.

Agrawal, A., and A. Goldfarb. 2008. "Restructuring Research: Communication Costs and the Democratization of University Innovation." *American Economic Review* 98 (4): 1578–90.

Boudreau, K., T. Brady, I. Ganguli, P. Gaule, E. C. Guinan, A. Hollenberg, and K. R. Lakhani. 2014. "A Field Experiment on Search Costs and the Formation of Scientific Collaborations." Working Paper, Harvard University.

Catalini, C., C. Fons-Rosen, and P. Gaulé. 2014. "Air Travel Costs and Scientific Collaborations." Unpublished Manuscript, MIT Sloan School of Management.

Cummings, J. N., and S. Kiesler. 2005. "Collaborative Research across Disciplinary and Organizational Boundaries." *Social Studies of Science* 35 (5): 703–22.

deB. Beaver, Donald. 2004. "Does Collaborative Research Have Greater Epistemic Authority?" *Scientometrics* 60 (3): 399–408.

Freeman, Richard B. 2010. "Globalization of Scientific and Engineering Talent: International Mobility of Students, Workers, and Ideas and the World Economy." *Economics of Innovation and New Technology* 19 (5): 393–406.

Freeman, R. B., and W. Huang. 2014. "Collaborating with People Like Me: Ethnic Co-authorship within the US." NBER Working Paper no. 19905, Cambridge, MA.

Guerrero Bote, Vicente P., Carlos Olmeda-Gómez, and Félix de Moya-Anegón. 2013. "Quantifying the Benefits of International Scientific Collaboration." *Journal of the American Society for Information Science and Technology* 64 (2): 392–404.

Jones, Benjamin F. 2011. "As Science Evolves, How Can Science Policy?" In *Innovation Policy and the Economy*, vol. 11, edited by Josh Lerner and Scott Stern, 103–31. Chicago: University of Chicago Press.

Jones, Benjamin F., S. Wuchty, and B. Uzzi. 2008. "Multi-University Research Teams: Shifting Impact, Geography, and Stratification in Science." *Science* 322 (5905): 1259–62.

Katz, J. S., and D. Hicks. 1997. "How Much Is a Collaboration Worth? A Calibrated Bibliometric Model." *Scientometrics* 40 (3): 541–54.

Lancho Barrantes, Bárbara S., Vicente P. Guerrero Bote, Zaida Chinchilla Rodríguez, and Félix de Moya Anegón. 2012. "Citation Flows in the Zones of Influence of Scientific Collaborations." *Journal of the American Society for Information Science and Technology* 63 (3): 481–89.

Lawani, S. M. 1986. "Some Bibliometric Correlates of Quality in Scientific Research." *Scientometrics* 9 (1–2): 13–25.

National Science Board. 2012. Science and Engineering Indicators 2012. http://www.nsf.gov/statistics/seind12/.

Olson, G. M., and J. S. Olson. 2000. "Distance Matters." *Human-Computer Interaction* 15 (2): 139–78.

———. 2003. "Mitigating the Effects of Distance on Collaborative Intellectual Work." *Economics of Innovation and New Technology* 12 (1): 27–42.

Rigby, John. 2009. "Comparing the Scientific Quality Achieved by Funding Instruments for Single Grant Holders and for Collaborative Networks within a Research System: Some Observations." *Scientometrics* 78 (1): 145–64.

Sauermann, H., and M. Roach. 2013. "Increasing Web Survey Response Rates in Innovation Research: An Experimental Study of Static and Dynamic Contact Design Features." *Research Policy* 42 (1): 273–86.

Scellato, G., C. Franzoni, and P. Stephan. 2012. "Mobile Scientists and International Networks." NBER Working Paper no. 18613, Cambridge, MA.

Simkin, M. V., and V. P. Roychowdhury. 2003. "Read before You Cite!" *Complex Systems* 14:269–74.

Thomson Reuters. 2012. Web of Science. http://wokinfo.com/.

Wuchty, S., B. F. Jones, and B. Uzzi. 2007. "The Increasing Dominance of Teams in Production of Knowledge." *Science* 316 (5827): 1036–39.

The (Changing) Knowledge Production Function
Evidence from the MIT Department of Biology for 1970–2000

Annamaria Conti and Christopher C. Liu

2.1 Introduction

Knowledge has been recognized as a major contributor to technological change and to economic growth (Romer 1990). In the knowledge production function, one of the most important inputs is knowledge created by university researchers. Indeed, a report by the National Science Board (2008) has revealed that university researchers are responsible for more than 70 percent of all scientific articles. Moreover, scholars have shown that academic knowledge is responsible for a large percentage of industrial innovations (Jaffe 1989; Mansfield 1995).

Academic knowledge has increasingly become a collective phenomenon. Seminal studies have documented the increase in the size of scientific collaborations, with special focus on the evolution of the geographic dispersion of team members (e.g., Adams et al. 2005; Wuchty, Jones, and Uzzi 2007). Even though university scientists collaborate more and more across research institutions, the scientific laboratory remains the major locus of knowledge production (Stephan 2012a). These laboratories are largely populated by graduate students and postdocs, whose contributions to their laboratory's knowledge stock have been recognized in a number of studies

Annamaria Conti is assistant professor of strategic management at the Scheller College of Business, Georgia Institute of Technology. Christopher C. Liu is assistant professor of strategic management at the Rotman School of Management, University of Toronto.

We are indebted to Nathan E. Gates, Adam Jaffe, Ben Jones, Paula Stephan, Marie Thursby, Fabian Waldinger, and seminar participants at the NBER Changing Frontier conferences (October 2012 and August 2013) for their valuable comments. We give particular thanks to the MIT Department of Biology. For acknowledgments, sources of research support, and disclosure of the author's or authors' material financial relationships, if any, please see http://www.nber.org/chapters/c13036.ack.

(see, for instance, Stephan 2012a; Conti, Denas, and Visentin 2014). These research trainees have coauthored an important percentage of their laboratory's papers and, moreover, have produced a considerable share of the articles published in highly ranked journals (Black and Stephan 2010).

In this study, we use a unique database that allows us to examine the productivity, training duration, and the collaborative behavior of graduate students and postdocs as well as the extent to which these aspects have changed over time. We interpret the patterns we find in light of two paradigms: the increased burden of knowledge that successive generations of scientists face (Jones 2009, 2010a) and the limited availability of permanent academic positions (Stephan 1996; Freeman et al. 2001).

Our data encompass the complete set of laboratories in the MIT Department of Biology, observed from 1970 to 2000. This department has been a major locus of basic and applied discoveries in the life sciences for the latter half of the twentieth century. Through the time frame of our data set, the scientists working at the MIT Department of Biology made discoveries as varied as the molecular mechanisms underpinning recombinant DNA (e.g., the discovery of splicing and introns), cell death, aging, and the progression of cancer. This work has resulted in six Nobel Laureates and forty-three members of the National Academy of Sciences between 1966 and 2000. MIT's Department of Biology has roughly doubled in size, from twenty-seven laboratories in 1966 to forty-nine laboratories in the year 2000. Given this department's elite status, the findings in this chapter may be difficult to extend beyond other elite North American laboratories. With this caveat in mind, we follow in the footsteps of other scholars and trade analytical depth with a focus on an elite setting (Azoulay, Zivin, and Wang 2010; Zuckerman 1977).

We collected a detailed set of information on the graduate students and postdocs who populated these laboratories, including their publication output. For the purposes of this study, we use this information to analyze the evolution over time of four fundamental aspects of their productivity: (a) training duration, (b) time to a first publication, (c) productivity over the training period, and (d) collaboration with other scientists.

We identified four main trends that are common to graduate students and postdocs. First, training periods have increased for later cohorts of graduate students and postdocs. Second, recent cohorts tend to publish their first article later than the earlier cohorts. Third, they produce fewer first-author publications. Finally, collaborations with other scientists, as measured by the number of coauthors on a paper, have increased. This increase is driven by collaborations with scientists outside of a trainee's laboratory.

The remainder of this study is organized as follows. Section 2.2 describes the empirical setting. Section 2.3 presents the scientific productivity trends for graduate students and postdocs. Section 2.4 concludes and discusses policy implications.

Table 2.1	**Personnel composition of Professor Baltimore's laboratory**
Professor:	David Baltimore
Visiting scientists:	Samuel Latt and Richard Van Etten
Postdoctoral associates:	Brygida Berse, Mark Feinberg, Michael Lenardo, Jing-Po Li, Shiv Pillai, Louis Staudt, and Xiao-Hong Sun
Postdoctoral fellows:	Raul Andino, Patrick Baeuerle, Andre Bernards, Lynn Corcoran, Sunyoung Kim, Towia Libermann, Ricardo Martinez, Mark Muesing, Cornelis Murre, Jacqueline Pierce, Stephen Smale, Didier Trono, Anna Voronova, and Astar Winoto
Technical assistants:	Ann Gifford, Carolyn Gorka, Patrick McCaw, Michael Paskind, and Gabrielle Rieckhof
Graduate students:	George Daley, Peter Jackson, Marjorie Oettinger, David Schatz, and Dan Silver
Undergraduate student:	Anna Kuang

2.2 Empirical Setting

Our core data source is the MIT Department of Biology's series of annual reports. The primary purpose of the annual report was to document, on a yearly basis, the department's internal activities. This information was then distributed to each member of the department, allowing individuals to be cognizant of their peers' scientific activities. To serve this purpose, the reports included both a roster of laboratory members that comprised the department, as well as technical summaries of ongoing projects. From 1966 to 1989, technical summaries were at the project level, which included both laboratory members affiliated with the project as well as a short project summary. The size of the annual report grew in accordance with the size of the department. After the annual report reached 629 pages in 1987, project summaries were condensed to two pages per laboratory, regardless of its size. In the year 2001, annual reports were no longer published, and our data set ceases at this point.

The annual report documents a roster of each laboratory's members. We know the names of every individual in each laboratory as well as the individual's personnel type (e.g., postdoc, graduate student, technician). As a result, we know the characteristics of the department, its laboratories, and its individual members over the duration of our data set. Table 2.1 provides an example of the roster data available for any given laboratory-year. We know of no other data source that provides as detailed a view into the organization of scientific work as this one.

We supplemented this departmental personnel roster with a number of other data sources. To examine scientific outputs, we hand collected each laboratory head's (i.e., principal investigator [PI]) paper output from Medline. We then matched each laboratory's extracted publication-author list with our personnel roster to examine the extent to which individual laboratory members contributed to the scientific output. In instances where matching

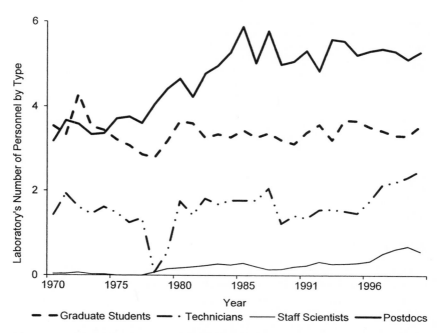

Fig. 2.1 Number of laboratory's personnel by type

was ambiguous (e.g., Liu), we examined the article directly. It is exceedingly rare for laboratory members to publish scientific papers without their PI listed as an author. Hence we do not believe we are missing any publications.

Overall, our data set comprises 1,494 laboratory-years and 20,324 laboratory member-years that span the period 1966–2000. Within this data set, there are 120 unique professors and 6,938 laboratory members who collectively produced 7,553 journal publications (in Medline).

We restrict our analysis to the years 1970–2000 as there was ambiguity in personnel categories prior to 1970. We begin with a description of the laboratories and their changes over time. We then turn our attention to examine the laboratory members with a particular emphasis on the two dominant personnel types, postdocs and graduate students, who comprise more than half of our personnel roster.

Within our data set, the average laboratory has ten members of whom five are postdocs, three are graduate students, and two are technicians. Staff scientists are rare, but their prevalence has increased over time. As shown in figure 2.1, laboratories have grown in size through the latter part of the twentieth century, and this increase has been driven by an increase in the number of postdoctoral scientists. There is no change in the number of graduate students or technicians over time, although the number of salaried staff (i.e., technicians and staff scientists) appears to have increased in the late 1990s.

Fig. 2.2 Number of laboratory's publications and impact factor-weighted publications

Figure 2.2 presents trends in scientific output for our laboratories. As shown, the average number of articles has steadily increased over time, from an average of four articles per laboratory-year in the 1970s to six articles per laboratory-year in the 1990s. We observe a very similar trend in the number of impact factor-weighted publications.

We focus our analysis of laboratory members on graduate students and postdocs for a number of reasons. First, these individuals make large contributions to a PI's publication output (Conti and Liu forthcoming). Their purpose is to directly produce scientific publications, rather than to play a supporting role (e.g., technicians). Second, these two types are the most prevalent personnel categories within the roster. Together they make up more than half of the laboratory. Third, these two personnel types have been the focus of recent interest in the literature because of their contributions to knowledge and technology production (e.g., Dasgupta and David 1994; Waldinger 2010). Lastly, we note that graduate students and postdocs are easily and unambiguously identified from one another, suggesting that the distinction in these roles may be salient (e.g., Azoulay, Liu, and Stuart 2014).

Our sample is composed of 991 graduate students and 2,427 postdocs. Figures 2.3A and 2.3B provide descriptive results of the distribution of graduate students and postdocs by their publication count. Interestingly, a significant proportion of them (about 35 percent) did not publish any articles during

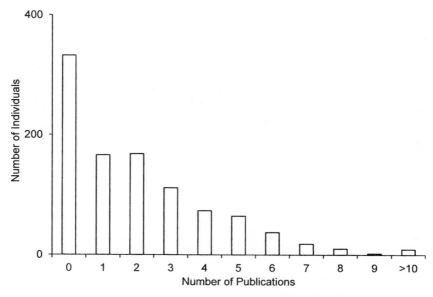

Fig. 2.3A Distribution of graduate students by their number of papers

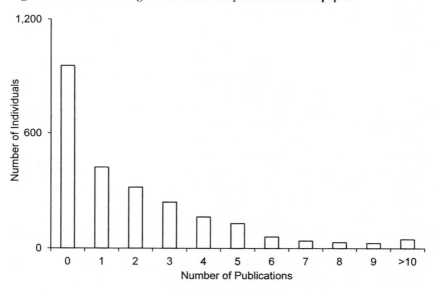

Fig. 2.3B Distribution of postdocs by their number of papers

their training period. Conditioned upon having published, the mean number of papers is about three articles for both graduate students and postdocs.

2.3 Trends in Scientific Productivity of Graduate Students and Postdocs

This section explores the trends in four major dimensions of graduate student and postdoc scientific productivity. First, we look at training duration. Second, we investigate the timing to a first publication. Third, we examine scientific output. Finally, we explore collaboration patterns.

In analyzing these trends, we should keep in mind that while both postdocs and graduate students are formally considered laboratory trainees, they fundamentally differ in a number of aspects. Postdocs are more experienced than graduate students and have accumulated a greater wealth of knowledge and skills. As a consequence, matching between postdocs and PIs is based upon prior ability and experience, rather than the future expectation of productivity as in the case of graduate students (Stephan 2012a).

2.3.1 Training Duration

We begin this section by presenting descriptive statistics for the average training duration of postdoc and graduate students over our sample period. We then investigate whether the length of training has changed over time. Figures 2.4A and 2.4B show the distribution of graduate students and postdocs by their training duration. That the training period for graduate students is longer than postdoctoral training is clearly evident. Indeed, the majority of graduate students in our sample completed their training between five and seven years, while postdocs tended to spend between two and four years in a PI's laboratory.[1]

Figure 2.5 documents training periods for graduate students (dotted line) and postdocs (solid line) over the period 1970–1995. We exclude the years 1996 through 2000 since students who enrolled in these years might not have completed their training by the end of 2000, when our data set is right-censored. Consistent with previous studies,[2] we find that training periods for recent cohorts of students are approximately one year longer than those for the earliest cohorts. The training period increases from three to approximately four years for postdocs and from five to six years for graduate students over our data set.

There are at least three reasons that can explain these trends. The first reason is that as knowledge accumulates, earlier trainee cohorts face a greater

---

1. It is possible for postdocs to have worked in more than one PI's laboratory before they are offered a faculty position. However, from discussions with MIT PIs as well as from an examination of a CV sample, it is evident that, at least for the period we examine, this is rarely the case for MIT postdocs.

2. See, for instance, the findings by Tilghman (1998), Jones (2009), Jones and Weinberg (2011), and Freeman et al. (2001).

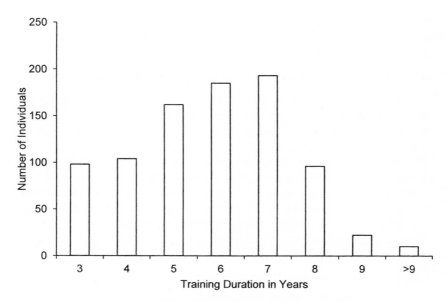

Fig. 2.4A Distribution of graduate students by their training duration

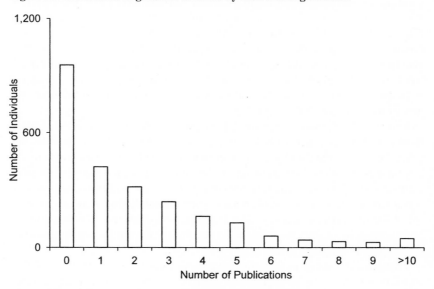

Fig. 2.4B Distribution of postdocs by their training duration

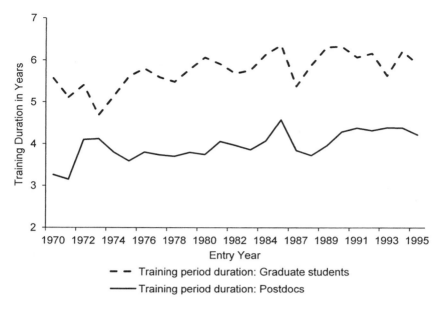

Fig. 2.5 Training duration for graduate students and postdocs over time

educational burden than do the older cohorts (Jones 2009, 2010a). Second, it is also possible that the recent cohorts of postdocs and graduate students tend to stay longer in their positions because of the increased mismatch between the trainees' supply and the availability of permanent academic positions (Stephan 1996; Freeman et al. 2001). Finally, one cannot exclude the possibility that the increased pressure on PIs to publish and apply for grants has led them to impose longer training periods on their students (Freeman et al. 2001).

To more formally assess the evolution of training periods over time, we estimate Poisson regression models, with robust standard errors, in which we relate the training duration of graduate students and postdocs to whether these trainees had enrolled during the following periods: (a) 1970–1979, (b) 1980–1989, and (c) 1990–1995. The distribution of students across enrollment periods is reported in table 2.2.

The equation we estimate is:

(1) $\quad y_i = \exp(\beta_1 D1980\text{–}1989 + \beta_2 D1990\text{–}1995 + v_i + \theta_i + \varepsilon_i),$

where y_i is training duration, measured in number of years. Moreover, D1980–1989 is an indicator variable that equals one if trainee i enrolled during 1980–1989 and equals zero otherwise. D1990–1995 equals one if trainee i enrolled during 1990–1995 and, similarly, equals zero otherwise. We omit the 1970–1979 indicator variable and use it as a reference. Hence,

Table 2.2   Distribution of graduate students and postdocs by enrollment period

	Graduate students	Postdocs
1970–1979	289	560
1980–1989	334	868
1990–1995	247	565
1996–2000	121	434

Table 2.3   Regression results for graduate student and postdoc training duration

	Graduate students		Postdocs	
	Coeff.	Coeff.	Coeff.	Coeff.
D1980–1989	0.103***	0.075**	0.111***	0.065*
	(0.028)	(0.033)	(0.032)	(0.036)
D1990–1995	0.128***	0.055	0.209***	0.143***
	(0.029)	(0.039)	(0.034)	(0.041)
Field FE	Yes	Yes	Yes	Yes
PI FE		Yes		Yes
R^2	0.01	0.04	0.01	0.03
N		870		1993

Note: We estimated Poisson models. Robust standard errors are in parentheses. For these analyses we only consider trainees who had enrolled before 1996.
***Significant at the 1 percent level.
**Significant at the 5 percent level.
*Significant at the 10 percent level.

the coefficients of β_1 and β_2 should be interpreted as the change in training duration relative to the duration of trainees enrolled in 1970–1979. When investigating training duration, it is important to consider the scientific field in which a laboratory operates. Different scientific fields use different tools and it is likely that trends in training durations vary across fields (Galison 1997). To account for field effects, we include a series of indicator variables, v_i, corresponding to the modal experimental organism used in each laboratory. Specifically, we generated indicators for protein biochemists, bacteriologists, unicellular systems (e.g., HeLa cells), genetic systems (e.g., yeast), rodents, and others (e.g., frogs). Finally, we include a set of PI dummies, θ_i, to control for variations in duration trends across laboratory heads (i.e., laboratory fixed effects).[3]

Table 2.3 presents the regression results for graduate student and postdoc training duration. For each trainee category, we first include biology field

3. In an alternative specification, we substituted the PI dummies with the PI five-year, presample stock of publications to compare cohorts of trainees from supervisors with similar characteristics.

fixed effects (column [1]) and, subsequently, we add PI fixed effects (column [2]). We begin by describing the results for graduate students and then for postdocs.

As table 2.3 shows, in the baseline model the dummies D1980–1989 and D1990–1995 have a positive and statistically significant coefficient. These results confirm the descriptive evidence that later cohorts of students take longer to complete their PhD than earlier cohorts (cohorts who enrolled during the 1970–1979 period). In the second column, we add PI effects and the magnitude of the coefficients declines, together with their statistical significance. This last result suggests that PI characteristics are a source of positive correlation between period dummies and training duration.

We find similar results for postdocs. The coefficients of the 1980–1989 and 1990–1995 period dummies are positive and statistically significant in both model specifications, although the magnitude and significance is, again, reduced with the inclusion of PI fixed effects.

To summarize, the results in this section suggest that training periods have increased in recent years for both graduate students and postdocs. While we cannot precisely disentangle the mechanisms behind these trends, we believe that increasing challenges imposed on recent trainees, in terms of increased educational burden or reduced availability of permanent academic positions, play an important role.

2.3.2 Time to a First Publication

A singular advantage to our data set is the ability to discern the year in which graduate students or postdocs *enter* their training, and begin to be at "risk" for being an author on an article. In this section, we use this aspect of the data set to focus on the time it takes trainees to publish their *first* article. We considered the time interval between a trainee's enrollment and first publication as the time it takes to acquire the knowledge to develop publishable findings. This interval then becomes a measure of trainee distance to the existing knowledge frontier. Figure 2.6 presents Kaplan-Meier estimates of the time to a first publication for postdocs and graduate students. As shown, the probability of publishing a paper in each training year appears to be higher for postdocs than for graduate students. This holds true even when we focus exclusively on first-author publications, which we take as a proxy for those projects to which trainees have given their greatest contribution.[4]

Once more, we are interested in the evolution of time to a first publication over our sample period, for both graduate students and postdocs. If the knowledge burden for the more recent cohorts is larger than that for the oldest ones, then we should expect that the time it takes to publish a first

---

4. For the sake of brevity, we do not show the results for first-author publications, but they are available upon request.

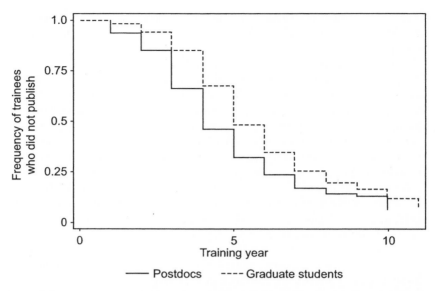

Fig. 2.6 Kaplan-Meier estimates of the time to a first publication: Graduate students and postdocs

article has increased for the most recent cohorts. There are other reasons to expect such a trend. One of these could be a lengthening of the review process at scientific journals. While this is a documented trend in the economic field (Ellison 2002), there are grounds for believing that this phenomenon is not confined to economic journals. As one example, statistics available for the *EMBO* journal reveal an increase over time in the number of days from submission to final decision.[5]

Figures 2.7 and 2.8 display Kaplan-Meier estimates of the time it takes to publish a first article, distinguishing between the following periods: (a) 1970–1979, (b) 1980–1989, and (c) 1990–2000. They provide evidence that the probability of publishing a paper at any given period is higher for the oldest cohorts than for the more recent ones. These trends seem to be more accentuated for postdocs than for graduate students. Moreover, for graduate students, they are more evident in first-author publications than they are in other publications.

Do these trends persist once we take into account field or PI characteristics, which are likely to be a source of correlation between enrollment periods and time to a first publication? Formally, we estimate a series of Cox proportional hazard models in which the hazard of publishing a first article is conditioned on a number of control variables (including period, field, and PI indicators) as above.

5. Statistics are available from http://www.nature.com/emboj/about/process.html.

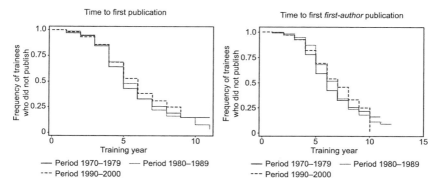

Fig. 2.7 Kaplan-Meier estimates of the time to a first publication: Graduate students over time

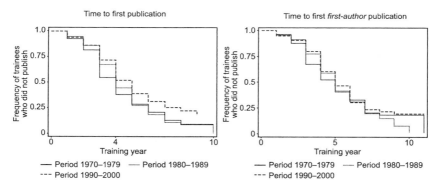

Fig. 2.8 Kaplan-Meier estimates of the time to a first publication: Postdocs over time

Specifically, we estimate the following equation:

$$(2) \qquad h(t|x_i) = h_0(t)\exp(x_{i}\beta_x),$$

where $h(t|x_i)$ is the hazard of publishing a first article, $h_0(t)$ is the baseline hazard (i.e., the hazard when all covariates are equal to zero), and x_i is a matrix of covariates. As in our previous equation, x_i includes period indicator variables as well as field and PI dummies. This time we also include in the sample trainees who had enrolled after 1995. Hence, the last period indicator variable equals one for trainees who had enrolled during 1990–2000 and zero otherwise. The results for graduate students are presented in table 2.4, while those for postdocs are in table 2.5. Standard errors are clustered around PI.

We begin by presenting the results for graduate students, distinguishing between the time to a first publication and the time to an initial first-author publication in table 2.4. Estimates are presented in terms of their effect on

Table 2.4   Hazard models for the time to a first publication: Graduate students over time

	Any publication		First-author publications	
	Hazard ratios	Hazard ratios	Hazard ratios	Hazard ratios
D1980–1989	0.969	0.768**	0.888	0.718***
	(0.121)	(0.095)	(0.110)	(0.091)
D1990–2000	0.837	0.650***	0.780*	0.613***
	(0.110)	(0.103)	(0.099)	(0.081)
Field FE	Yes	Yes	Yes	Yes
PI FE		Yes		Yes
Log likelihood	−4,121	−4,042	−3,467	−3,380
N		991		

Note: We estimate Cox proportional hazards models with standard errors clustered around PI. We report hazard ratios.
***Significant at the 1 percent level.
**Significant at the 5 percent level.
*Significant at the 10 percent level.

Table 2.5   Hazard models for the time to a first publication: Postdocs over time

	Any publication		First-author publications	
	Hazard ratios	Hazard ratios	Hazard ratios	Hazard ratios
D1980–1989	0.850***	0.788***	0.862	0.795**
	(0.061)	(0.056)	(0.083)	(0.075)
D1990–2000	0.665***	0.615***	0.658***	0.602***
	(0.061)	(0.061)	(0.071)	(0.062)
Field FE	Yes	Yes	Yes	Yes
PI FE		Yes		Yes
Log likelihood	−10,583	−10,478	−8,626	−8,517
N		2,427		

Note: We estimate Cox proportional hazards models with standard errors clustered around PI. We report hazard ratios.
***Significant at the 1 percent level.
**Significant at the 5 percent level.
*Significant at the 10 percent level.

the odds of publishing a first paper. Hence, a coefficient smaller (larger) than one reflects a negative (positive) effect. When we only include field fixed effects, the coefficients of the 1980–1989 and 1990–2000 period dummies are smaller than one, as expected, but not statistically significant. With the inclusion of PI fixed effects, the odds ratio of publishing a first paper decreases in magnitude (relative to the excluded reference category, 1970–1979) and is now statistically significant. This result, which is our preferred specification, indicates that trends in the time to a first publication vary across PIs.

When we examine first-author publications, we find additional evidence that the time to a first publication has increased for later cohorts of graduate students relative to earlier ones. Indeed, the coefficients of both period dummies are smaller than one and the coefficient for the 1990–2000 indicator is statistically significant. The coefficient magnitudes suggest that the hazard of publishing an initial first-author paper, for graduate students who enrolled in the 1980–1989 period, is 0.9 times the hazard of those who enrolled in the 1970–1979 period. It declines to 0.8 times for graduate students who enrolled during 1990–2000. As before, once we introduce PI fixed effects the significance of the coefficients improves and the magnitude declines.

In the case of postdocs, both the time to a first publication and that to an initial first-author publication appear to have increased for later cohorts relative to earlier ones. Across multiple regression specifications, the hazard of publishing a first paper is lower for postdocs who started in the 1980–1989 period, than for postdocs who enrolled during 1970–1979. And this downward, temporal shift in the hazard of publishing continues for those postdocs who started during 1990–2000. Moreover, the coefficients tend to be statistically significant with and without PI fixed effects.[6]

Taken together, we provide evidence that the time to an initial first-author publication has increased for both graduate students and postdocs and this trend line shows no evidence of leveling off. Moreover, in the case of postdocs, results indicate that the time to a first publication has increased even for non-first-author articles. As a complement to our prior results, that overall training periods have increased over time, increasing times to first publication suggest that, at least in part, recent cohorts of trainees require extra training time to "ramp-up" to the productive training periods.

2.3.3 Publication Trends

In this section, we turn our attention to trends in the overall publication output of graduate students and postdocs. The question we want to explore is whether recent cohorts of graduate students and postdocs have become less productive than older ones. Indeed, if one posits that recent cohorts of scientists face a larger learning burden or that the reviewing process at scientific journals has increased over time, then we should observe a declining trend in the publication output of graduate students and postdocs.

To investigate this hypothesis, we estimate count regression models in which we relate publication outputs that graduate students and postdocs had produced during their training as a function of whether their enrollment year falls within the 1970–1979, 1980–1989, or 1990–1995 periods. We adopt a Poisson specification with robust standard errors. We measure publication

---

6. In column (3) the coefficient for the 1980–1989 period dummy is not significant. However, a test of joint significance of period dummies rejects the null hypothesis that they are (jointly) equal to zero with a p-value of 0.00.

Table 2.6   **Regression results for graduate student publications**

	No. publications		No. first-author publications		Probability of publishing a first-author publication	
	Coeff.	Coeff.	Coeff.	Coeff.	Coeff.	Coeff.
D1980–1989	0.022	−0.103	−0.114	−0.241**	−0.009	0.071
	(0.083)	(0.103)	(0.087)	(0.104)	(0.038)	(0.048)
D1990–1995	−0.071	−0.257**	−0.221**	−0.499***	−0.084**	−0.203***
	(0.090)	(0.120)	(0.096)	(0.130)	(0.041)	(0.059)
Duration	Yes	Yes	Yes	Yes	Yes	Yes
Field FE	Yes	Yes	Yes	Yes	Yes	Yes
PI FE		Yes		Yes		Yes
R^2	0.05	0.15	0.03	0.12	0.105	0.28
N				870		

Note: Standard errors are in parentheses. For the Poisson models we use robust standard errors, while for the linear probability model we cluster standard errors around PI. For these analyses we only consider trainees who had enrolled before 1996.
***Significant at the 1 percent level.
**Significant at the 5 percent level.
*Significant at the 10 percent level.

output by counting the number of publications from the moment a trainee joins a PI laboratory until two years after the trainee was last observed in the laboratory. In this way, we account for the fact that there are lags between the moment a research project is completed and the moment its results are published. As for the analysis of training durations, we exclude the latest years because graduate students and postdocs who enrolled in these years might not have completed their training by the end of our sample period.

The equation we estimate is:

$$(3) \quad y_i = \exp(\beta_1 D1980\text{–}1989 + \beta_2 D1990\text{–}1995 + \beta_3 Duration_i + \nu_i + \theta_i + \varepsilon_i),$$

where y_i is either the total count of trainee i's publications or the count of their first-author publications. D1980–1989 is an indicator variable that equals one if trainee i enrolled during 1980–1989 and equals zero otherwise. D1990–1995 equals one if trainee i enrolled during 1990–1995 and, similarly, equals zero otherwise. Duration$_i$ is defined as the number of years a trainee has spent in a laboratory. Finally, ν_i and θ_i are field and PI fixed effects, respectively.

The results for graduate students are displayed in table 2.6, while those for postdocs are presented in table 2.7. When we consider the total publication count (column [1]), we find that graduate students who enrolled in more recent periods are no less productive than their colleagues who enrolled during 1970–1979. In fact, none of the coefficients for the 1989–1990 and 1990–1995 period dummies are statistically significant. Once we include

Table 2.7   **Regression results for postdoc publications**

	No. publications		No. first-author publications		Probability of publishing a first-author publication	
	Coeff.	Coeff.	Coeff.	Coeff.	Coeff.	Coeff.
D1980–1989	−0.160**	−0.250***	−0.174**	−0.255***	−0.018	0.019
	(0.067)	(0.071)	(0.068)	(0.074)	(0.026)	(0.029)
D1990–1995	−0.173**	−0.314***	−0.238***	−0.384***	−0.064**	−0.076**
	(0.075)	(0.086)	(0.076)	(0.089)	(0.028)	(0.035)
Duration	Yes	Yes	Yes	Yes	Yes	Yes
PI FE	Yes	Yes	Yes	Yes	Yes	Yes
Entry Year FE		Yes		Yes		Yes
R^2	0.10	0.18	0.08	0.14	0.22	0.23
N				1993		

Note: Standard errors are in parentheses. For the Poisson models we use robust standard errors, while for the linear probability model we cluster standard errors around PI. For these analyses we only consider trainees who had enrolled before 1996.

***Significant at the 1 percent level.
**Significant at the 5 percent level.
*Significant at the 10 percent level.

supervisor fixed effects, the coefficient of the dummy for student enrollment during 1990–1995 becomes statistically significant and has a negative sign. While this last result suggests that there are some supervisor characteristics that are correlated with productivity trends, we cannot conclude that there is a general declining tendency in the graduate student paper count. In support of this conjecture, descriptive evidence reported in figure 2.9 does not reveal a decreasing trend for the annual publication count. In regressions not reported here (but available upon request), we find very similar results when we use the impact-factor weighted publication count as the output measure.

We show different findings when analyzing first-author publications. In this case, both period dummies have a negative coefficient and the coefficient for the 1990–1995 period variable is significant, regardless of whether we include PI fixed effects. One might wonder whether this effect is driven by the fact that fewer graduate students are publishing first-author papers in recent years. To investigate this possibility, we estimate a linear probability model in which the dependent variable is an indicator that takes a value of one if graduate students have published at least one article during their training. The results are displayed in the last column of table 2.6. The coefficient for the 1990–1995 period dummy is negative and statistically significant, independent of the regression specification. These results suggest that at least part of the declining output trend is explained by a lower publishing probability for the most recent cohorts. Overall, we find that later graduate student cohorts

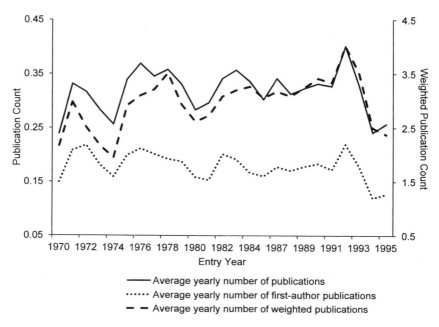

Fig. 2.9 Publication output of graduate student cohorts

Note: Counts normalized by duration.

produce fewer first-author articles than earlier ones and, this time, regression results seem to be supported by descriptive evidence reported in figure 2.9.

When we turn our attention to postdocs (table 2.7), we find strong evidence that the postdoc cohorts enrolled during 1980–1989 and 1990–1995 produce fewer articles than cohorts enrolled during 1970–1979. This result holds true regardless of whether we look at total or first-author publication counts. Indeed, the coefficients of our period dummies are negative and statistically significant, with and without PI fixed effects. When we analyze the probability of publishing at least one first-author paper, we find that part of the declining trend for the first-author paper count is explained by a lower publishing probability for the most recent cohorts. Overall, these findings are consistent with the descriptive trends presented in figure 2.10, which shows an over-time decline in publication outputs by postdoc students.

In analyses not presented here for the sake of brevity, we attempted to analyze whether the decline in the number of first-author graduate student publications was correlated with larger time intervals between papers, for subsequent publications. Thus we estimated hazard models for publishing a second first-author paper, conditioned on having published an initial one, and for publishing a third first-author paper, conditioned on having published a second. Because we have annual data, we cannot analyze the time

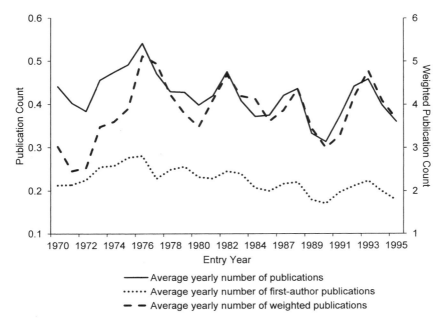

Fig. 2.10 Publication output of postdoc cohorts
Note: Counts normalized by duration.

interval between two papers published in the same year. With this caveat in mind, we find that the time intervals between first-author publications, subsequent to the first, are not larger for the most recent graduate student cohorts. This seems to suggest that the decline in the number of first-author papers for graduate students could be explained by the fact that trainees take longer to publish a first article or they publish fewer articles per year. Similar results were obtained when we estimated the hazard that postdoc students publish a paper or a first-author paper, conditioned on an initial publication.

To summarize, the results from this section lead us to infer that when we measure graduate student productivity by their first-author publication count, later cohorts appear to be less productive than earlier ones. As for postdocs, recent cohorts appear to be less productive in terms of both first-author and total paper counts.

2.3.4 Collaboration Trends

We have analyzed changes in both the training periods and scientific productivity of postdoc and graduate students from the 1970s through the 1990s. Although an array of mechanisms may explain these trend lines, a final question we pose is whether trainees have reacted to these challenges by working in larger teams, in a similar fashion to other researchers.

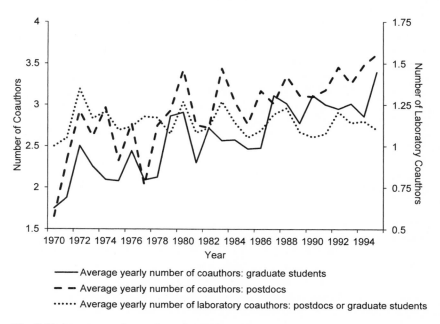

Fig. 2.11 Average yearly number of coauthors per paper

The benefits of teamwork have been extensively discussed in the economics literature and include output gains derived from labor specialization (Becker and Murphy 1992) and from the circulation of new ideas among team members (Adams et al. 2005). In the economics of science, scholars have found that scientists increasingly work in teams (Wuchty, Jones, and Uzzi 2007)[7] and that team size has expanded over time (Adams et al. 2005), largely due to an intensification of multiuniversity collaborations (Jones, Wuchty, and Uzzi 2008).

Figure 2.11 reports trends over time in the average number of coauthors per paper, distinguishing between postdocs and graduate students. In line with previous studies, we observe that for both trainee categories the average number of coauthors per paper has increased over time from approximately 1.5 at the beginning of the 1970s to approximately 3.5 by the second half of the 1990s. Interestingly enough, we also observe that the increased collaboration size was mainly driven by an increase in the number of outside laboratory coauthors.[8]

---

7. See also Agrawal and Goldfarb (2008) and Forman and Van Zeebroeck (2012).

8. In this case, we do not report regression results. The reason is that regression analyses must be conditioned on the sample of trainees who published at least one paper. The resulting sample size is quite limited and, thus, if we add PI and field fixed effects the regression estimates become imprecise. Nevertheless, the signs of the interest coefficients are the ones expected. Particularly, in the regression for the average number of coauthors on a paper, the coefficients of the dummies for the most recent decades are positive. Moreover, in the regression for the number of laboratory coauthors, the magnitude of the decade indicators' coefficients is almost zero, reflecting the trend that appears in figure 2.11.

Overall, this suggests that trainees, similar to other scientists across a broad range of disciplines, are increasingly working in teams and these teams tend to encompass authors from outside the focal trainees' laboratories.

2.4 Conclusions and Policy Implications

2.4.1 Summary

Knowledge production is widely considered to be one of the main determinants of economic growth. Within this domain, scientific knowledge production centered at universities, which results in codified outputs designed largely for dissemination and replication, is particularly important.

This study focuses on the contributions to academic knowledge by postdocs and graduate students. Using data from the MIT Department of Biology from 1970 to 2000, we looked at the evolution of four fundamental aspects of their productivity: (a) training duration, (b) time to a first publication, (c) productivity over the training period, and (d) collaboration with other scientists.

We identified four main trends that are common to both graduate students and postdocs. First, training periods have increased for later cohorts of research trainees. Second, recent cohorts tend to publish their initial first-author article later than the earlier cohorts. Third, they produce fewer first-author publications. Finally, collaborations with other scientists, as measured by the number of coauthors on a paper, have increased over time. This increase is driven by collaborations with scientists outside of a trainee's laboratory.

2.4.2 Interpreting the Results

What are the mechanisms that drive our results? Our findings are consistent with Jones's educational burden story (Jones 2009, 2010a), which states that as knowledge accumulates, future generations of scientists require greater effort (or more time) to absorb and build upon this accumulated knowledge base. To offset this trend, one possibility is for individuals to specialize, narrowing their field of expertise. A second consequence of specialization may be the need to broaden patterns of collaboration with other scientists. Our first three results—longer training periods, longer time to publish, lower productivity for later trainee cohorts—could be interpreted as an indication that the knowledge burden has increased, particularly for recent trainees. The final result regarding increased trainee collaboration provides an indication that these cohorts have become more specialized, although other possibilities abound.

While the educational burden story is a compelling explanation, we nevertheless think that other mechanisms might also be responsible for our results. One of these mechanisms is the mismatch between the supply of trainees and the availability of posttraining academic positions that scholars have discussed in recent decades (Stephan 2012b; Freeman et al. 2001). Data from the NSF-NIH Graduate Students and Postdoctorates in Science and Engineering survey shows that enrollment into PhD life science programs has increased by 80 percent between 1972 and 2005.[9] While we do not have information on the availability of posttraining positions, it is plausible that selection into (desirable) postdoctoral positions has become harder over time. Lastly, we also should note that longer training periods certainly benefit and are encouraged by PIs. Specifically, many PIs are reluctant to allow their most productive laboratory members (i.e., high-tenure trainees) to depart. In fact, a PI's compensation is, increasingly, linked to a tournament model in which seminal laboratory member (i.e., trainee) contributions are essential (Freeman et al. 2001).

If market frictions were to be responsible for longer training periods, should we also expect them to explain the lower productivity of recent trainee cohorts and their increased propensity to work in collaboration with other scientists? Is it plausible to posit that market disequilibria last for decades? Why is the market not redirecting the excess supply of trainees to other fields?

To answer the first question, one might consider that the excess supply of scientists has led to an increase in academic journal submissions, without a corresponding increase in the number of publications. If there is an excess supply of submissions, then the direct consequence is that publishing becomes more difficult, which might explain the lower productivity of recent trainee cohorts. Moreover, specialization and collaboration become ways of dealing with market disequilibria and one wonders whether the reduction in recent cohort productivity could have been even more accentuated had recent trainees not worked with other scientists. This mechanism is not necessarily in contrast with the educational burden explanation; rather, it offers a complementary perspective. In fact, market imbalances might act as a stimulus for scientists to expand the knowledge frontier in order to publish, thus increasing the burden on future generations.

While the mechanisms we have highlighted seem to be plausible, one cannot exclude the possibility that the mismatch between the supply of trainees and the availability of academic positions has led the brightest students to shy away from careers in the life sciences. Thus, the increase in training periods and the reduced productivity of the most recent cohorts is a reflection of their lower quality skills. With our current data set, we cannot disentangle these possibilities.

To answer the second and third questions regarding the duration of market imbalances, we should refer to studies by Freeman et al. (2001) and Stephan (2012a) and mention that, increasingly, PhD programs in life science, among others, tend to be populated by foreign students. Indeed, while some domes-

9. Data is available from https://webcaspar.nsf.gov/.

tic students might be discouraged from continuing their studies in the life sciences PhD programs, American PhD programs remain attractive to foreign students not only because of their prestige, but also because salary differentials between foreign countries and the United States are typically large. To verify that the proportion of foreign graduate students in the MIT Department of Biology has increased over time, we examined our trainees' first and last names. We then codified those who had a Chinese last name as well as those with an Italian or French first and last name.[10] We found that the proportion of Asian, Italian, or French trainees has increased from 17 percent in 1970 to 27 percent in 1995. While these figures are only suggestive, given that we cannot distinguish between foreign or native-born students, they provide some indication that foreign trainees have recently become an increasingly large proportion of the trainee population.

2.4.3 Policy Implications

Ultimately, this chapter has served to document the mechanisms underlying two important trends in the scientific community: the increasing duration of scientist trainees and an increasing propensity for collaborative activity (e.g., Agrawal, McHale, and Oettl [chapter 3, this volume]; Tilghman 1998). Additionally, we have provided evidence of a decline in the scientific output of recent trainees. What implications do these trends have for the scientific community?

First, we note the remarkable consistency linking changes in graduate students and postdocs training with an array of outcomes (i.e., training duration, time to first publication, and productivity). It is very possible that the act of doing science has become more difficult over time. To be productive, there is more knowledge that must be learned. It is possible that few policy changes can offset these trends, and that longer training durations are a necessary byproduct of scientific advances in the twentieth century.

Second, institutional constraints may be at play. Even for established scientists (i.e., PIs), increasing difficulties in acquiring funding may cascade through the scientific system in multiple ways. To increase efficiency, PIs may hold on to their productive students longer. Faced with increasing uncertainty over funding, universities may hire only the most experienced and productive postdocs. And lastly, each of these processes may feed back on one another over time.

Third, regardless of the reasons for the observed trends, it is important to note that the costs of science have increased (Jones 2010b). These costs are paid by the individual, who must endure longer training and uncertain future prospects, as well as by society at large, which does not recuperate the returns from its investment. As previous scholars have highlighted (Jones 2010b; Stephan 2012b), costs can be reduced by ensuring that graduate

---

10. Given the authors' backgrounds, we found it easiest to codify these student ethnicities.

students and postdocs receive adequate pedagogical support during their training period. This, in turn, improves the efficiency of trainee learning and may serve to offset increases in learning burdens. Moreover, decision makers could cap the trainee teaching load, thereby ensuring that the majority of their time is dedicated to research.

It is also worth mentioning that, as the pre-PI career path for life scientists has become incredibly long, talented scientists may increasingly choose to opt out. Our data illustrate that total trainee duration has crested ten years and this evidence is not unique to the MIT Department of Biology and to elite institutions (Stephan 2012a). Longer training duration raises the opportunity costs of a scientific career and makes other occupations more attractive. Thus, if employment in other fields entails shorter training periods, lower uncertainty and higher salaries, we may increasingly see a shift from the careers where graduate training is the passkey to the profession, such as the life sciences, toward other, equally rewarding careers (e.g., engineering).

Given fecundity differentials across the sexes, increasing training durations may affect women more severely than men, further exacerbating issues of female participation in the sciences (Ding, Murray, and Stuart 2006). As training increasingly comes to dominate individuals in their thirties, work-life balance issues, including considerations such as family constraints and career uncertainty (Kaminski and Geisler 2006) may come to dominate. Certainly, longer training durations do not help ease these concerns.

We conclude with a final important issue that has attracted the attention of recent scholars, namely the allocation of research credit in collaborations (Bikard, Murray, and Gans 2013). Working in teams entails a trade-off. On the one hand, teamwork seems to produce more knowledge breakthroughs than solo work (Singh and Fleming 2010). On the other, it involves costs, some of which are related to the assessment of the team members' contributions (Dasgupta and David 1994). This trade-off is especially relevant for trainees given that access to tenure-track positions requires that they be able to prove their ability to conduct impactful independent research.

References

Adams, J. D., G. C. Black, J. R. Clemmons, and P. E. Stephan. 2005. "Scientific Teams and Institutional Collaborations: Evidence from US Universities, 1981–1999." *Research Policy* 34 (3): 259–85.

Agrawal, A., and A. Goldfarb. 2008. "Restructuring Research: Communication Costs and the Democratization of University Innovation." *American Economic Review* 98 (4): 1578–90.

Azoulay, P., C. C. Liu, and T. Stuart. 2014. "Social Influence Given (Partially) Deliberate Matching: Career Imprints in the Creation of Academic Entrepreneurs." Working Paper. http://pazoulay.scripts.mit.edu/docs/imprinting.pdf.

Azoulay, P., J. G. Zivin, and J. Wang. 2010. "Superstar Extinction." *Quarterly Journal of Economics* 125 (2): 549–89.

Becker, G. S., and K. M. Murphy. 1992. "The Division of Labor, Coordination Costs, and Knowledge." *Quarterly Journal of Economics* 107 (4): 1137–60.

Bikard, M., F. Murray, and J. Gans. 2013. "Exploring Tradeoffs in the Organization of Scientific Work: Collaboration and Scientific Reward." *NBER Working Paper* no. 18958, Cambridge, MA.

Black, G. C., and P. E. Stephan. 2010. "The Economics of University Science and the Role of Foreign Graduate Students and Postdoctoral Scholars." In *American Universities in a Global Market*, edited by Charles T. Clotfelter, 129–61. Chicago: University of Chicago Press.

Conti, A., O. Denas, and F. Visentin. 2014. "Knowledge Specialization in PhD Students Groups." *IEEE Transactions on Engineering Management* 61 (1): 52–67.

Conti, A., and C. C. Liu. 2014. "Bringing the Lab Back in: Personnel Composition and Scientific Output at the MIT Department of Biology." *Research Policy*.

Dasgupta, P., and P. A. David. 1994. "Toward a New Economics of Science." *Research Policy* 23:487–521.

Ding, W. W., F. Murray, and T. E. Stuart. 2006. "Gender Differences in Patenting in the Academic Life Sciences." *Science* 313 (5787): 665–67.

Ellison, G. 2002. "The Slowdown of the Economics Publishing Process." *Journal of Political Economy* 110 (5): 947–93.

Forman, C., and N. Van Zeebroeck. 2012. "From Wires to Partners: How the Internet has Fostered R&D Collaborations within Firms." *Management Science* 58 (8): 1549–68.

Freeman, R., E. Weinstein, E. Marincola, J. Rosenbaum, and F. Solomon. 2001. "Competition and Careers in Biosciences." *Science* 294:2293–94.

Galison, P. 1997. *Image and Logic: A Material Culture of Microphysics.* Chicago: University of Chicago Press.

Jaffe, B. A. 1989. "Real Effects of Academic Research." *American Economic Review* 79 (5): 957–70.

Jones, B. F. 2009. "The Burden of Knowledge and the 'Death of the Renaissance Man': Is Innovation Getting Harder?" *Review of Economics and Statistics* 76 (1): 283–317.

———. 2010a. "Age and Great Invention." *Review of Economics and Statistics* 92 (1): 1–14.

———. 2010b. "As Science Evolves, How Can Science Policy?" In *Innovation Policy and the Economy*, vol. 11, edited by Josh Lerner and Scott Stern, 103–31. Chicago: University of Chicago Press.

Jones, B. F., and B. A. Weinberg. 2011. "Age Dynamics in Scientific Creativity." *Proceedings of the National Academy of Sciences* 108 (47): 1–5.

Jones, B. F., S. Wuchty, and B. Uzzi. 2008. "Multi-University Research Teams: Shifting Impact, Geography, and Stratification in Science." *Science* 322 (5905): 1259–62.

Kaminski, D., and C. Geisler. 2006. "Survival Analysis of Faculty Retention in Science and Engineering by Gender." *Science* 335 (6070): 864–66.

Mansfield, E. 1995. "Research Underlying Industrial Innovations: Sources, Characteristics, and Financing." *Review of Economics and Statistics* 77 (1): 55–65.

National Science Board. 2008. Science and Engineering Indicators. http://www.nsf.gov/statistics/seind/.

Romer, P. M. 1990. "Endogenous Technological Change." *Journal of Political Economy* 98 (5): 71–102.

Singh, J., and L. Fleming. 2010. "Lone Inventors as Sources of Breakthroughs: Myth or Reality?" *Management Science* 56 (1): 41–56.

Stephan, P. 1996. "The Economics of Science." *Journal of Economic Literature* 34 (3): 1199–235.

———. 2012a. *How Economics Shapes Science*. Cambridge, MA: Harvard University Press.

———. 2012b. "Perverse Incentives." *Nature* 484:29–31.

Tilghman, S. 1998. "Trends in the Early Careers of Life Sciences." Report by the Committee on Dimensions, Causes, and Implications of Recent Trends in the Careers of Life Scientists, National Research Council. Washington, DC: National Academy Press.

Waldinger, F. 2010. "Quality Matters: The Expulsion of Professors and the Consequences for PhD Student Outcomes in Nazi Germany." *Journal of Political Economy* 118 (4): 787–831.

Wuchty, S., B. F. Jones, and B. Uzzi. 2007. "The Increasing Dominance of Teams in Production of Knowledge." *Science* 316 (5827): 1036–39.

Zuckerman, H. 1977. *Scientific Elite: Nobel Laureates in the United States*. New York: Free Press.

Collaboration, Stars, and the Changing Organization of Science
Evidence from Evolutionary Biology

Ajay Agrawal, John McHale, and Alexander Oettl

3.1 Introduction

The spatial organization of science is undergoing a fundamental transformation. New patterns of institutional participation, division of labor, and star scientist centrality are emerging. Given the essentially combinatory nature of invention and innovation, changes in organization that affect access to knowledge and ease of collaboration to produce new knowledge are potentially of great importance to aggregate technological progress and economic growth.[1]

In this chapter, we document and discuss significant changes in the spatial organization of science over recent decades in the field of evolutionary biology. Specifically, we identify two trends that appear contradictory at first glance. First, we find that the concentration of scientific output at the institution level is falling. More institutions are participating in scientific research over time and relatively more activity is migrating to lower-ranked institu-

Ajay Agrawal is the Peter Munk Professor of Entrepreneurship and associate professor of strategic management at the Rotman School of Management of the University of Toronto and a research associate of the National Bureau of Economic Research. John McHale is Established Professor and Head of Economics at the National University of Ireland, Galway. Alexander Oettl is assistant professor of strategic management at the Georgia Institute of Technology.

This research was funded by the Centre for Innovation and Entrepreneurship at the Rotman School of Management, University of Toronto and the Social Sciences and Humanities Research Council of Canada. We thank Adam Jaffe, Ben Jones, Julia Lane, and seminar participants at the NBER preconference workshop on "The Changing Frontier: Rethinking Science and Innovation Policy" as well as at the University of Toronto for valuable feedback. Errors remain our own. For acknowledgments, sources of research support, and disclosure of the authors' material financial relationships, if any, please see http://www.nber.org/chapters/c13038.ack.

1. For influential work that emphasizes the role of combining ideas in the generation of new knowledge, see Romer (1990), Jones (1995), Weitzman (1998), and Mokyr (2002).

tions, broadening the base of science. At the same time, however, we find that the concentration of scientific output at the individual level is increasing. Publications and citations have always been highly skewed toward star performers. However, the relative importance of stars has increased in recent decades.

What could explain these seemingly contradictory trends? Collaboration offers a possible explanation. An increase in collaborative activity could broaden the base of science at the department level by raising the relative amount of participation by previously lesser-involved research institutions and at the same time increase the concentration of output at the individual level by disproportionately benefiting highly productive scientists, perhaps through more efficient matching with collaborators, thus enabling more finely grained specialization. Stars may disproportionately benefit from better matching because they have a larger pool of potential distant collaborators to choose from. We report descriptive evidence that is consistent with these conjectures.

Specifically, we report evidence that the level of collaboration has increased significantly over time. Furthermore, the average distance between collaborators has grown in terms of both physical distance and the rank separation of collaborating institutions, consistent with the conjecture that collaboration plays a role in the expanding base of science. We further show that the base of institutional participation has grown, including the entry of institutions from emerging economies. Moreover, we show that star scientists have an increasingly larger pool of potential collaborators relative to nonstars, consistent with the assertion that they may disproportionately benefit from lower communication costs due to better matching opportunities.

Why might collaboration be increasing? We see two central forces that increase the returns to collaboration although also work in opposing directions on the returns to colocation. The first is the rising "burden of knowledge" (Jones 2009). The increasing depth of knowledge required to work at the scientific frontier is leading to increasing returns to specialization. This in turn raises the returns to collaboration, given the need to combine ideas and skills to produce new ideas. Furthermore, to the extent that colocation lowers the cost of collaboration, the rising burden of knowledge increases the returns to colocation. Agrawal, Goldfarb, and Teodoridis (2013) report evidence consistent with the knowledge burden hypothesis. Utilizing the sudden and unexpected release of previously hidden knowledge caused by the collapse of the Soviet Union as a natural experiment, the authors find that an outward shift in the knowledge frontier does indeed cause an increase in collaboration.

The second force is the improvement in collaboration-supporting technologies that reduce the barriers created by distance such as e-mail, low-cost conferencing, and file-sharing technologies (Agrawal and Goldfarb 2008; Kim, Morse, and Zingales 2009). All else equal, these advances allow for a greater physical dispersal of collaborating scientists. So, although the declining cost of communication may decrease the returns to colocation, it increases the returns to collaboration. Therefore, despite the potential conflict between these two forces with respect to their impact on the relative returns to colocation, they both increase the returns to collaboration.

We develop a simple model that provides a potential unified explanation for these facts. The key idea behind the model is that stars have a larger set of potential collaborators to choose from—perhaps due to having more former graduate students—and thus have the potential to gain disproportionately from improvements in collaboration technology. Moreover, some of these cross-institutional collaborations may occur with scientists from lesser-ranked institutions, consistent with a broadening institutional base in the production of science.

Drawing on parallel work on the causal impact of star scientists on departmental performance (Agrawal, McHale, and Oettl 2013), we speculate on the efficiency of the emerging spatial distribution of scientific activity. Recognizing the existence of knowledge, reputational, and consumption externalities associated with the location decisions of star scientists, we make no presumption that the resulting spatial distribution of stars is efficient. We find that stars attract other stars and also that the recruitment of a star can have positive effects on the productivity of colocated incumbents working in areas related to the star. These effects appear to be particularly strong when recruitment takes place at non-top-ranked institutions.

However, while strong forces may lead to star agglomeration due to localized externalities, lower-ranked institutions may have strong incentives to compete for stars as a core part of strategies aimed at climbing the institutional rankings. We document significant movement up the rankings for a select set of institutions that begin outside the top-ranked institutions. Congestion effects may also exist from star colocation due to clashing egos and increasing returns to "vertical collaboration" across skill sets located at different institutions. Overall, we find a tendency toward reduced concentration of the field's best scientists at its top-ranked institutions. In addition, the increasing propensity to collaborate across institutions, particularly across institutions of significantly different rank, further diminishes the concentration of knowledge production. Thus, fears of excessive concentration of stars due to positive sorting might be overblown, although research on the normative implications of the observed changes in the organization of science is still at an early stage.

We structure the rest of the chapter as follows. In section 3.2, we explain the construction of our evolutionary biology data at the institutional and individual levels. In section 3.3, we report evidence of the broadening institutional and international base of scientific activity. We describe the increasing concentration of individual productivity and, in particular, the rising importance of stars in section 3.4. In section 3.5, we present data on the overall rise of collaborative activity and the change in collaboration patterns. In section 3.6, we develop a model that offers a potential unified explanation of the facts documented in previous sections. Finally, in section 3.7 we provide a more speculative discussion of possible normative implications of these participation, concentration, and collaboration patterns, with an emphasis on the role of the location of stars and the increasing propensity to collaborate across institutions.

3.2 Data

Our study focuses on the field of evolutionary biology, a subfield of biology concerned with the processes that generate diversity of life on earth. Although some debate exists among historians of science and practicing evolutionary biologists over the key early contributors to this discipline, the general consensus remains that *On the Origin of Species by Means of Natural Selection*, authored by Charles Darwin in 1859, is the foundational text of this field. As in most fields of science, research in evolutionary biology consists of both theoretical and experimental contributions. In addition to specializing in particular topic areas, empiricists often specialize in working with particular organisms such as *Macrotrachela quadricornifera* (rotifer), *Drosophila melanogaster* (fruit fly), and *Gasterosteus aculeatus* (three-spined stickleback fish). The returns to species specialization result from, for example, the upfront costs of learning how to work with a particular species (including, in many cases, learning where to find them and how to catch and care for them to facilitate reproduction in order to observe, for instance, the variation in genotypes and phenotypes of offspring over multiple generations) as well as setting up the infrastructure in a lab or in nature to study them.

3.2.1 Defining Evolutionary Biology

Defining knowledge in evolutionary biology is not straightforward. On the input side, evolutionary biology, as in many areas of science, draws from many fields, such as statistics, molecular biology, chemistry, genetics, and population ecology. Furthermore, on the output side, some of the most influential papers are published in general interest as opposed to field-specific journals. Therefore, identifying the set of papers that comprise the corpus of the field is complicated because although every paper in the *Journal of Evolutionary Biology* is probably relevant, most papers in *Science* and *Nature* are not, although a significant fraction of the field's most important papers are published in those latter two journals.

Therefore, we follow a three-step process for defining "evolutionary biology papers." First, using bibliometric data from the ISI Web of Science, we collect data on all articles published during the twenty-nine-year period of 1980 through 2008 in the journals associated with the four main societies that focus on the study of evolutionary biology: the Society for the Study of Evolution, the Society for Systematic Biology, the Society for Molecular Biology and Evolution, and the European Society of Evolutionary Biology. Their respective journals are: *Evolution, Systematic Biology, Molecular Biology and Evolution*, and *Journal of Evolutionary Biology*. We focus on these four society journals because every article published within them is relevant to evolutionary biologists. In other words, unlike general interest journals such as *Science, Nature*, and *Cell*, which include papers from evolutionary biology but also research from many other fields, these four journals focus specifically on our field of interest. This process yields 15,256 articles.

Second, we collect all articles that are referenced at least once by these 15,526 society journal articles. There are 149,497 unique articles that are referenced at least once by the set of 15,256 evolutionary biology society articles. This set of 149,497 articles includes, for example, papers that are important to the field but are published outside the four society journals, such as key evolutionary biology papers published in *Science* that are cited, likely multiple times, by articles in the four society journals. We call this set of 149,497 papers the corpus of influence because each of these articles has had impact on at least one "pure" evolutionary biology article.

Third, we citation-weight the corpus of influence. We do this by counting the references to each of the 149,497 articles from the original 15,256 society journal articles. There are 501,952 references from the 15,256 society journal articles. So, on average, articles in the corpus are cited 3.4 times. Unsurprisingly, the distribution of citations is highly skewed. The minimum number of citations is one (by construction), the median is one, and the maximum is 906.[2] For most of the analyses in this chapter, we use counts of citation-weighted publications. When we do so, we use the 149,497 articles weighted by the 501,952 society article references.

3.2.2 Identifying Authors

We follow several steps to attribute the 149,497 articles in the corpus of influence to individual authors. The reason this process requires several steps is that authors are not uniquely identified and therefore name disambiguation is necessary. In other words, when we encounter multiple papers authored by James Smith, we need to determine whether each is written by the same James Smith or if instead these are different people with the same name. This process is made more challenging because until recently the ISI Web of Science only listed the first initial, a middle initial (if present), and the last name for each author. Is J Smith the same person as JA Smith? Name disambiguation is particularly important for properly assessing researcher productivity over time and changing collaboration patterns.

2. This paper is "The Neighbor-Joining Method—A New Method for Reconstructing Phylogenetic Trees" published in *Molecular Biology and Evolution* (1987) by Saitou Naruya (University of Tokyo) and Masatoshi Nei (University of Texas).

To address this issue, we employ heuristics developed by Tang and Walsh (2010). The heuristic utilizes backward citations of focal papers to estimate the likelihood of the named author being a particular person. For example, if two papers reference a higher number of the same papers (weighted by how many times the paper has been cited, i.e., how popular or obscure it is), then the likelihood of those two papers belonging to the same author is higher. We attribute two papers to the same author if both papers cite two or more rare papers (fewer than fifty citations) in both papers. We repeat this process for all papers that list nonunique author names (i.e., same first initial and last name). We exclude scientists who do not have more than two publications linked to their name.

Overall, 171,428 authors are listed on the 149,497 articles. We drop 140,240 names because they do not have more than two publications linked to their name. Employing the process described above, we assign the remaining 31,188 author names to 32,955 unique authors (a single name may map to more than one person). We conduct our analyses using these 32,955 authors. It is important to note that this is the total number of scientists in our sample over the twenty-nine-year period, but that the number of active scientists varies from year to year. Unsurprisingly, the output produced by these authors is highly skewed. Considering the overall period of our study, the minimum number of publications per author is three (by construction), the median is four, the mean is 7.5, and the maximum is 210 (Professor Rick Shine at the University of Sydney).

We use citation-weighted paper counts per year as our primary measure of author output. We treat as equal every paper on which a scientist is listed as an author. In other words, we do not distinguish between a paper on which a scientist is one of two authors from one on which they are one of three authors. An alternative approach is to use fractional paper counts where half a paper unit is attributed to the focal author in the former case and a third in the latter. Although we report results using the former approach, we conduct our analysis using both approaches. The results are qualitatively similar.

In certain analyses, we refer to the top 100 (or 200, or 50) scientists. When we do so, we determine ranking by the accumulated stock of citation-weighted output over the preceding years. When we refer to "stars," we are referring to scientists in the 90th percentile in a given year in terms of their accumulated stock of citation-weighted paper output over the preceding years. We provide a more detailed explanation of how we identify stars and related features of the data in our companion paper that focuses on stars and that uses the same data (Agrawal, McHale, and Oettl 2013).

3.2.3 Identifying Scientist Locations

Using the unique author identifiers we generate in the process described above for each evolutionary biology paper, we then attribute each scientist to a particular institution for every year they are active. A scientist is active from the year they publish their first paper to the year they publish their last paper. Here again, we must overcome a data deficiency inherent within the ISI Web of Science data; until recently, the Web of Science did not link institutions listed on an article to the authors. Instead, we impute author location using reprint information that provides a one-to-one mapping between the reprint author and the scientist's affiliation. In addition, we take advantage of the fact that almost 57 percent of evolutionary biology papers are produced with only a single institution listing. Thus, we are able to directly attribute the location of all authors on these papers to the focal institution. This method of location attribution is more effective for evolutionary biology than for many other science disciplines since articles in this field are generally produced by smaller-sized teams relative to other disciplines in the natural sciences (3.32 average number of authors per paper).

Overall, we are able to attribute 78.9 percent of the 32,955 unique authors to an institution. We drop institutions that do not produce at least one publication in each of the twenty-nine years under study. This results in the identification of 255 institutions that actively produce new knowledge in the field of evolutionary biology throughout our study period. Although we refer to these as "departments," they are actually the set of authors at an institution (e.g., Georgia Tech) who publish at least three articles that we categorize as being part of the corpus of influence in evolutionary biology during the study period. In other words, these individuals may not all formally belong to the same department within the institution. Again, the output of departments is highly skewed. Over the twenty-nine-year period, the minimum number of publications per institution per year is one (by construction), the median is eleven, the mean is 17.7, and the maximum is 181 (Harvard University in 2005).

3.3 Participation: A Broadening Base

The first trend in the organization of evolutionary biology we document is a decline in the skew of the distribution of output across institutions. This may reflect: (a) an increasing emphasis in knowledge production across previously lesser-producing institutions that are now more concerned about rankings and thus increasingly emphasizing research output as a factor in promotion and tenure, (b) changing preferences of faculty who have spent more time than their predecessors developing specialized research expertise, and/or (c) mounting political pressure to distribute government funding more evenly across institutions and political jurisdictions. In addition, we find a dramatic increase in research activity in emerging economies, possibly reflecting a broader movement toward higher value-added activities as part of the economic development process.

In figures 3.1, 3.2, and 3.3, we report evidence of the broadening base of science in terms of the declining department-level concentration of sci-

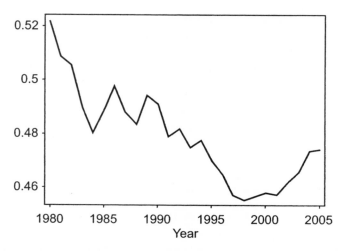

Fig. 3.1 Gini coefficients by year for the distribution of scientists across departments

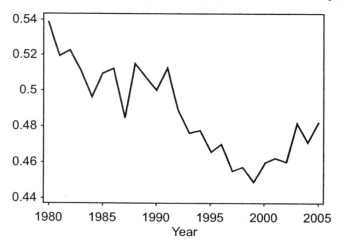

Fig. 3.2 Gini coefficients by year for the distribution of publications across departments

entists, publications, and citations, respectively. Specifically, we plot Gini coefficients to illustrate the distribution of scientists (publications, citations) across departments by year. The pattern of falling concentration is pronounced for the period between 1980 and 2000, although there is some indication of a turnaround in this pattern after 2000.

In figure 3.4, we plot department-level Lorenz curves for publications and citations. These curves illustrate the overall shift in the distribution over time. For example, the top 20 percent of departments produce 60 percent of all publications in 1980, but only 50 percent in 2000. Similarly, the

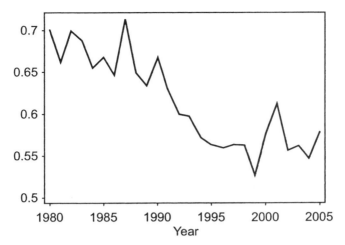

Fig. 3.3 Gini coefficients by year for the distribution of citations received across departments

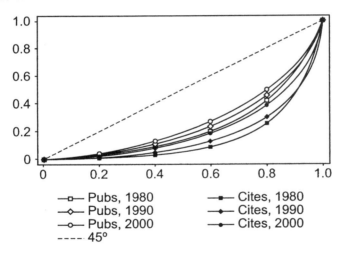

Fig. 3.4 Lorenz curves by department for publications and citation-weighted publications

top 20 percent produce 75 percent of all citation-weighted publications in 1980, but only 60 percent in 2000. It is important to note that we use a balanced panel for these analyses, including only the 255 institutions publishing in evolutionary biology throughout the period under study. In other words, we do not allow for entry of new institutions part way through the study period. Since most institutions that are ever meaningful contributors to this field are active throughout our study period, this is not a serious restriction.

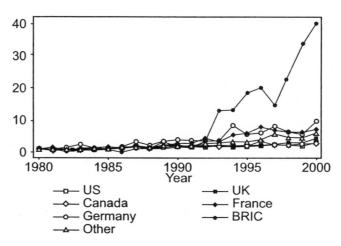

Fig. 3.5 Publication count by country by year normalized relative to output in 1980

However, we relax the no-entry restriction for the data we use in the next graph where we plot output by country because several institutions in previously low-income countries are not active in the early years, but have since become increasingly important in the overall production of knowledge. We plot the increasing importance of institutions based in emerging markets in figure 3.5. The growth rate of publications from institutions based in BRIC countries (Brazil, Russia, India, and China) begins to rise dramatically from the early 1990s onward and increases fortyfold by 2000. However, in absolute terms, the BRIC countries are still minor knowledge producers in this field relative to the leading nations, such as the United States, the United Kingdom, France, Germany, and Canada.

These decentralization findings from university-based research in evolutionary biology are consistent with prior findings on the decentralization of innovative activity more broadly (Rosenbloom and Spencer 1996; Bresnahan and Greenstein 1999). Also, more recently, in a study of innovation in information and communication technologies (ICTs) over almost the identical period as our study (1976–2010), Ozcan and Greenstein (2013) examine US patent data and find that although the top twenty-five firms account for 72 percent of the entire patent stock and 59 percent of new patents in 1976, they account for only 55 percent and 50 percent, respectively, by 2010. The decline is even more dramatic when they restrict the sample to the ownership of high-quality patents (82 percent down to 62 percent). They interpret their results as supporting the view that decentralization is resulting from "more widespread access to the fundamental knowledge and building blocks for innovative activity" (5).

Overall, we interpret our data as reflecting a decline in the concentration of output at the department level. In other words, the top institutions

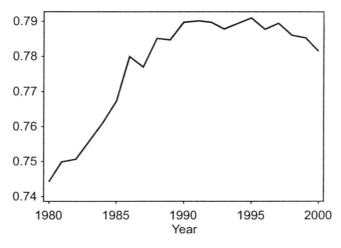

Fig. 3.6 Gini coefficients for the distribution of citation-weighted publications across individuals

are producing a decreasing fraction of the overall output, and previously lesser-producing institutions are now contributing a higher portion of overall output. However, this is not the case at the individual level. We turn to this unit of analysis next.

3.4 Concentration: The Increasing Importance of Stars

With greater democratization in knowledge production across departments, is science becoming a less elite activity, with a falling centrality of stars as they compete with scientists from an ever-widening base? One might expect the broadening base of science at the department level to be accompanied by a reduction in the concentration of output at the individual level. However, we find evidence of the opposite.

We again plot Gini coefficients by year using citation-weighted publications, but this time at the individual level. These data, illustrated in figure 3.6, indicate a significant increase in concentration during the 1980s and then relative stability during the following decade. Then, in figure 3.7, we plot individual-level Lorenz curves for 1980, 1990, and 2000 with the same data to illustrate how the full distribution shifts over time. Again, we see individual-level output increasing over time. For example, the top 20 percent of scientists produce 70 percent of output in 1980 but 80 percent by 2000, with most of the shift occurring in the first decade. Furthermore, in figure 3.8, we illustrate the increasing spread between the top-performing scientists and the rest by comparing the number of citation-weighted publications required to be in the top 50, which increases fivefold, to the average number

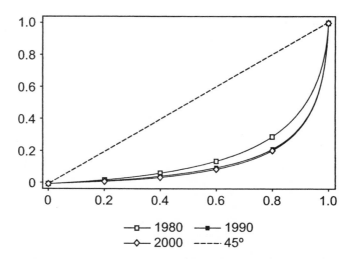

Fig. 3.7 Lorenz curves by individual for citation-weighted publications

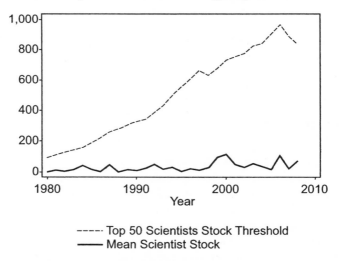

Fig. 3.8 Publication stock of 50th-ranked scientist

of citation-weighted publications, where the increase over the same time period is negligible.

How might we reconcile decreasing concentration at the department level but increasing concentration at the individual level? The answer may lie in the changing patterns of collaboration. Recall that although the rising burden of knowledge and declining communication costs exert opposing forces on the returns to colocation, both increase the returns to collaboration. We turn to the topic of collaboration next.

3.5 Collaboration: Increasing across Distance and Rank

The trend toward increasing collaboration is a well-documented feature of the changing organization of science (e.g., see Wuchty, Jones, and Uzzi 2007). In fact, the rising role of collaboration is one of the most common themes across the chapters in this volume. For example, Branstetter, Li, and Veloso (chapter 5) state: "Our study suggests that the increase in US patents in China and India are to a great extent driven by MNCs (multinational corporations) from advanced economies and are *highly dependent on collaborations* with inventors in those advanced economies."[3] Forman, Goldfarb, and Greenstein (chapter 6) explain: "We show that these [geographic distribution of inventive activity] results are *largely driven by patents filed by distant collaborators* rather than by noncollaborative patents or by patents by nondistant collaborators." Stephan (chapter 10) argues: "Much of the equipment associated with these shifts in logic were, although expensive, still affordable at the lab or institutional level. Some, however, such as an NMR (nuclear magnetic resonance), carried sufficiently large price tags to *encourage, if not demand, collaboration across institutions.*" Conti and Liu (chapter 2) report: "Collaborations with other scientists, as measured by the number of coauthors on a paper, have increased. This increase is driven by *collaborations with scientists outside of a trainee's laboratory.*" Freeman, Ganguli, and Murciano-Goroff (chapter 1) discover through their survey: "*The major factor cited for all types of collaborations was 'unique knowledge, expertise, capabilities. . . .* Non-colocated and international teams were more likely to have a coauthor contributing data, material, or components—a pattern that has been increasing over time.*"

We document this phenomenon of increasing collaboration over time in our own setting in figure 3.9. Specifically, this figure illustrates the steady increase in the average number of authors on evolutionary biology papers, rising from 2.3 in 1980 to 3.8 in 2005. Moreover, this collaboration increasingly has been taking place across university boundaries (Jones, Wuchty, and Uzzi 2008). We illustrate this in figure 3.10, where we plot the average number of unique institutions represented on a paper over time. The figure shows that this number increases from 1.46 in 1980 to 2.45 in 2005.

We also observe a dramatic trend in the average rank difference between authors on coauthored papers (figure 3.11). For example, in 1980 the average distance in rank between collaborating institutions is approximately thirty (e.g., one collaborator is at an institution ranked number twenty and the coauthor is at an institution ranked number fifty), by 2005 the difference increases to approximately fifty-five. Furthermore, we find evidence of increasing distance between collaborators over time. We illustrate this in

---

3. The emphasis in this and the other quotes in this paragraph is our own, not that of the original authors.

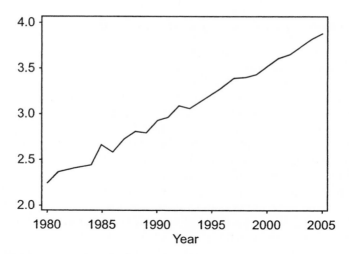

Fig. 3.9 Mean number of authors per paper

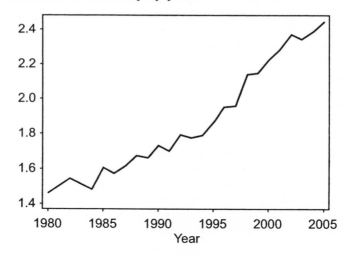

Fig. 3.10 Mean number of unique institutions per paper

figure 3.12, where the average distance between coauthors increases from 325 to 500 miles over the period 1980 to 2005.

Why might the falling cost of distant collaboration disproportionately benefit stars? Freeman, Ganguli, and Murciano-Goroff (chapter 1, this volume) present survey evidence indicating that, in general, a large fraction of collaborations occur between scientists who were previously colocated. We conjecture that one reason stars disproportionately benefit from a drop in the cost of distant collaboration is because they have a greater number of distant

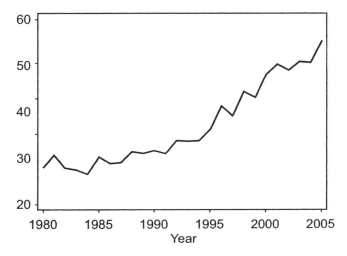

Fig. 3.11 **Mean difference in institution rank between coauthors**

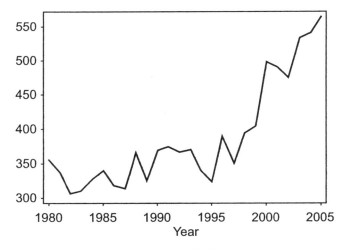

Fig. 3.12 **Mean distance between coauthors (miles)**

potential collaborators. For example, stars are likely to have more graduate students and postdoctoral students than nonstars, on average, and these students are likely to subsequently move to other institutions. To the extent that communication technologies like the Internet are most suitable for facilitating communication between individuals with an already established relationship as opposed to establishing new relationships (Gaspar and Glaeser 1998), then lowering communication costs will disproportionately benefit those individuals, such as stars, who have more previously colocated, but now distant poten-

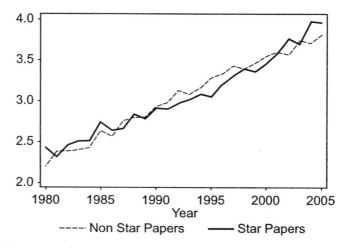

Fig. 3.13 Mean number of authors per paper (star versus nonstar)

tial coauthors. In other words, stars are able to employ this technology over a larger number of previously colocated but now distant potential collaborators.

This benefit to stars could accrue through two non-mutually exclusive channels. First, stars could disproportionately increase the number of individuals they collaborate with. Our descriptive evidence suggests that although stars do increase their propensity to collaborate over time, so do nonstars. We illustrate this in figures 3.13 and 3.14. First, we show that although the number of coauthors per paper increases over time, there is no meaningful difference between papers with and without stars. Second, we construct three measures of stars (top 50, top 100, and top 200 scientists) and plot the number of unique coauthors per year for stars versus nonstars. These data indicate that although the annual number of unique collaborators is increasing over time for star scientists, stars do not seem to increase their number of unique collaborators at a meaningfully faster rate than nonstars.

Second, stars may disproportionately benefit from the fall in communication costs because they are able to make better matches with coauthors since they have more potential collaborators to choose from. In other words, the best of the available pool of potential collaborators is better for stars than for nonstars. So, for example, if the falling cost of communications increases the returns to collaboration such that both a star and a nonstar increase their number of collaborations by one, then the average star may choose the best-suited collaborator from a pool of many previously colocated but now distant potential collaborators, while the nonstar can only choose from a pool of few. Even if stars and nonstars are choosing collaborators from pools with the same distribution in terms of quality or range of skills, stars likely will be able to choose a superior match simply due to the larger pool size to which they have access.

(a) Top 200 Stars

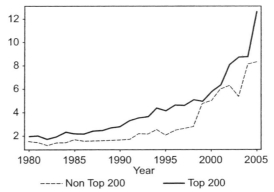

(b) Top 100 Stars

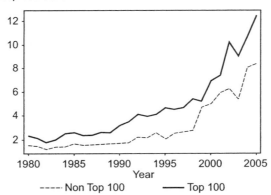

(c) Top 50 Stars

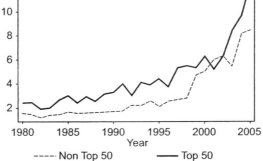

Fig. 3.14 Number of unique coauthors per year
Note: Panel (a) = top 200 stars; panel (b) = top 100 stars; and panel (c) = top 50 stars.

We construct a measure of the size of the pool of previously colocated but now distant collaborators by counting the cumulative number of individuals who coauthor with the focal scientist at least once while located at their home institution and then subsequently at least once while at another institution. We again construct three measures of stars (top 50, top 100, and top 200 scientists). In figure 3.15, we plot the potential distant coauthor pool size for stars versus nonstars (cumulative number of unique coauthors that were previously colocated but are now distant). It is important to note that this count is not simply the aggregation of the annual counts plotted in the prior figure. That is because in the prior figure repeated coauthorships are counted as distinct in each new year (although multiple coauthorships with the same individual in the same year are not double counted). However, in this plot only unique coauthorships that are unique in the absolute sense (cumulatively) are counted. Furthermore, in figure 3.15 we only count distant coauthors that were previously located whereas in the prior figure there were no distance or prior colocation restrictions in counting unique coauthors. These data indicate that the pool size of potential collaborators (such as graduate students and postdocs) grows significantly faster for stars than for nonstars. Furthermore, in figure 3.16, we plot the inverse of the ratio of the number of actual collaborations in a given year to the number of potential collaborators in the pool that year and compare the change in this ratio over time for stars versus nonstars. We interpret the ratio as a proxy for the degree of selectivity afforded to stars and nonstars. In other words, a higher ratio for stars versus nonstars indicates that stars collaborate with a smaller fraction of their pool of potential coauthors than nonstars. The figure thus suggests that the relative selectivity of stars versus nonstars in terms of choosing collaborators is increasing over time. While not conclusive, these descriptive data are consistent with the conjecture that stars disproportionately benefit from falling communication costs by way of an increased pool size of distant collaborations to choose from relative to nonstars.

3.6 Improved Collaboration Technology and the Distribution of Scientific Output: An Integrating Model

In this section, we develop a simple model to examine the effects of an improvement in collaboration technology on the distribution of scientific output. In particular, we examine how such an improvement both disproportionately affects stars and leads to more collaboration. The model's results are consistent with both an increased concentration of scientific output across individual scientists—that is, a star concentration effect—and also a broadening institutional base of science.

A key assumption is that relationships with previously colocated but now distant former coauthors, such as former graduate students and postdocs, are central to developing opportunities for subsequent collaboration. This is consistent with the survey evidence on collaboration reported by Freeman,

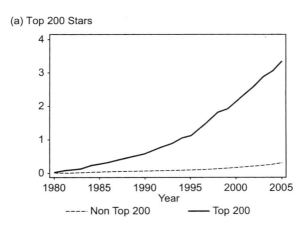

(a) Top 200 Stars

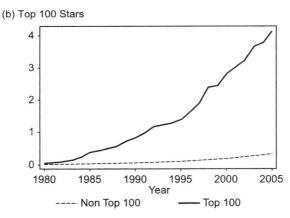

(b) Top 100 Stars

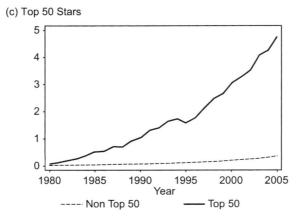

(c) Top 50 Stars

Fig. 3.15 Cumulative number of unique previously colocated but now distant coauthors
Note: Panel (a) = top 200 stars; panel (b) = top 100 stars; and panel (c) = top 50 stars.

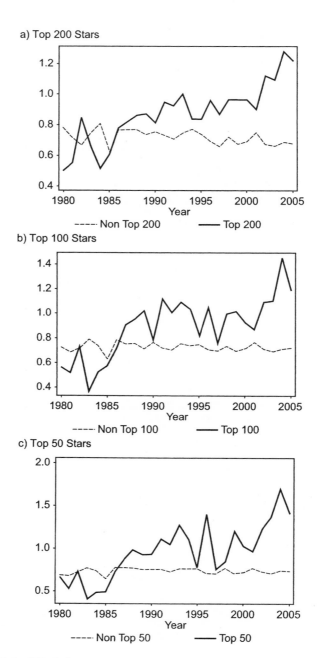

Fig. 3.16 Selectivity index

Note: Panel (a) = top 200 stars; panel (b) = top 100 stars; and panel (c) = top 50 stars. We construct the selectivity index as the inverse of the ratio of the number of unique collaborators in a given year over the cumulative number of previously colocated but now distant collaborators.

Ganguli, and Murciano-Goroff (chapter 1, this volume), which documents the extent to which such relationships account for the majority of collaborative partnerships. We also assume that the number of feasible collaborative relationships is limited due to the costs of collaboration. Furthermore, we assume that stars know a larger set of former graduate students and postdocs from which to choose their collaborative relationships. We do not need to assume that stars have better graduate students in general or engage in more collaborative relationships. We show that simply having a greater range of graduate students to choose from enables stars to gain disproportionately from an improvement in collaborative technologies, which we take to be due to improvements in communication technologies (e-mail, file-sharing technologies, etc.).

For a given scientist, we assume the value of a collaborative relationship, X, with a given former graduate student is uniformly distributed on the interval $[0, M]$. We assume that an improvement in collaboration technology increases the value of any relationship by a multiplicative factor. The increased value of collaboration could also reflect a greater need for collaboration due to the burden of knowledge effect. Thus, we can simply model an improvement in technology (or the greater need for collaboration) as an increase in M, effectively a stretching of the distribution to the right.

3.6.1 Basic Model

We assume initially that each scientist chooses the single-best relationship from her set of n former graduate students. We use the size of n as a proxy for the scientist's degree of stardom. For a given scientist, the expected value of the best available relationship is:

$$(1) \qquad E(X) = \int_0^M X \frac{n}{X} \left(\frac{X}{M}\right)^n dX = \frac{n}{1+n} M.$$

This result uses the distribution of the maximum value of n, which draws from the uniform distribution.[4]

The increase in expected value from a small increase in the available collaboration technology is then:

$$(2) \qquad \frac{\partial E(X)}{\partial M} = \frac{n}{1+n}.$$

The size of this increase is increasing in n,

$$(3) \qquad \frac{\partial^2 E(X)}{\partial M\, \partial n} = \frac{1}{(1+n)^2} > 0.$$

Thus, stars—those with a high n—gain disproportionately from the improvement in the collaboration technology.

---

4. The CDF for this extreme value distribution is: $F(X) = (X / M)^n$. The density function is then: $f(X) = (n / X)(X / M)^n$.

3.6.2 Extended Model

A limitation of the basic model is that it assumes a scientist will choose to collaborate with her best former graduate student no matter how low the value of that best collaboration. A more realistic assumption is that scientists have some threshold below which they will not collaborate, given the opportunity costs of collaboration (e.g., reduced time for sole authorship). Denoting this threshold as X^*, the expected value of a collaboration is now:

$$(4) \qquad E(X) = \int_{X^*}^{M} \frac{n}{X}\left(\frac{X}{M}\right)^n dX = \left[\left(\frac{n}{1+n}\right)M\right]\left[1 - \left(\frac{X^*}{M}\right)^{n+1}\right].$$

The expected value is lower than when the threshold is absent because best draws from the distribution that are below the threshold result in zero value. It is also increasing in M, so that improvements in the collaboration technology are again beneficial.

$$(5) \qquad \frac{\partial E(X)}{\partial M} = \left[\frac{n}{1+n}\right]\left[1 + n\left(\frac{X^*}{M}\right)^{n+1}\right] > 0.$$

We again ask if the technology improvement disproportionately benefits stars. This requires that the cross derivative with respect to n is positive. Making use of logarithmic differentiation, the cross derivative is:

$$(6) \qquad \frac{\partial^2 E(X)}{\partial M\,\partial n} = \frac{1}{(1+n)^2} + \left(\frac{n^2}{1+n}\right)\left(\frac{X^*}{M}\right)^{n+1}\left[\frac{2}{n} - \frac{1}{1+n} + \ln\left(\frac{X^*}{M}\right)\right].$$

This cross derivative is obviously quite a complex function of n, M, and X^*. However, it can be shown to be positive for all n given a low enough value of X^* relative to the starting value of M, so that $\partial E(X) / \partial M$ is then monotonically increasing in n. Figure 3.17 shows the cross derivative as a function of n for different values of X^* (conveniently scaled by the starting value of M): 0.1, 0.2, and 0.3. At high values of X^* / M, the cross derivative can be negative over an intermediate range of n but becomes positive for high enough values of n. We assume, however, that the threshold is sufficiently low such that the cross derivative is positive for all n.

An additional consequence of introducing a threshold for collaboration is that the probability of collaboration is now itself a function of M.

$$(7) \qquad \text{Prob}[X > X^*] = 1 - \left(\frac{X^*}{M}\right)^n.$$

This probability is also increasing in M, so that improvements in the collaborative technology lead to more as well as higher expected value collaboration:

$$(8) \qquad \frac{\partial \text{Prob}[X > X^*]}{\partial M} = \frac{n}{M}\left(\frac{X^*}{M}\right)^n > 0.$$

Summing up, the extended model demonstrates two effects of an improvement in collaborative technology that could impact the distribution of scien-

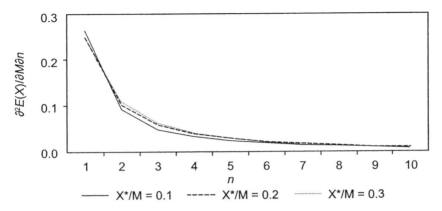

Fig. 3.17 **Relationship of the cross derivative to** n

tific output. First, provided that scientists do not set too high a threshold for engaging in collaboration, the benefit from the improvement in technology is increasing in n, so that stars—whom we assume to be disproportionately endowed with previously colocated but now distant former coauthors—benefit disproportionately. This is consistent with an increased concentration of scientific output at the individual level. Second, it will be beneficial for more scientists to engage in collaborative research. This is consistent with an expanding institutional base of science as more former students and postdocs—who will have dispersed across the institutional ranks—are involved in collaborative research.

3.7 Discussion: Normative Implications of Star Location

Our review of the basic trends in participation, concentration, and collaboration reveals the dramatically changing organization of scientific activity in the field of evolutionary biology. The emerging picture also points to the increasingly central role played by stars in collaboration and overall output. Moreover, stars, like the overall research community, appear to be increasingly collaborating across distance and institution rank. Overall, we see evidence of a developing cross-institutional division of scientific labor, with stars playing a leadership role in institution- and distance-spanning multiauthor research teams.

The rising centrality of stars raises questions about the efficient distribution of stars across institutions. We thus reflect on the efficiency of the emerging pattern of the division of labor, drawing on both the factual picture just documented and parallel work on the causal impact of star scientists at the departmental level (Waldinger 2012, 2013; Agrawal, McHale, and Oettl 2013). A key question is whether the emerging spatial distribution of stars is efficient from the perspective of maximizing the value of scientific output.

We do not presume that the distribution will be efficient, given the free location choices of individual scientists and the productivity, reputational, and consumption externalities associated with those choices. We note in particular that the reputational spillover from locating at top-ranked institutions could lead to an excessive positive sorting of stars at these institutions. Such inefficiency, if it exists, could be ameliorated by easier cross-institution collaboration, effectively making the location of stars less important to knowledge production. Even so, given the ongoing costs of distance-related collaboration, a concern still remains that there may be excessive concentration from a social welfare perspective.

In Agrawal, McHale, and Oettl (2013), we show that the arrival of a star, whom we define as a scientist whose output in terms of citation-weighted publications is above the 90th percentile of the citation-weighted stock of papers published up until year t_{-1}, leads to a significant increase in the productivity of colocated scientists. More specifically, we show this effect operates through two channels: knowledge and recruiting externalities. We show that the arrival of the star leads to an increase in the productivity of incumbents, those scientists already at the department prior to the arrival of the star, but only for those incumbents working on topics related to those of the star. We do not find any evidence of productivity gains by incumbents working in the field of evolutionary biology, but on topics unrelated to those of the star. These effects are robust to including controls for broader departmental and university expansion. Furthermore, they are robust to placebo tests for the timing of the effect; we find no evidence of a pretrend in terms of increasing productivity prior to the arrival of the star. Moreover, the results are also robust to using a plausibly exogenous instrument for star arrival.

The star's arrival also leads to a significant increase in subsequent joiner quality (recruits hired after the arrival of the star), which is most pronounced for related joiners but also occurs for unrelated joiners. These results also hold when subjected to the robustness tests described above. These recruiting results raise a concern about the possibility of reputation-driven positive sorting at top institutions, with stars attracting stars irrespective of productivity-increasing knowledge spillovers. This in turn raises a concern about lost opportunities for stars to seed focused and dynamic research clusters at lower-ranked institutions.

But are these opportunities actually lost? Given the apparent role of star recruitment in department building—which our evidence suggests would be particularly effective where the institution already has a cadre of incumbents working in related areas to the star and has a sufficient flow of new openings to take advantage of star-related recruitment externalities—an offsetting force to excessive concentration could come from the incentive of lower-ranked institutions to use star-focused strategies to ascend departmental rankings. We show how departmental rankings changed between 1980 and 2000 in figure 3.18. While these data imply a reasonably high degree of rank persistence, they also show that some institutions made significant movements up the rankings. Anecdotal evidence suggests that the recruitment of stars may have played an important role here.

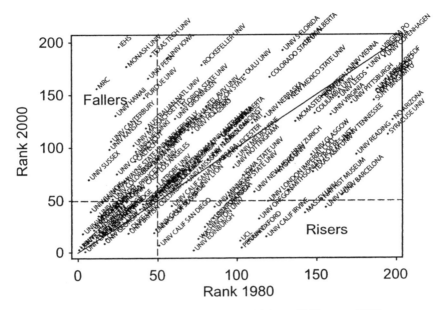

Fig. 3.18 Department-level rank in evolutionary biology: 1980 versus 2000

Moreover, stars may increasingly benefit institutions they do not join but where they have collaborative relationships. Azoulay, Graff Zivin, and Wang (2010) and Oettl (2012), who both use the unexpected death of star scientists to estimate their effect on the productivity of their peers, report evidence that stars significantly influence the productivity of their collaborators. Moreover, Agrawal and Goldfarb (2008) show that the greatest effect of universities connecting to Bitnet (an early version of the Internet) in terms of influencing cross-institution collaboration patterns was not between researchers at tier 1 institutions but rather tier 1–tier 2 collaborations. This is consistent with the data we report here on the increasing institution rank distance between collaborators. One interpretation of this result is that lowering communication costs particularly benefits vertical collaboration, suggesting an increasingly vertically disaggregated division of labor as communication costs fall. Perhaps, for example, declining communication costs increase the returns for individuals at top institutions specializing in leading major research initiatives, identifying key research questions, and writing grant applications, while their collaborators at lower-ranked institutions run experiments, collect and analyze data, and work together with all collaborators to interpret and write their results. The results reported by Kim, Morse, and Zingales (2009) are consistent with this when they document the rise of lesser-ranked universities.

To obtain more direct evidence of changes in star concentrations, we plot in figure 3.19 the share of the top 100 evolutionary biology scientists at the

a) Top 100 Scientists in Top 50 Institutions

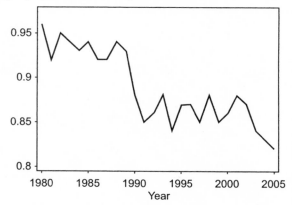

b) Top 100 Scientists in Top 25 Institutions

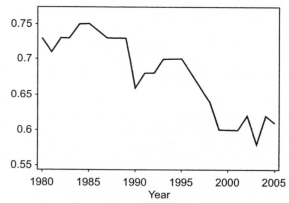

c) Top 100 Scientists in Top 10 Institutions

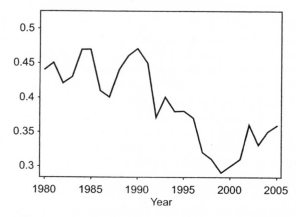

Fig. 3.19 Fraction of top 100-ranked researchers at top-ranked departments

top 50, top 25, and top 10 evolutionary biology departments. The basic pattern shows, if anything, a fall in the concentration of stars at top institutions, somewhat allaying fears of excessive concentration due to reputation-driven positive sorting.

Our examination of the efficiency of the emerging organization of activity in the field of evolutionary biology is unavoidably preliminary and speculative given current levels of knowledge. The broad pattern of increased spatial and cross-institution collaboration—often centered on a star—is pronounced in the data. However, despite institution-level evidence of reputation-based sorting, we do not observe the feared rise in concentration at top institutions. Given the importance of the spatial and institutional distribution of stars to the workings of collaborative science, we expect the normative implications of the changing spatial distribution of scientific activity—and its stars—to be an active area of future research on the organization of science.

References

Agrawal, A., and A. Goldfarb. 2008. "Restructuring Research: Communication Costs and the Democratization of University Innovation." *American Economic Review* 98 (4): 1578–90.

Agrawal, A. K., A. Goldfarb, and F. Teodoridis. 2013. "Does Knowledge Accumulation Increase the Returns to Collaboration? Evidence from the Collapse of the Soviet Union." NBER Working Paper no. 19694, Cambridge, MA.

Agrawal, A. K., J. McHale, and A. Oettl. 2013. "Why Stars Matter." NBER Working Paper no. 20012, Cambridge, MA.

Azoulay, P., J. Graff Zivin, and J. Wang. 2010. "Superstar Extinction." *Quarterly Journal of Economics* 125 (2): 549–89.

Bresnahan, T. F., and S. Greenstein. 1999. "Technological Competition and the Structure of the Computer Industry." *Journal of Industrial Economics* 47 (1): 1–40.

Gaspar, J., and E. L. Glaeser. 1998. "Information Technology and the Future of Cities." *Journal of Urban Economics* 43 (1): 136–56.

Jones, B. F. 2009. "The Burden of Knowledge and the 'Death of the Renaissance Man': Is Innovation Getting Harder?" *Review of Economic Studies* 76 (1): 283–317.

Jones, B. F., S. Wuchty, and B. Uzzi. 2008. "Multi-University Research Teams: Shifting Impact, Geography, and Stratification in Science." *Science* 322 (5905): 1259–62.

Jones, C. I. 1995. "R&D-Based Models of Economic Growth." *Journal of Political Economy* 103 (4): 759–84.

Kim, E. H., A. Morse, and L. Zingales. 2009. "Are Elite Universities Losing Their Competitive Edge?" *Journal of Financial Economics* 93 (3): 353–81.

Mokyr, J. 2002. *The Gifts of Athena: Historical Origins of the Knowledge Economy.* Princeton, NJ: Princeton University Press.

Oettl, A. 2012. "Reconceptualizing Stars: Scientist Helpfulness and Peer Performance." *Management Science* 58 (6): 1122–40.

Ozcan, Y., and S. Greenstein. 2013. "The (De)concentration of Sources of Inventive Ideas: Evidence from ICT Equipment." Working Paper, Northwestern University.

Romer, P. M. 1990. "Endogenous Technological Change." *Journal of Political Economy* 98 (5): S71–102.
Rosenbloom, R. S., and W. J. Spencer. 1996. *Engines of Innovation: US Industrial Research at the End of an Era.* Boston: Harvard Business Press.
Tang, L., and J. P. Walsh. 2010. "Bibliometric Fingerprints: Name Disambiguation Based on Approximate Structure Equivalence of Cognitive Maps." *Scientometrics* 84 (3): 763–84.
Waldinger, F. 2012. "Peer Effects in Science: Evidence from the Dismissal of Scientists in Nazi Germany." *Review of Economic Studies* 79 (2): 838–61.
———. 2013. "Bombs, Brains, and Science: The Role of Human and Physical Capital for the Creation of Scientific Knowledge." Working Paper, University of Warwick.
Weitzman, M. L. 1998. "Recombinant Growth." *Quarterly Journal of Economics* 113 (2): 331–60.
Wuchty, S., B. F. Jones, and B. Uzzi. 2007. "The Increasing Dominance of Teams in Production of Knowledge." *Science* 316 (5827): 1036–39.

Comment   Julia Lane

The authors address an interesting and important question about the way in which scientific collaboration has changed over time. They use a creatively constructed data set on evolutionary biology to show how scientific collaboration for the subset of authors they identify has changed. The results reported are consistent with other work in the book. Most interestingly, the geographic distance between coauthors has increased substantially, notably that the concentration of publications within an institution has decreased and that the institutional rank distance between coauthors has increased. They find that the concentration of publications at the individual level has increased. They also note the pool of potential coauthors has increased. The authors posit that these trends are the result of two factors: the burden of knowledge and collaboration-supporting technologies.

Their work thus provides potential new areas that could be examined in future research. One is whether evolutionary biology is unique among scientific disciplines: it would be extremely useful to know whether similar changes in collaboration are found for such subsets of authors in "big science," like astrophysics, and smaller scale sciences, like chemistry. It would also be useful to examine across different disciplines whether observed changes in collaboration are due to specialization of innovative labor. It would also be very useful to understand the role of technology in driving geographically dispersed collaboration. It is possible, and anecdotal evidence suggests, that increasingly technology-intensive science that requires large-scale complex equipment is a driving force behind changes in collaboration.

Julia Lane is an Institute Fellow at the American Institutes for Research.
For acknowledgments, sources of research support, and disclosure of the author's material financial relationships, if any, please see http://www.nber.org/chapters/c13039.ack.

The research also raises interesting questions about whether a field, even though it might be named the same over time, is indeed still the same. Does evolutionary biology still mean the same thing now as it did twenty-nine years ago? Or has it now become more interdisciplinary and been influenced by the convergence of physics, chemistry, and biology? In other words, how much are changes in collaboration due to changes in the very structure of science itself? There is a great deal of interest in using natural language processing techniques to study the evolution of scientific disciplines; future research could examine the role of changes in the topics covered under different disciplinary nomenclatures on collaboration (Herr 2009; Talley et al. 2011).

The findings also suggest interesting potential research into other areas—particularly the role of monetary incentives in changing collaboration. Over the period that is studied, US science agencies changed the way in which science was funded. Although the big interdisciplinary funding initiatives—such as funding for nanotechnology and the human genome sequence—are well known and fundamentally changed both the nature and scale of scientific endeavor in a number of fields, there have been many such smaller scale initiatives. In addition, during the period of study, funding abruptly doubled for the NIH and then leveled off—and so the sharp change in funding could be used to examine the role of monetary incentives.

The authors also posit some interesting hypotheses about the role of graduate students and postdoctoral fellows in the changes in collaborations they observe among their subset of authors. We know that the composition and size of scientific teams is not only dramatically different across research fields, but also changes substantially in response to monetary incentives (Stephan 2007, 2012). Subsequent research on the role of team structure in collaboration could examine both how the structure of teams in evolutionary biology changed over the period—and whether and how subsequent placements of team members in industry or in academia evolved over the time period in question.

Much can also be learned from the extensive hard work that was done to structure the data set. The chapter is partly about authors and partly about institutions, which makes it very difficult to develop a representative frame. The authors have done an enormous amount of work to develop a frame that is based on a very specific selection of publications; this is the link asset that includes information about both.

Future research could be very useful to inform us about the generalizability of the authors' results. In particular, it would be extremely useful to learn whether the known skewness of productivity of scientists, of institutions, and of the salience of publications yields an analysis that is representative of all three dimensions. There has been a rich vein of research in other fields, like labor economics, to understand how the link between workers and firms (jobs) can become a frame of study in its own right (Abowd and Vilhuber 2011; Abowd, Haltiwanger, and Lane 2004; Burgess, Lane, and Stevens 2000);

similar research could usefully be undertaken using data sets constructed in the way described in this chapter. I outline a possible research agenda below.

The Link Asset: Publications. The authors use an imaginative approach to construct a frame from which to draw their sample. In particular, the frame described in this chapter is derived from publications in four journals over twenty-nine years whose focus is evolutionary biology. It would be extremely useful for other researchers to determine whether frames constructed like this are representative of research fields. Are such frames constant over time, both with respect to coverage and content? In particular, how does the emergence of new journals that might siphon off contributors affect the representativeness of the frame? For example, in economics, would a frame based on the publications *Journal of Political Economy*, *Quarterly Journal of Economics*, *American Economic Review*, and *Econometrics*, for example, be equally representative over time, both in levels, and as a proportion of all journals in which economists publish? New research in computer science is beginning to enable us to determine what is missing—by using the same capture/recapture techniques that are used to generate censuses of wildlife, we can determine the coverage of journals over time (Khabsa, Treeratpitu, and Giles 2012; Giles and Khabsa 2014).

The authors use an equally imaginative approach to weighting their sample, and their approach should stimulate an interesting line of research. In particular, they use citations to weight publications, rather than the absolute measure of publications themselves. Returning to my earlier analogy about what we have learned from researching firms, if the focus of the research is firms, then firms can be used as the unit of analysis. If the focus of research is employment, then it is more appropriate to use employment-weighted firms. In the former case, the small start-up has the same weighting as does General Motors. In the latter case, General Motors' behavior will dominate the analysis. Similarly, in this chapter, the results will be dominated by the activities of a few highly productive (and cited) researchers; an important line of analysis for future research would be to examine how the results change when all individual researchers are weighted equally.

The Article Authors. The sample design used in the chapter suggests even more useful research that could be undertaken to build a richer understanding of how the empirical approach can be expanded to understand the activities of active researchers. For example, the original draw from the four journals identifies 171,428 authors; 140,240 of these are dropped because there are no more than two publications linked to their names. Additional research could examine the sensitivity of the results to, for example, distance measures (how many authors are dropped because it is not possible to identify nonunique author names, and understanding the selection bias associated with nonunique Asian author names); time variability (how does the sample selection change over time); and institutional variation (how many institutions are dropped). There are equally interesting questions about the importance of using fractionally weighted output measures rather than full output measures. In the case of this chapter, the authors decide to weight author output measures by the full publications; since coauthorship has increased overtime, this decision weights output in later time periods more heavily than previous time periods. Much useful research could examine how different weighting choices affect the econometric results. At the least, both weighted and unweighted results should be presented and discussed throughout.

The chapter also identifies an important possible set of sources for collaboration—namely the possibility of faculty coauthoring papers with their postdoctoral fellow and graduate students. This builds on a rich literature that has suggested that within-group specialization is an important feature of modern science (Black and Stephan 2010; Conti, Denas, and Visentin, forthcoming). It would be extraordinarily valuable to build a database that included information about the links between the authors (including the postdoc/graduate student/principal investigator relationship) to establish this hypothesis more generally.

The Institutions. Another important potential line of research that is identified in this chapter is the role of institutional locations, about which so little is known. The structure of the sample identifies some interesting features of collaboration. Just over half (57 percent) of the papers have a single institution listed, and so all authors are located with this institution, and 79 percent of authors are attributed to an institution. Future research could examine how a sample constructed like this might influence our understanding of the role of institutions—particularly focusing on how the structure of institutional affiliations changes over time, both domestically and internationally—and how a selection decision based on institutional affiliations affects the number of coauthors included in the analytical data set.

Finally, the authors put together a very interesting model about the role of graduate and postdoctoral students in generating increasing collaborations. This framework can be used to add to the three alternative hypotheses that are also bruited in the literature. In principle, an empirical exercise could test the model by including data on the pool of available graduate students, data on the role of funding agencies in incentivizing collaboration, the role of technological complementarity, as well as the role of previous collocation.

In sum, this chapter presents some very provocative results, which are very congruent with companion chapters in the book. The challenge to the research community is to extend their interesting results generated from a painstakingly assembled and idiosyncratic sample and determine the generalizability of the approach to both evolutionary biology and other areas of science.

References

Abowd, J., J. Haltiwanger, and J. Lane. 2004. "Integrated Longitudinal Employee
-Employer Data for the United States." *American Economic Review* 94:224–29.

Abowd, J. M., and L. Vilhuber. 2011. "National Estimates of Gross Employment and Job Flows from the Quarterly Workforce Indicators with Demographic and Industry Detail." *Journal of Econometrics* 161:82–99.

Black, G., and P. E. Stephan. 2010. "The Economics of University Science and the Role of Foreign Graduate Students and Postdoctoral Scholars." In *American Universities in a Global Market*, edited by C. Clotfelter, 129–61. Chicago: University of Chicago Press.

Burgess, S., J. Lane, and D. Stevens. 2000. "Job Flows, Worker Flows, and Churning." *Journal of Labor Economics* 18 (3): 473–502.

Conti, A., O. Denas, and F. Visentin. Forthcoming. "Knowledge Specialization in PhD Student Groups." *IEEE Transactions on Engineering Management*. http://papers.ssrn.com/sol3/papers.cfm?abstract_id=2210462.

Giles, L., and M. Khabsa. 2014. "The Number of Scholarly Documents on the Web." *PLoS One*. doi: 10.1371/journal.pone.0093949.

Herr, B. W., E. M. Talley, G. A. P. C. Burns, D. Newman, and G. LaRowe. 2009. "The NIH Visual Browser: An Interactive Visualization of Biomedical Research." Information Visualization, 13th Annual Conference, Barcelona, July. Institute of Electronics and Electronic Engineers. http://ieeexplore.ieee.org/xpl/abstractAuthors.jsp?tp=&arnumber=5190811&url=http%3A%2F%2Fieeexplore.ieee.org%2Fxpls%2Fabs_all.jsp%3Farnumber%3D5190811.

Khabsa, M., P. Treeratpitu, and L. Giles. 2012. "AckSeer: A Repository and Search Engine for Automatically Extracted Acknowledgments from Digital Libraries." Proceedings of the 12th ACM/IEEE-CS Joint Conference on Digital Libraries. http://clgiles.ist.psu.edu/pubs/JCDL2012-AckSeer.pdf.

Stephan, P. 2007. "Wrapping It Up in a Person: The Mobility Patterns of New PhDs." In *Innovation Policy and the Economy*, vol. 7, edited by A. Jaffe, J. Lerner, and S. Stern. Cambridge, MA: MIT Press.

———. 2012. *How Economics Shapes Science*. Cambridge, MA: Harvard University Press.

Talley, Edmund M., David Newman, David Mimno, Bruce W. Herr II, Hanna M. Wallach, Gully Burns, Miriam Leenders, and Andrew McCallum. 2011. "Database of NIH Grants Using Machine-Learned Categories and Graphical Clustering." *Nature Methods* 8:443–44. doi:10.1038/nmeth.1619.

Credit History
The Changing Nature of
Scientific Credit

Joshua S. Gans and Fiona Murray

4.1 Introduction

We are interested in the institutional conditions confronting scientists engaged in cumulative innovation and the ways they have changed since Vannevar Bush famously evoked the image of an endless frontier of scientific progress. While many scholars have explored changes in outputs, that is, in the rate and direction of the scientific frontier, we examine changes in the production of scientific knowledge. In seeking to explain how the organization of knowledge production has changed, scholars have focused on two critical factors: the vast number of new technologies that enable scientific progress (Mokyr 2002; Agrawal and Goldfarb 2008) and the increasing burden of knowledge needed for cumulative progress at the frontier (Jones 2009). We propose a third, previously overlooked, factor grounded in the formal institutions and norms of science: the assessment and allocation of credit. The institutions of credit are central to the incentive system of open science (Merton 1957; Dasgupta and David 1994). While not as easy to observe as the growing array of new equipment that fill laboratories today, and harder to conceive of than the notion that to contribute to the frontier

Joshua S. Gans is professor of strategic management and holder of the Jeffrey S. Skoll Chair of Technical Innovation and Entrepreneurship at the Rotman School of Management, University of Toronto, and a research associate of the National Bureau of Economic Research. Fiona Murray is the Associate Dean of Innovation at the MIT Sloan School of Management as well as the Alvin J. Siteman (1948) Professor of Entrepreneurship and a research associate of the National Bureau of Economic Research.

We thank Suzanne Scotchmer, Ben Jones, an anonymous referee, and participants at the NBER Changing Frontier conference for helpful comments. Responsibility for all views expressed lies with the authors. The latest version of this chapter is available at research.joshuagans.com. For acknowledgments, sources of research support, and disclosure of the authors' material financial relationships, if any, please see http://www.nber.org/chapters/c13042.ack.

today means knowing more than those contributing fifty years ago, we argue that credit has also changed in a range of critical dimensions in the years since the "endless frontier."

We start our essay by focusing briefly on economic and sociological perspectives on the nature of scientific credit. In section 4.2, we then develop our perspective on the core organizational choices made by scientists as a way of motivating the central importance of scientific credit in the ways in which knowledge production is organized, using an example from number theory in mathematics to highlight these choices. Section 4.3 elaborates our "credit history"—how the institutions and norms of scientific credit have changed over the past fifty years. We do so by exploring three debates that have animated the scientific community over the past fifty years. Building on the qualitative insights from the past fifty years, section 4.4 lays out a formal model that places credit allocation alongside the changing technical costs and knowledge burden of research to explore the relative importance of these three factors. Section 4.5 considers predictions for how science is likely to change going forward and implications of the continually changing nature of scientific credit for the scientific community. Section 4.6 concludes.

4.2 Credit and the Organization of Science

4.2.1 Choices in the Organization of Science

The post–Vannevar Bush years have seen growing interest among scholars in disciplines spanning the history, economics, and sociology of science in understanding the production and accumulation of scientific knowledge (rather than simply a focus on scientific knowledge itself). While economists such as Arrow (1962), Nelson (1959), and Rosenberg (1990) have famously pursued questions regarding the rate and direction of scientific progress, sociologists including de Solla Price, Garfield, and Merton were among those highlighting the social and institutional nature of scientific progress at the frontier. More recently, the economics of science has been concerned with the institutions that shape knowledge accumulation, among them, decisions regarding the production and disclosure of knowledge (Azoulay, Graff Zivin, and Manso 2010; David 2008; Furman and Stern 2011; Mokyr 2002; Murray and Stern 2007; Murray et al. 2009).

As they work at the knowledge frontier, scientists (in a variety of organizational settings) also make a range of meaningful organizational choices, although these have been less widely examined by scholars of science. Particularly for those researchers working within academia, there is considerable flexibility with regard to a range of choices. Indeed, one can think of academic laboratories as small enterprises in which the faculty scientist is effectively a chief executive officer (CEO) with significant autonomy. Among their autonomous organizational choices, two are critical: First, the scope of the project—whether to undertake a small research project and then disclose or to undertake a more substantial but possibly longer project and disclose at a later time. Second, whether and with whom to collaborate with, who to have as a coauthor, and so forth. We place these choices at the front and center in our analysis of the organization of knowledge production and its transformation over the past fifty years.

A useful way to understand scientists' organizational choices is to build on the conceptual approach developed by Green and Scotchmer (1995) in the context of cumulative innovation in the private sector and by Aghion, Dewatripont, and Stein (2008) with regard to scientific research. Accordingly, research follows a particular "line" that sets an intellectual trajectory for progress and along which research can be understood as taking place in discrete stages or "chunks." At each stage, scientists (or those who fund them) have the freedom to determine specific organizational arrangements and control rights and rewards within the constraints of the broader institutional context. With respect to the organizational arrangements made by scientists working within academia, we argue that three elements are critical: First, they must determine a sequence of cumulative projects that follow along the line they are pursuing; that is, they set a particular intellectual trajectory and map out two or more projects along that line. Second, they must determine the optimal way to approach these projects with respect to collaborative choices. Third, they must determine their disclosure choices for these projects. Taken together, these three elements lead to three distinctive organizational outcomes for any two steps along a research trajectory (and can thus be generalized along a much more significant path):

- Integration. Under integration, scientists may choose to undertake both projects in a line themselves (i.e., within their laboratory with no external collaborators) and only then publish both steps.
- Collaboration. Under a collaboration strategy, scientists bring in collaborators (from other laboratories presumably with complementary skills) to complete both the projects in a line and publish a paper describing both steps with coauthors.
- Publication. Rather than collaborate or integrate, a scientist may choose to publish the first stage in the line and then simply wait to be cited (in the market for ideas) by follow-on researchers who pursue the second stage of the line at some later point.

While conceptual in our exposition, these three organizational alternatives reflect the very real choices made by academic scientists throughout the course of their careers as independent investigators. They are sharply illustrated in the recent case of discoveries in number theory.

On April 14, 2013, Dr. Yitang Zhang, a previously unknown lecturer at the University of New Hampshire, submitted a paper to the *Annals of Mathematics* that purported to prove that there were infinitely many pairs of consecutive prime numbers with a gap of, at most, seventy million. This was the first such bound established and one of the most significant steps toward proving a long-standing conjecture that there exist an infinite number of twin primes (that is, a bound of size 2, the smallest possible). The paper was accepted for publication on May 14, 2013.

Zhang's contribution reflects a strategy that we would describe as "publication"—it was sole authored but he chose to publish as soon as he had established an advance rather than take the next step, that is, follow an integration strategy. What is interesting is what has happened since that time: in subsequent months, other researchers showed that the seventy million bound established by Zhang was capable of significant refinement. By July, it had been reduced from seventy million to just 5,414. With each advance, Zhang's contribution became more significant.

What is salient for the purposes of our analysis is the way in which follow-on researchers chose to chart the continued research trajectory. And as the bound fell to 400,000, there were contributions from a number of individual researchers who raced to publish even the smallest improvements; that is, they followed a publication strategy. However, in June 2013, Field's medalist Terence Tao proposed a change in organization. He set up the bounded primes problem as a polymath project. Polymath projects are online collaborative endeavours in mathematics using many researchers to solve unsolved problems in a short amount of time. There had been seven such projects over the previous two years, all with some measure of success. The important feature of the polymath project is that all conjectures, failed routes, and advances are made public and transparent. For the bounded primes problem, in just a month the bound fell from just below 400,000 to its current level. The end result was a many-authored paper with this final result and proofs of varying efficiently, that is, a modern form of collaboration.

The bounded primes example vividly demonstrates the range of organizational choices that can be pursued, as well as the changes in organizational modes scientists pursue today. It highlights the importance of thinking more deeply about organizational choices and credit in science. The simple dichotomy between sole authored and collaborative works does not capture the richness of the scientific knowledge production process. Here we argue that much can be gained by explicitly considering publication (and citation) as an organizational model for cumulative scientific endeavours alongside integration (sole authorship and secrecy) as well as collaboration.

4.2.2 Organizational Choices and Institutions of Credit

Our organizational perspective highlights the factors that influence scientists as well as the central role of credit in the organization of science along research trajectories. Without the consideration of credit, the reward structure that lies at the heart of scientific work is ignored and our explanations of knowledge production are inadequate. Our argument is as follows: selecting whether to integrate, collaborate, or publish (and rely on the citation market) depends at least in part on the ways in which scientists' believe that they will be rewarded for each of these alternatives. Specifically, a scientist choosing among these options must consider the cost of pursuing each project along the line as well as the time it will take to accumulate the relevant specialized knowledge—the traditional factors thought to shape the organizational calculus made by researchers from one laboratory to another. Nonetheless, these explanations are incomplete. While the costs and benefits of the necessary technology and specialization are critical, scientists must also consider the benefits and costs in terms of the level of credit they will receive under different organizational arrangements.

Under *integration* a scientist receives all the credit for a substantial amount of research progress along the line, but must balance this against the potential costs of acquiring the specialized knowledge and accessing relevant technology. In contrast, the attractiveness of *collaboration* depends upon the trade-offs between the benefits of additional resources (expertise and technology) brought to the project by coauthors and the possible costs of how credit is allocated and shared between scientists and their collaborators (see Bikard, Murray, and Gans [forthcoming] for an empirical elaboration of this issue). Last, under the *publication* choice, citation markets provide an alternative form of credit—in the form of citation and acknowledgment that may itself be valued by researchers and those who evaluate them—that must be considered as a scientists may then receive credit for the first-stage project in the form of publication and credit in the form of citation recognition from the second stage researchers.

The trade-offs driving scientists' organizational choices emphasize the importance of credit as a more institutionally grounded, but nonetheless important, countervailing set of costs and benefits that balance the technical costs and benefits of pursuing particular organizational strategies along a given research line. The role of credit as a central institutional feature shaping the organization of science came to the attention of economists upon publication of Dasgupta and David's influential 1994 paper "Towards a New Economics of Science." In it, they highlighted the importance of reexamining the organizational structures as well as the institutions and policies of science. The paper argues that science "is a system that remains an intricate and rather delicate piece of social and institutional machinery" (489), and emphasizes the importance of the norms and general "institutions" governing the production of knowledge. In focusing on the less tangible features of scientific work, Dasgupta and David build upon a long sociological tradition examining the institutional arrangements found in academia.

Distinguished sociologist of science Robert Merton identified the central role of credit and the informal norms regarding credit, describing them as the "psychosocial processes affect[ing] the allocation of rewards to scientists for their contributions—an allocation which in turn affects the flow of ideas and findings through the communication networks of science" (Merton 1968, 56). Merton also argued that credit could be "mis-allocated" under some conditions noting, famously, that "[e]minent scientists get proportionately great credit for their contributions to science while relatively unknown scientists tend to get disproportionately little credit for comparable contributions" (Merton 1968, 443)—a feature of scientific credit Merton dubbed the "Matthew Effect."

While the Matthew Effect emphasizes a specific instantiation of credit and its (mis-) allocation, Merton's other work and that of subsequent sociologists have explored a variety of ways in which the scientific reward system and credit serve key elements in the institutional life of scientists. For example, in shaping the career trajectories of scientists (most especially tenure), credit was traditionally allocated through processes that take place within a small "inner circle" of scientists who adjudicate claims for priority and, therefore, credit on the basis of close personal relationships (Crane 1969). Beyond credit allocation within closed social contexts, Hägstrom (1969) recognized that citations to prior publications by colleagues also provided an additional reward to researchers as part of an exchange relationship, whereby credit and recognition are placed at the center of a system for knowledge disclosure with information provided in return for credit in the form of citations (see Murray [2010] for an exploration of this process in the context of patent rights). Cole and Cole (1973) further elaborated our understanding of credit and rewards by considering the different types of rewards scientists' accrue: professorships in leading departments, honorific titles, and wide citation being among the most salient.

Most recently, historians of science working in these institutional traditions have sought to use the idea of credit as a way of explaining key historical events in the scientific community. Perhaps the best example are the lengths to which Galileo went to ensure that his novel telescope (and his unique access to its design and production) became a means of receiving both financial and reputational credit from a variety of patrons across Europe, carefully balancing his maintaining control of the telescope to garner further credit with sharing with others so as to ensure that his scientific claims could be validated and thus given appropriate credit (Biagioli 2003, 2006).

Dasgupta and David (1994) can be credited with incorporating this line of thinking from the sociology of science into an economic framework. Their approach has been to examine the reward system as a central element in science and argue that "an individual's reputation for 'contributions' acknowledged within their collegiate reference groups is the fundamental 'currency' in the reward structure that governs the community of academic scientists" (498). This reference to credit as currency highlights the notion of credit and currencies of credit as central both to the norms of science and to its economic foundations. From the perspective of a social planner, the importance of priority in credit speeds up discovery along research lines and ensures their disclosure. From the perspective of a scientist organizing to pursue a given research line, issues of credit allocation shape their organizational choices. However, what remains to be understood, and what serves as the central focus of the remainder of this chapter, is the way in which credit and the changing nature of credit more precisely and more generally shapes the organization of science.

The notion that credit is a critical factor driving the organization of science is in counterpoint to prior approaches taken in the academic literature. Traditionally, scholars have examined two main determinants of the organization of research: the technology of knowledge production and the burden of knowledge. With regard to the influence of technological change, a significant body of knowledge has argued for the important (albeit complex) role of new technologies in facilitating the pursuit of scientific progress (see Mokyr 2002, 2010). From Boyle's air pump (Shapin and Schaffer 1985) to Volta's pile (Pancaldi 2005), new technologies have enabled scientists to pursue more complex and distinctive research lines.

In the post–Vannevar Bush period, technology has been of particular importance in areas such as biology and physics (Knorr-Cetina 1999). In biology, the invention and automation of the polymerase chain reaction (PCR), DNA sequencing and DNA synthesis have, among other technologies, opened a wealth of new biological research lines and, at least in part, been the driving force behind new modes of organization (Huang and Murray 2009). Likewise in physics, the development of new, more powerful telescopes and massive particle colliders (each with their attendant computing power) have enabled the exploration of new knowledge frontiers, while at the same time changing the lives of physicists and their ways of collaborating and organizing of research (Galison 1997). Beyond the specific technologies of knowledge production, recent work (e.g., Agrawal and Goldfarb 2008; Ding et al. 2010) has highlighted the ways in which the coming of the Internet shaped the organization of research and the extent and nature of collaboration versus integration. In particular, data (from engineering research) show that faculty in middle-ranking universities have seen the greatest organizational change, becoming more likely to be engaged in top-tier collaborations than prior to Bitnet introduction.

An alternative, or perhaps more accurately a complementary, perspective on the changing organization of science is articulated by Ben Jones who outlined the importance of the burden of knowledge on a researcher's organizational choices (Jones, Wuchty, and Uzzi 2008; Jones 2009). His line of argument focuses on the growing length of scientific training as scientists seek to accumulate an ever-growing body of knowledge in order to make contributions at the frontier. As a corollary to the increasing requirement for training, scientists are accordingly becoming narrower in their expertise and more highly specialized—an effect he refers to as the death of the Renaissance man (Jones 2009). According to the burden of knowledge argument, the combined need for more and more specialized knowledge leads researchers into pursuing their chosen research lines through higher levels of collaboration. In support of this perspective are data on the rise in the number of authors on scientific publication across all fields (Wuchty, Jones, and Uzzi 2007).

Our argument is that changing the technology and specialization costs (and benefits) of scientific knowledge production takes us only part way toward understanding the organizational choices of researchers at any given moment in time. Moreover, it fails to account for the fact that, while technology and specialization has surely changed over the past fifty years, so too has the nature of credit. While less observable in laboratories than the changing equipment and less immediate than the growing shoulders of giants upon which all scholars must stand, credit too has likely changed. In particular, with the expansion of the scientific community (both within the United States and more globally), simple informal networks and scientific inner circles are less likely to be effective at adjudicating questions of credit. Likewise, the observable rise in collaboration challenges traditional institutions of credit and its allocation. Thus, credit becomes central to the calculus of academic scientists. In what follows, we offer a limited "credit history" of the past fifty years as a window into the changing credit allocation process and the exchange rates in the currencies of credit.

4.3 Credit History

The intangible nature of credit and its allocation makes it challenging to trace and demonstrate. To overcome this invisibility, we develop a short history of the debates around credit and the scientific reward system as told by scientists themselves in the pages of their journals: a window into a narrative we refer to as a credit history. This history is basically gleaned from the issues that animated the editorial pages of major research journals—*Science* and *Nature* as well as some medical journals, and the *Chronicle of Higher Education* (CHE). While not comprehensive, our explication of the credit arguments that animated scientists serve as a window into the challenges that they confront as they wrestle with their autonomous credit system.

Two debates of particular import can be traced through the historical record that are salient to the link between credit and the organization of scientific knowledge production: first, authorship conventions (including ordering and ghost authorship), which link to trade-offs around credit and collaboration; second, salami slicing, which speaks to the role of credit in choices of integration versus citation markets.

4.3.1 Authorship Conventions and Credit

The most important debates that animate scientists as they consider the role of credit are those that explicitly link authorship and credit allocation

(Cawkell 1976). Gaeta[1] (1999) coined the term authorship "law and order" to connote not only the rules of authorship, that is, the "law," but also the specific role of ordering of authorship (see Gans and Murray [2013] for a more comprehensive theoretical treatment of author ordering and credit and Engers et al. [1999] for a theoretical treatment). At its core the debate raises the possibility that while some genuine changes in collaboration may be taking place (see Price and Beaver 1966) driven by specialization, thus accounting for a rise in average authorship, gratuitous authorship may also be increasing, particularly from the 1970s onward (Alberts and Shine 1994).

To highlight this possibility, Broad provides an example in his 1981 article on the topic. He notes: "The fellowship application for the American College of Physicians asks a candidate to list percent participation in studies in which he is a listed author. Though seemingly a workable solution, the accuracy of the resulting judgments has been called into question. In at least one instance, when a whole research team applied for fellowships, their total participation came to 300 percent" (Broad 1981, 1138).

Changing authorship norms (specifically norms adding further authors) places an increased burden on the reward system of science, particularly in the evaluation of young faculty at key career milestones. Over a fifty-year period, when the number of researchers (and their specialization) has increased, the evaluation of individual biographies (i.e., published contributions) has grown increasingly complex. Even in 1981, an editor of the *New England Journal of Medicine* noted that "You have to know the journals, and what impact they have. . . . You have to know the institutions, the people, the meetings. . . . It's a ticklish matter" (quoted in Broad [1981], 1139). Today, evaluative choices for promotion, tenure, grant making, and a wealth of other forms of scientific credit rely on publishing records, even while those records are increasingly murky and hard to interpret (see Simcoe and Waguespack [2011] for an analysis of the impact of missing authors). Likewise, scientists themselves must make organizational choices over collaboration, integration, or publishing in the shadow of a complex credit allocation process: one that is beset with indeterminacy over credit and authorship norms (Häussler and Sauermann 2013).

In responding to this debate, a number of scientific journals have taken the lead in asking authors to carefully document their contributions to scholarly publications. Most notably, on New Year's Day 2010, Bruce Alberts, editor-in-chief at *Science* magazine, published an editorial promoting scientific standards and focusing on authorship issues. In it, he described a change in policy to discourage "honorary authorship" in which:

> before acceptance, each author will be required to identify his or her contribution to the research (see www.sciencemag.org/about/authors). Sci-

---

1. http://www.ncbi.nlm.nih.gov/pubmed/10230981.

ence's policy is specifically designed to support the authorship requirements presented in *On Being a Scientist: Third Edition*, published by the
US National Academy of Sciences. That report emphasizes the importance of an intellectual contribution for authorship and states that "Just
providing the laboratory space for a project or furnishing a sample used
in the research is not sufficient to be included as an author." (Alberts
2010, 12)

This view was echoed in a 2012 column in *Nature*, in which the author
argued that "[w]hen it comes to apportioning credit, science could learn from
the movies" (Frische 2012, 475).

In other instances, faculty themselves have developed internal "lab norms"
to adjudicate authorship claims. As recently described in *Science*, Harvard
psychology professor Stephen Kosslyn has developed his own points system,
which he describes in detail on the website for his laboratory. "Anyone who
works with him on a project that results in a paper can earn up to 1,000
points, based on the extent of their contribution to six different phases of
the project: idea, design, implementation, conducting the experiment, data
analysis, and writing. The first and last phases—idea and writing—get the
most weight. Those who make a certain cut-off are granted authorship, and
their score determines their order on the list" (Venkatramen 2010). While
this example is unusual in its specificity, local norms at the field level are
continually evolving and fields such as high energy physics have moved to
norms that rely on alphabetical ordering of the many hundreds of authors
that are often part of a paper that relies on massively costly, centralized
technology for the production of knowledge.

Not only an issue for credit, authorship conventions—rather than authorship that simply reflects changes in underlying research organization—also
raise questions of responsibility and liability for research findings and for the
potential "false science," including fraud (see Furman, Jensen, and Murray
[2012] and Azoulay et al. [2013] for a broader analysis of retractions in this
context). To combat this challenge and the liabilities (as well as the credit) that
arises with publishing, *Science*, in the same editorial noted above, also outlines a policy in which senior authors record that they have personally examined original data and attest to its appropriate presentation (Alberts 2010).

4.3.2 Salami Slicing and Credit

A second major theme that emerges in our credit history takes the colour-
ful label "salami slicing." It describes an ongoing debate regarding the
"size" of the least publishable unit (LPU) along a research line. This debate
emphasizes the organizational choices between integration (leading to a
larger published slice) on the one hand and publication (of a smaller slice
of the research line) with follow-on citation by other researchers (or by self-
citation by the research team). The question posed by scientists is whether
or not to publish the small step embodied in project 1 and wait for citation

by another researcher pursuing project 2 (or pursing self-citation) or to complete projects 1 and 2 before publishing a larger slice of research, thus making a larger contribution.

There is clear statistical evidence from the 1970s onward to support the claim that publication length, at least in the biological and physical sciences, is shortening (although Card and DellaVigna [2013] show evidence that in contrast, publications in economics are longer). While not conclusive evidence of the rise of LPUs, anecdotal evidence supports the publishing dilemma of young faculty and the link between LPUs and credit. One dean of science described the dilemma in an article in CHE as follows: "In order to appear to have more publications on their CVs, young scholars are often advised to break their research down into pieces and publish those pieces in multiple articles—i.e., LPUs. . . . Having a couple of LPUs will ensure that the bean counters cannot assail her record. We both know that there are those among us who would easily ignore her aggressive pursuit of grants and a single brilliant paper in *Cell* if her four years here did not include the magic two papers."[2]

Far from being a new issue, discussions over LPUs and the link between publication strategies and credit can be traced back through the editorial pages of *Science* at least to 1981 (Broad 1981). In a provocative article, the careers of young scientists in the 1980s—who typically had between fifty and one hundred publications at the time of promotion—were contrasted with scientists from the late 1950s such as James Watson who, when being evaluated by peers had only eighteen papers (albeit one that described the structure of DNA). Broad notes the emergence of the LPU and argues that "the increases stem not from a sharp rise in productivity but rather from changes in the way people publish. Coauthorship is on the rise, as is multiple publication of the same data" (1137). He also notes the challenge for credit allocation arguing that, in combination, LPU practices and changing coauthorship obfuscate the effort made by young scholars, making evaluation and credit much more challenging.

More recently, in 2005, the journal *Nature Materials* explored the impact on the sustainability of scientific publishing of what it referred to as "fragmenting single coherent bodies of research into as many publications as possible—the practice of scientific salami slicing."[3] Scientists also speculate that salami slicing potentially leads to a much greater likelihood that publications will be plagiarized (at least in part), be overlapping, or in other ways cross the boundary into false science. It also has important implications for the effectiveness and capacity of scientists to engage in meaningful peer review and credit allocation. As the editors of *Nature Materials* outlined, poor practices associated with salami slicing "deny referees and editors the

2. http://chronicle.com/article/In-Defense-of-the-Least/44761.
3. http://www.nature.com/nmat/journal/v4/n1/full/nmat1305.html.

opportunity of assessing the true extent of its contribution to the broader body of research" raising the question of credit allocation for researchers selecting between integration and publishing.

While incomplete in their coverage, these two examples of credit history highlight both the central importance and the critical challenges associated with the role of credit in the organization of science. Far from being a wrinkle in the changing tapestry of science, scientists are making a range of organizational choices in the shadow of the complex and changing beliefs about the nature and meaning of scientific credit. In what follows, we bring clarity to these issues by developing a formal model that links credit with technology and specialization.

4.4 Formal Model

The combined elements of technology, knowledge burden, and credit institutions together shape the observable outputs of science—the number of publications, the number of authors, the rate of progress and its direction. More importantly, they inform the underlying organization of science: decisions made by scientists about "laboratory life" as it pertains to any particular research line the laboratory is pursuing. With this in mind, we provide a formal model of the drivers of the (optimal) choice of organizational form for cumulative science and, in particular, how this choice is driven by institutions to allocate credit to individual scientists for their role in knowledge production.

4.4.1 Basic Set-Up

We consider an environment whereby knowledge is created by cumulative scientific research. Specifically, we focus on a pioneer scientist's (P) decisions with respect to an initial scientific project, 1. Following Green and Scotchmer (1995) (and also Bresnahan 2011), we assume that the (opportunity) cost to the scientist of pursuing 1 is c_1. The stand-alone (expected) quality of project 1 is x.[4]

A follow-on project that builds on 1, project 2, may also be possible and can be conducted by P or another follow-on scientist, F. For a scientist, i, with in-depth knowledge of project 1, the probability that they perceive the project 2 opportunity is p_i. To acquire the necessary in-depth knowledge of project 1 costs a scientist, C_i, so long as they have access to project 1's knowledge in the first place. The idea here is that, while project 1 knowledge may be disclosed (say through publication or communication), understanding it

---

4. It is useful to note that the publication and associated citation plays a similar role to ex post licensing in the Green and Scotchmer (1995) model except that the key parameters are market determined rather than determined through bilateral negotiation. That said, Green and Scotchmer (1995) do bring some of those factors into play when they consider how a planner might set patent length as well as antitrust policy.

in a way that leads to follow-on research takes additional effort (for instance, by undertaking replication studies). That said, an alternative interpretation of C_i is as a communication cost. If project 2 is possible, it comes with an expected quality, y, and research (opportunity) cost, c_{2i}.

As noted in our discussion above, there are three ways research into projects 1 and 2 can be organized. First, under *integration*, P conducts both projects before publishing the results under sole authorship. Second, under *collaboration*, P collaborates with another scientist, F, over both projects. In this situation, each focuses on aspects of the project they can do at least cost but pool their skills and communication in understanding the implications of project 1 for the project 2 opportunity. Third, under *publication*, P publishes project 1 results and then F conducts research into project 2, citing back to P's initial contribution. Under both collaboration and publication, the market awards P and F with attribution regarding each scientist's contributions. A key focus here will be on how that attribution takes place.

Integration. In this option, P pursues both projects. Of key importance is that the entire quality of research, should it take place, is attributed to P. Thus, P's expected payoff is:

(1) $v_P^{Int} = \max[x - c_1 + \max[p_F(y - c_{2P}) - C_P, 0], 0]$.

Note that it is entirely possible that project 1 has no stand-alone value (i.e., $x = 0$) and its value rests solely on its ability to lead to research in project 2.

Collaboration. Under collaboration, P identifies F ex ante, and they choose to pursue both projects jointly.[5] The first consequence of this is that the costs of understanding the implications of project 1 for project 2 are shared across scientists. To this end, we assume that these costs are C_{PF} and can be allocated to P and F through internal bargaining, with P's share being s. Similarly, we assume that the consequent probability that project 2's opportunity is perceived is p_{PF}.

The second consequence of collaboration is that coauthorship is formally given to both P and F on projects 1 and 2. Of course, one can imagine a scenario whereby this is only done with respect to project 2 but, as explained below, this does not necessarily lead to different conclusions regarding whether collaboration is chosen. If both projects are successful, the research quality of their collaborative effort is $x + y$. However, the market—comprised of scientific peers—will award each with personal attribution of that output. We assume that the attribution going to scientist i is α_i rather than simply equal sharing on the basis that the market may have some reason to assign differential weights to each scientist. Otherwise we assume that attribution has to be the same in equilibrium; that is, $\alpha_P = \alpha_F = \alpha$. Importantly, we make no assumption that $\alpha_P + \alpha_F = 1$. Indeed, (as observed in the example

---

5. There is an issue associated with whether F can be simply identified or not. As we note below, publication can work without this condition.

from medical research above) credit could be greater than 1, although we assume that the total quality from the projects can be no greater than $x + y$.

Under collaboration, the expected payoffs to each scientist are:

$$v_P^{Col} = \max[\alpha_P x - c_1 + \max[p_{PF}\alpha_P y - sC_{PF}, 0], 0],$$ (2)

$$v_F^{Col} = \max[\alpha_F x + \max[p_{PF}(\alpha_F y - c_{2F}) - (1 - s)C_{PF}, 0], 0].$$ (3)

This reflects the notion that P has the lowest cost associated with conducting project 1 and that P and F choose the scientist with the lowest cost to conduct project 2. The allocation of the costs, C_{PF}, is assumed to be determined internally. To keep things simple, it will be assumed that all of the internal bargaining rests with P and so s is the minimal amount (if it exists) that will ensure that F collaborates.

To see what s will be, let us assume that it is jointly profitable for project 1 to be investigated and, individually, profitable for project 2 to proceed (i.e., $\alpha_F y \geq c_{2F}$). In this case, from equation (3), the minimal s that allows F to earn a positive return is:

$$\hat{s} = 1 - \frac{\alpha_F x + p_{PF}(\alpha_F y - c_{2F})}{C_{PF}}.$$ (4)

Thus, for P, its expected return is:

$$v_P^{Col} = (\alpha_P + \alpha_F)(x + p_{PF}y) - c_1 - C_{PF} - p_{PF}c_{2F}.$$ (5)

Note that the total surplus from collaboration is:

$$\max[x - c_1 + \max[p_{PF}(y - c_{2F}) - C_{PF}, 0], 0].$$ (6)

Importantly, while the market can potentially award P and F a "free lunch" if $\alpha_P + \alpha_F > 1$, total surplus only involves the "real" variables.

Publication. The third organizational option for cumulative science is for P to research and publish the results of project 1 and then for another scientist, F, to investigate this project outcome and (potentially) perceive and research project 2. For F, should P publish their research from project 1, they will have a choice as to whether to conduct an in-depth investigation of that research and, if that provides an opportunity, research and publish project 2. It is assumed that if project 2 is published that it includes a citation to P's research in project 1. The market will then partially attribute credit for some of project 2 to P as a share β_P of y and attribute β_F of y to F.

Given this, F's expected payoff following a publication by P is:

$$v_F^{Pub} = \max[p_F(\beta_F y - c_{2F}) - C_F, 0].$$ (7)

If F's expected payoff is negative, we assume here that publication is infeasible as, if they are given the choice, P would prefer integration to publication. However, if F's expected payoff is positive and research into project 2 goes ahead, P's expected payoff is:

Table 4.1   **Value attributed to each scientist**

	P	F
Integration	$1(x + y)$	0
Collaboration	$\alpha_P(x + y)$	$\alpha_F(x + y)$
Publication	$x + \beta_P y$	$\beta_F y$

(8) $$v_P^{Pub} = \max[x - c_1 + p_F\beta_P y, 0].$$

In this case, social surplus from publication is:

(9) $$\max[x - c_1 + p_F(y - c_{2F}) - C_F, 0].$$

Again, we assume that even if $\beta_P + \beta_F \neq 1$ the social surplus from project 2 if it is successful is y.

Table 4.1 summarizes the value attributed to each scientist under each organizational mode.

4.4.2 Equilibrium Choices

We now turn to consider P's organizational choice for research. While the specification above allows for the possibility that under some organizational forms both stages of research will be completed while under others one or neither will be completed, this model can be more easily exposited if we assume that, regardless of organizational mode, both stages of research are completed.

To that end we assume the following:

(A1) $$x - c_1 + p_F(y - c_{2P}) \geq C_P$$
(both projects are undertaken under integration);

(A2) $$(\alpha_P + \alpha_F)(x + p_{PF}y) - c_1 - c_{2F} \geq C_{PF}$$
(both projects are undertaken under collaboration); and

(A3) $$x - c_1 + p_F\beta_P y \geq 0 \quad \text{and} \quad p_F(\beta_F y - c_{2F}) \geq C_F$$
(both projects are undertaken under publication).

Thus,

(10) $$v_P^{Int} = x - c_1 + p_P(y - c_{2P}) - C_P$$

(11) $$v_P^{Col} = (\alpha_P + \alpha_F)(x + p_{PF}y) - c_1 - C_{PF} - p_{PF}c_{2F}$$

(12) $$v_P^{Pub} = x - c_1 + p_F\beta_P y$$

(13) $$v_F^{Pub} = p_F(\beta_F y - c_{2F}) - C_F.$$

In addition, we assume that:

(A4) $$c_{2F} \leq c_{2P}.$$

That is, F has a comparative advantage in conducting stage 2 (e.g., P has a higher opportunity cost of their time). Under these assumptions, P will choose $\max\{v_P^{Int}, v_P^{Col}, v_P^{Pub}\}$. It is instructive to consider the pairwise choices between these organizational forms.

Collaboration versus Integration. Collaboration will be chosen by P if:

(14) $$v_P^{Col} \geq v_P^{Int}$$

$$\Rightarrow (\alpha_P + \alpha_F - 1)x + ((\alpha_P + \alpha_F)p_{PF} - p_P)y$$

$$\geq p_{PF}c_{2F} - p_P c_{2P} + C_{PF} - C_P.$$

By contrast, collaboration is socially superior to integration if:

(15) $$x + p_{PF}(y - c_{2F}) - C_{PF} \geq x - c_1 + p_P(y - c_{2P}) - C_P$$

$$\Rightarrow (p_{PF} - p_P)y$$

$$\geq p_{PF}c_{2F} - p_P c_{2P} + C_{PF} - C_P.$$

It is clear that in our modeling set up, the social choice and P's choice will coincide if and only if $\alpha_P + \alpha_F = 1$, a result that sharply contrasts to scientists' own discussions and informal evidence which, as we outlined, suggest that many scientists believe that $\alpha_P + \alpha_F > 1$. Based on our model, if collaboration allows knowledge transfer costs (C_{PF}) to be shared between both scientists, overweighting the collaboration rewards would encourage overcollaboration.

Publication versus Integration. Publication will be chosen by P if:

(16) $$v_P^{Pub} \geq v_P^{Int} \Rightarrow p_F\beta_P y \geq p_P(y - c_{2P}) - C_P.$$

In this case, publication is socially preferable to integration if:

(17) $$x - c_1 + p_F(y - c_{2F}) - C_F \geq x - c_1 + p_P(y - c_{2P}) - C_P$$

$$\Rightarrow (p_P - p_F)y + p_F c_{2F} + C_F$$

$$\leq p_P c_{2P} + C_P.$$

Note, however, that because we assume that, under publication, F will choose to conduct project 2, $p_F\beta_F y \geq p_F c_{2F} + C_F$. Thus, if publication is chosen by P we know that:

(18) $$(p_P - p_F\beta_P)y - p_P c_{2P} - C_P \leq p_F\beta_F y - p_F c_{2F} - C_F$$

$$\Rightarrow (p_P - p_F(\beta_P + \beta_F))y + p_F c_{2F} + C_F$$

$$\leq p_P c_{2P} + C_P.$$

This is a necessary condition for publication to be chosen by P. Thus, if $\beta_P + \beta_F \leq 1$, if publication is chosen in equilibrium then it is socially preferable to integration. However, if $\beta_P + \beta_F > 1$, it is possible that publication will be chosen in equilibrium when it is not socially preferable to integration. Specifically, equation (18) may hold when equation (17) does not.

Publication versus Collaboration. Publication will be chosen by P over collaboration if:

$$(19) \qquad v_P^{Pub} \geq v_P^{Col}$$

$$\Rightarrow (1 - \alpha_P - \alpha_F)x + ((\alpha_P + \alpha_F)p_{PF} - p_F\beta_P)y$$

$$\leq p_{PF}c_{2F} + C_{PF}.$$

In this case, publication is socially preferable to collaboration if:

$$(20) \qquad (p_{PF} - p_F)y + p_Fc_{2F} + C_F \leq p_{PF}c_{2F} + C_{PF}.$$

As above recall that, $p_F\beta_F y \geq p_Fc_{2F} + C_F$. Thus, if publication is chosen in equilibrium by P, then

$$(21) \quad (1 - \alpha_P - \alpha_F)x + (\alpha_P + \alpha_F)p_{PF}y - p_F\beta_P y \leq p_{PF}c_{2F}$$

$$+ C_F \Rightarrow (1 - \alpha_P - \alpha_F)x + (\alpha_P + \alpha_F)p_{PF}y$$

$$- p_F\beta_P y \underbrace{-p_F\beta_F y + p_Fc_{2F} + C_F}_{<0} \leq p_{PF}c_{2F} + C_F.$$

This is a necessary condition for publication to be chosen by P. Thus, if $\alpha_F + \alpha_P = 1$ and $\beta_P + \beta_F \leq 1$ then, if publication is chosen by P, it will also be socially preferred to collaboration. However, if $\beta_P + \beta_F > 1$, then it is possible that publication will be chosen in equilibrium when it is not socially preferable to collaboration. Specifically, equation (21) may hold even when equation (20) does not hold.

4.4.3 Optimal Attribution

The above analysis suggests that setting $\alpha_F + \alpha_P = 1$ and $\beta_P + \beta_F = 1$ may have some desirable properties. However, it also demonstrated that inefficient outcomes can arise involving each of the three evaluated organizational choices. Consequently, we consider what levels of $(\alpha_F, \alpha_P, \beta_F, \beta_P)$ might generate an efficient outcome, the idea being to imagine that these parameters were chosen by a planner and to evaluate their properties exploring how this is in accordance (or discordance) with what actually happens in science. To this end, we follow Green and Scotchmer (1995) considering situations where the follow-on scientist, F, may earn no surplus as a convenient means of avoiding having to deal with the potential range of parameters that may give rise to an efficient outcome. The idea here being that since P's choice determines the outcome, it makes sense to ensure that as much of the surplus goes to P as possible.

PROPOSITION 1. *There exists* $(\alpha_P, \alpha_F, \beta_P, \beta_F)$ *such that* $\alpha_F + \alpha_P = 1$ *and* $\beta_P + \beta_F = 1$ *that results in an equilibrium choice for* P *that is efficient.*

The proposition is easily proved by finding the $\hat{\beta}_F$ such that $v_F^{Pub} = 0$ and letting $\hat{\beta}_P = 1 - \hat{\beta}_F$ and substituting $\alpha_F + \alpha_P = 1$ so that:

$$v_P^{Col} \geq v_P^{Int} \Rightarrow (p_{PF} - p_P)y \geq p_{PF}c_{2F} - p_P c_{2P} + C_{PF} - C_P$$

$$v_P^{Pub} \geq v_P^{Int} \Rightarrow (p_P - p_F)y + p_F c_{2F} + C_F \leq p_P c_{2P} + C_P$$

$$v_P^{Pub} \geq v_P^{Col} \Rightarrow (p_{PF} - p_F)y + p_F c_{2F} + C_F \leq p_{PF}c_{2F} + C_{PF}.$$

These are identical to conditions in equations (15), (17), and (20). This demonstrates that the choices among each of the organizational forms are driven by the same conditions as the socially optimal choices.

Specifically, note that:

$$(22) \qquad (\hat{\beta}_P, \hat{\beta}_F) = \left(\frac{p_F(y - c_{2F}) - C_F}{p_F y}, \frac{p_F c_{2F} + C_F}{p_F y} \right),$$

such that so long as they sum to 1 it is arbitrary what α_F and α_P are. The reason for this is quite intuitive: with collaboration, we allowed P and F to negotiate cost sharing, but given the structure this allowed them to transfer utility. Thus, the decision was driven wholly by the joint market reward to collaboration rather than the precise division. No such instrument existed for publication and hence, the market rewards needed to determine division as well as overall value in order to generate a socially optimal outcome.

It is useful to consider how Proposition 1 might change if, in fact, s (the share of costs accruing to P under collaboration) was fixed and nonnegotiable.

PROPOSITION 2. *When s is fixed, there exists $(\alpha_P, \alpha_F, \beta_P, \beta_F)$ such that $\alpha_F + \alpha_P = 1$ and $\beta_P + \beta_F = 1$ that results in an equilibrium choice for P that is efficient.*

The proof proceeds using the same method as Proposition 1. In this case, the range of β_i remains as in equation (22). By contrast, the market weights for collaboration become:

$$(23) \quad (\hat{\alpha}_P, \hat{\alpha}_F) = \left(\frac{x + p_{PF}(y - c_{2F}) - (1 - s)C_{PF}}{x + p_{PF}y}, \frac{p_{PF}c_{2F} + (1 - s)C_{PF}}{x + p_{PF}y} \right).$$

In this case, it can easily be demonstrated that the payoffs realized are the same as in Proposition 1.

4.4.4 A Note on Social Surplus

Thus far we have been somewhat loose in our consideration of socially optimal outcomes. The propositions focus on efficiency, which is the expected difference between research quality realized and the costs incurred under the chosen organizational form. However, this is distinct from social efficiency that would take into account the broader impact of the research. For a very significant medical breakthrough, for example, the social surplus from that research may vastly exceed the assessed quality of that research accruing to scientists. In this case, the costs realized by these scientists would be very small relative to social outcomes. Thus, to generate a socially efficient outcome would require giving scientists kudos well above 1 even for integrated research lines.

Given this, how should we interpret Propositions 1 and 2? The first reasonable assumption is that a research paper has an assessed quality that is independent of the organizational form of the team that generated it. This is what the parameters x and y capture. As a result, we interpret $x + y$ as the maximum kudos that can be given to a research line that is sole authored. Given this, the shares, $(\alpha_P, \alpha_F, \beta_P, \beta_F)$, represent the relative shares given to collaboration or publication compared with that for a sole-authored paper. Thus, Propositions 1 and 2 ask if it is efficient to give collaboration or publication a different weight to integrated paper kudos that is fixed. We find that such a distinct weight is not warranted. However, that is a distinct weight relative to the level of kudos under integration. If that kudos could be aligned with social value, then that could generate a socially optimal outcome. For the moment, an interpretation as a relative weight is as far as we can go here.

4.5 Some Implications

Having constructed a model of organizational choice for cumulative science, we now turn to consider a number of issues raised in our credit history and discuss what insights the model gives us into how these may be reasonably resolved. We must emphasize that this analysis is suggestive rather than conclusive. Specifically, we do not know what determines the allocation of credit in science. Our formal model tells us what that allocation might look like if it was indeed efficient but, in fact, the processes by which these actually arise have likely been changing over time and are subject to various informational limitations that will lead to allocation being an inferred outcome rather than a precise one.

4.5.1 Collaborative Bias

The first element of our credit history explored the relationship between credit and decisions to collaborate. In particular, we illustrated the concerns scientists have expressed over rising collaboration and the attendant difficulties in credit allocation. This is particularly troublesome if there are trends that lead to overcollaboration. Specifically, it has been claimed that the market weights on a collaborative research project are greater than the weights that would arise if that project were not collaborative. This goes beyond the potential higher quality of such projects to whether, in fact, the market does and should divide the quality of collaborative projects by the number of authors when assigning attribution of credit. In other words, what are the implications of an inflated assessment of $\alpha_P + \alpha_F$ on the overall organizational form chosen?

As it turns out, if $\alpha_P + \alpha_F$ is set exogenously (assumed here to be > 1), then only one market weight remains to be determined: β_P. This is because β_F is determined as the value that leaves $v_P^{Pub} = 0$ and so is unchanged from Propositions 1 and 2. The issue becomes that β_P must do two things. It must continue to balance P's incentives with respect to publication versus integration. And it must now rebalance P's incentives with respect to publication and collaboration. It is easy to see that that latter task means that the optimal attribution, $\hat{\beta}_P$, will lie above the levels in Propositions 1 and 2. Thus, as a result of a bias to market weights on collaboration, not only will we observe socially suboptimal overcollaboration, but also overpublication as well. The point here is that these decisions interact and so any analysis of patterns must take into account all of the organizational form options facing scientists.

4.5.2 Blurred Contributions and Salami Slicing

Both with regard to the link between collaboration and credit, and in considering salami slicing, scientists have argued that a major challenge lies in the increasingly blurred relationship between actual observable effort in knowledge production and credit allocation. Among other things, this blurring arises because of the rise and increased geographic dispersion of the scientific community. More specifically, this likely means that the roles of individual scientists in collaborative endeavors have become increasingly blurred. Propositions 1 and 2 both suggested that optimal attribution would depend on factors specific to the project but importantly specific to the scientists themselves and their roles (as pioneer and follower, respectively).

If these factors are less known in more recent science compared to that in the past, what impact might this have on the choice of organizational form for cumulative science? The challenge here is to consider whether blurring will systematically bias organizational choices. To that end, we focus here on the new prominence given to citation counts. Basically, the value of a given research output is increasingly measured by the number of citations it receives above and beyond other factors. Within the context of our model here, that means that there is a systematic increase in β_P if there is follow-on research, while there is no necessary trade-off between the level of β_P and the level of β_F.

The direct impact of this, as predicted by the model, would appear to be an increasing bias toward publication as an organizational choice at the expense of collaboration and integration that may be more efficient. In other words, this result provides a direct link between blurring of credit, a rise in the use of citations, and the move to salami slicing that has been so widely documented and discussed among scientists.

4.5.3 The Matthew Effect

A third issue that has animated scholars of science more than scientists themselves is the Matthew Effect, which, as noted earlier, argues that more famous scientists (in terms of their past achievements or positions at elite institutions) receive more credit—in citations and kudos in collaborative projects. The issue is whether such credit is proportionate to their actual contribution (which may be high for the same reasons they are famous) or disproportionate. There is, to our knowledge, no formal economics model that derives the Matthew Effect in its disproportionate form as an equilibrium phenomenon.[6] It is instructive, therefore, to discuss this in the context of the model presented here. Specifically, if the market weights to collaboration and publication are determined optimally, what does this say about the Matthew Effect?

To consider this, let us focus on the pioneer scientist (P). There are several parameters that relate to P's ability to contribute to project 2. There is p_P, the probability that P has an insight that perceives the opportunity of project 2. There is C_P, the costs associated with understanding project 1 that will generate that insight. There is c_{P2}, P's costs associated with carrying out project 2. Finally, in a situation where it is exogenous, there is s, P's contribution to insight in a collaborative venture. The only time a contributive driver for P impacts a market weight is for s, when it is exogenous in a collaborative venture. In this case, if s is higher, P will receive more of the weight in kudos associated with collaboration. However, the implication is that F will receive too *little* weight in such collaborations and so collaboration will be infeasible. This will lead to more publication/integration than is efficient. However, in both those organizational forms there is no Matthew Effect distortion realized. The conclusion here is that for a pioneer scientist, the Matthew Effect is not a clear prediction of this model.

When it comes to a famous scientist's role as a follow-on researcher (F), there is more impact. If that scientist finds it less costly to engage in project 2 research (c_{F2}), then the market weights F receives under publication and collaboration both rise. Otherwise, drivers that are specific to F only impact the weight a scientist receives under publication; specifically, the higher are p_F and C_F, the more diminished is the market kudos flowing back to P for a citation. Note here, however, that while it is often said that the Matthew Effect works to provide a famous scientist with more citations, here it is operating to deny kudos flowing back to earlier researchers.

That said, the market weights for publication and collaboration are, in reality, given by market assessments of the underlying drivers as in equations (22) and (23). If, because of fame, these assessments are distorted upward for one scientist, this may have an impact on the relative choices of organizational forms. In particular, a market bias in favor of P that is understood

6. There are, however, evolutionary and other models where the Matthew Effect arises. For instance, Price (1976), David (1994), and Simkin and Roychowdhury (2005a, 2005b, 2006, 2007). However, the issue is that when scientists expect a Matthew Effect but are interested in extracting out a signal of actual contribution, it is difficult to sustain signals with a bias in equilibrium. Of course, this may be a flaw with the standard economic approach to equilibrium rather than a statement about the plausibility of the Matthew Effect in science itself.

by both scientists may render project 2 under publication infeasible for F (as their expected surplus was zero based on real variables and with a diminished market weight it will be negative). This would rule out publication as a choice for P. In addition, unless they can internally negotiate taking into account such biases, a diminished weight for F will render their participation in collaboration infeasible. Thus, a disproportionate weight on a famous P would have the effect of pushing organizational choice outcomes toward integration. Ironically, if this were done, the Matthew Effect would not be observable at all as it only arises under nonintegrated organizational forms.

4.6 Conclusions and Future Directions

In October 2013, François Englert and Peter Higgs were awarded the Nobel Prize for what is known as the "Higgs mechanism" that fills empty space that absorbs forces that were missing from observations. Robert Brout, who collaborated with Englert, had passed away but would have otherwise shared in the prize. As it turns out, that was just the beginning. There was a set of follow-on research that took the initial idea and built up the theory. Some of this was done very quickly but also independent of the initial contributions. Then the research program was "completed" by a very large team operating the Large Hadron Collider in Switzerland. This example shows not only how credit can be allocated, but also how follow-on contributions enable the value of initial contributions. Not surprisingly, this has given rise to debate as to whether Nobel prizes should include teams of research— both formal or part of a research line. Of course, in this case, the credit went to the pioneers, or stage 1, researchers.

Our chapter suggests that the experience in particle physics is commonplace and also that how credit is allocated will impact upon whether research lines are undertaken by formal teams or through a publication mechanism. Our formal model suggests that such choices have become important and have efficiency consequences. In addition, we suggest that the allocation of credit needs to reflect the balance of incentives in the organizational choices facing scientists.

However, this is just the beginning. What we have not addressed is how credit allocation arises in reality. While one can imagine institutions that may allocate credit to foster efficiency, it is not clear how existing credit allocations mechanisms including citations, collaboration, name ordering, careers, tenure, prestige, prizes, and phenomena naming actually function and what ends they promote. In another paper (Gans and Murray 2013) we consider how a market might evaluate the contribution of individual researchers when there is "team output" and show that the processes and mechanisms for credit allocation interact with market inferences and, hence, overall efficiency. But broader themes still remain including the impact of quantitative counts of credit (i.e., citation-based metrics) and the ongoing issue of the

Matthew Effect. These theoretical and empirical developments remain a task for follow-on research and the continuation of scientific investigation that we have continued here in this chapter.

References

Aghion, P., M. Dewatripont, and J. Stein. 2008. "Academic Freedom, Private-Sector Focus and the Process of Innovation." *RAND Journal of Economics* 39 (3): 617–35.
Agrawal, A., and Avi Goldfarb. 2008. "Restructuring Research: Communication Costs and the Democratization of University Innovation." *American Economic Review* 98 (4): 1578–90.
Alberts, Bruce. 2010. "Promoting Scientific Standards." *Science* 327 (5961):12.
Alberts, Bruce, and Kenneth Shine. 1994. "Scientists and the Integrity of Research." *Science* 266 (5191): 1660–61.
Arrow, K. 1962. "Economic Welfare and the Allocation of Resources for Invention." In *The Rate and Direction of Inventive Activity*, edited by R. Nelson, 609–25. Princeton, NJ: Princeton University Press.
Azoulay, P., J. Furman, J. Krieger, and F. Murray. 2013. "Retractions." NBER Working Paper no. 18499, Cambridge, MA.
Azoulay, P., J. Graff Zivin, and G. Manzo. 2010. "Incentives and Creativity: Evidence from the Academic Life Sciences." *RAND Journal of Economics* 42 (3): 527–54.
Biagioli, M. 2003. "Rights or Rewards: Changing Frameworks of Scientific Authorship." In *Scientific Authorship: Credit and Intellectual Property in Science*, edited by Mario Biagioli and Peter Galison, 253–79. New York: Routledge.
———. 2006. *Galileo's Instruments of Credit: Telescopes, Instruments, Secrecy*. Chicago: University of Chicago Press.
Bikard, M., F. Murray, and J. S. Gans. Forthcoming. "Exploring Trade-Offs in the Organization of Scientific Work: Collaboration and Scientific Reward." *Management Science*.
Broad, W. 1981. "The Publishing Game: Getting More for Less." *Science* 211 (4487): 1137–39.
Card, David, and Stefano DellaVigna. 2013. "Nine Facts about Top Journals in Economics." *Journal of Economic Literature* 51 (1): 144–61.
Cawkell, A. E. 1976. "Citations, Obsolescence, Enduring Articles, and Multiple Authorships." *Journal of Documentation* 32 (1): 53–58.
Cole, Jonathan R., and Stephen Cole. 1973. *Social Stratification in Science*. Chicago: University of Chicago Press.
Crane, Diana. 1969. "Social Structure in a Group of Scientists: A Test of the 'Invisible College' Hypothesis." *American Sociological Review* 34:335–52.
Dasgupta, P., and P. A. David. 1994. "Towards a New Economics of Science." *Research Policy* 23:487–521.
David, P. A. 1994. "Positive Feedbacks and Research Productivity in Science: Reopening Another Black Box." In *Economics and Technology*, edited by O. Grandstrand. Amsterdam: Elsevier.
———. 2008. "The Historical Origins of 'Open Science': An Essay on Patronage, Reputation and Common Agency Contracting in the Scientific Revolution." *Capitalism and Society* 3 (2): article 5. doi:10.2202/1932-0213.1040.

Ding, W. W., S. G. Levin, P. E. Stephan, and A. E. Winkler. 2010. "The Impact of Information Technology on Academic Scientists' Productivity and Collaboration Patterns." *Management Science* 56 (9): 1439–61.

Engers, M., J. S. Gans, S. Grant, and S. P. King. 1999. "First Author Conditions." *Journal of Political Economy* 107 (4): 859–83.

Frische, Sebastian. 2012. "It Is Time for Full Disclosure of Author Contributions." *Nature* 489 (7417): 475.

Furman, J., K. Jensen, and F. Murray. 2012. "Governing Knowledge Production in the Scientific Community: Quantifying the Impact of Retractions." *Research Policy* 41:276–90.

Furman, J., and S. Stern. 2011. "Climbing atop the Shoulders of Giants: The Impact of Institutions on Cumulative Research." *American Economic Review* 101 (5): 1933–63.

Gaeta, T. J. 1999. "Authorship: 'Law' and Order." *Academic Emergency Medicine* 6 (4): 297–301.

Galison, Peter. 1997. *Image and Logic: A Material Culture of Microphysics*. Chicago: Chicago University Press.

Gans, J. S., and F. Murray. 2013. "Markets for Scientific Attribution." Unpublished Manuscript, Massachusetts Institute of Technology.

Green, J., and S. Scotchmer. 1995. "On the Division of Profit in Sequential Innovation." *RAND Journal of Economics* 26 (1): 20–33.

Hagström, Warren O. 1965. *The Scientific Community*. New York: Basic Books.

Häussler, C., and H. Sauermann. 2013. "Credit Where Credit is Due? The Impact of Project Contribution and Social Factors on Authorship and Inventorship." *Research Policy* 42 (3): 688–703.

Huang, Kenneth, and F. Murray. 2009. "Does Patent Strategy Shape the Long-Run Supply of Public Knowledge? Evidence from the Human Genome." *Academy of Management Journal* 52 (6): 1193–221.

Jones, B. 2009. "The Burden of Knowledge and the 'Death of the Renaissance Man': Is Innovation Getting Harder?" *Review of Economic Studies* 76 (1): 283–317.

Jones, B., S. Wuchty, and B. Uzzi. 2008. "Multi-University Research Teams: Shifting Impact, Geography, and Stratification in Science." *Science* 322 (5905): 1259.

Knorr-Cetina, K. 1999. *Epistemic Cultures: How the Sciences Make Knowledge*. Cambridge, MA: Harvard University Press.

Merton, Robert. 1957. "Priorities in Scientific Discovery." *American Sociological Review* 22 (6): 635–59.

———. 1968. "The Matthew Effect in Science: The Reward and Communication Systems of Science are Considered." *Science* 159 (3810): 56.

Mokyr, Joel. 2002. *The Gifts of Athena: Historical Origins of the Knowledge Economy*. Princeton, NJ: Princeton University Press.

———. 2010. *The Enlightened Economy: An Economic History of Britain, 1700–1850*. New Haven, CT: Yale University Press.

Murray, F. 2010. "The Oncomouse that Roared: Hybrid Exchange Strategies as a Source of Productive Tension at the Boundary of Overlapping Institutions." *American Journal of Sociology* 116 (2): 341–88.

Murray, F., P. Aghion, M. Dewatripont, J. Kolev, and S. Stern. 2009. "Of Mice and Academics: Examining the Effect of Openness on Innovation." NBER Working Paper no. 14819, Cambridge, MA.

Murray, F., and S. Stern. 2007. "Do Formal Intellectual Property Rights Hinder the Free Flow of Scientific Knowledge? An Empirical Test of the Anti-Commons Hypothesis." *Journal of Economic Behavior and Organization* 63 (4): 648–87.

Nelson, R. R. 1959. "The Simple Economics of Basic Scientific Research." *Journal of Political Economy* 67 (3): 297–306.

Pancaldi, G. 2005. *Volta: Science and Culture in the Age of Enlightenment*. Princeton, NJ: Princeton University Press.

Price, D. J. 1976. "A General Theory of Bibliometric and Other Cumulative Advantage Processes." *Journal of the American Society for Information Science* 27:292–306.

Price, D. J., and D. B. Beaver. 1966. "Collaboration in an Invisible College." *American Psychologist* 21 (11): 1011–18.

Rosenberg, N. 1990. "Why Do Firms Do Basic Research (with Their Own Money)?" *Research Policy* 19 (2): 165–74.

Shapin, Steven, and Simon Schaffer. 1985. *Leviathan and the Air-Pump: Hobbes, Boyle and the Experimental Life*. Princeton, NJ: Princeton University Press.

Simcoe, T. S., and D. M. Waguespack. 2011. "Status, Quality, and Attention: What's in a (Missing) Name?" *Management Science* 57 (2): 274–90.

Simkin, M. V., and V. P. Roychowdhury. 2005a. "Copied Citations Create Renowned Papers?" *Annals of Improbable Research* 1 (1): 24–27.

———. 2005b. "Stochastic Modeling of Citation Slips." *Scientometrics* 62:367–84.

———. 2006. "Theory of Aces: Fame by Chance or Merit?" *Journal of Mathematical Sociology* 30:33–42.

———. 2007. "A Mathematical Theory of Citing." *Journal of the American Society for Information Science and Technology* 58 (11): 1661–73.

Venkatramen, Vijaysree. 2010. "Conventions of Authorship." *Science Careers*. http://sciencecareers.sciencemag.org/career_magazine/previous_issues/articles/2010_04_16/caredit.a1000039.

Wuchty, S., B. Jones, and B. Uzzi. 2007. "The Increasing Dominance of Teams in Production of Knowledge." *Science* 316 (5827): 1036.

II

The Geography of Innovation

5

The Rise of International Coinvention

Lee Branstetter, Guangwei Li, and Francisco Veloso

5.1 Introduction

For decades, international economists and development economists have worked with models that posit a kind of ladder of economic development. Countries begin the development process as largely agricultural economies. As they accumulate skill, capital, and technology, economies move into more complex manufacturing and service activities. Finally, after decades of development and steady increases in income, countries begin to create new-to-the-world technology. However, this is something that emerges at the end of the development process in the standard models (Vernon 1966; Krugman 1979; Grossman and Helpman 1990, 1991).

Despite many years of impressive growth, China, and especially India, are still in the early stages of the conventional development process—this is evidenced by their still low levels of per-capita output and income. China

Lee Branstetter is professor of economics and public policy at Carnegie Mellon University, a nonresident senior fellow at the Peterson Institute for International Economics, and a research associate of the National Bureau of Economic Research. Guangwei Li is a PhD student in public policy and management, and in strategy, entrepreneurship, and technological change at Carnegie Mellon University. Francisco Veloso is the Dean of the Católica–Lisbon School of Business and Economics.

We gratefully acknowledge the financial support of the US National Science Foundation (SciSIP grant no. 0830233), the Portuguese National Science Foundation (PTDC/ESC/71080/2006), the Carnegie Mellon Center for the Future of Work, the Kauffman Foundation, and the World Bank. We received helpful comments from Bronwyn Hall and from seminar audiences at the NBER, Carnegie Mellon, the 2012 Asia Pacific Innovation Conference, and the Atlanta Conference on Science and Innovation Policy 2013, and we appreciate the research assistance and intellectual input of Matej Drev. The views expressed in this chapter are those of the authors, and we retain sole responsibility for any errors. For acknowledgments, sources of research support, and disclosure of the authors' material financial relationships, if any, please see http://www.nber.org/chapters/c13028.ack.

and India lag far behind the industrial West, of course, but they also lag behind other developing countries, such as Brazil, South Africa, and Malaysia. However, India and China are already innovating, as is evidenced by the rapidly rising number of patents granted by the US and European patent offices to inventors residing in India and China. While the absolute number of patents remains low, the rates of growth have been exponential. A rapidly growing number of patent counts are not the only indicator of rising innovation in these emerging markets; India and China are also hosting an expanding number of research and development (R&D) centers sponsored by the world's technologically elite firms (Basant and Mani 2012; Freeman 2006). Does this trend contradict conventional wisdom? Should we abandon our conventional economic models, or at least presume that they may not apply to these dynamic Asian giants? Respected experts in international economics have suggested as much, calling upon advocates of more traditional models to "wake up and smell the ginseng" (Puga and Trefler 2010). The growing role of emerging economies in global innovation has also raised significant concerns among leaders in government, industry, and academia in the industrial West (National Research Council 2007, Royal Society 2010). Is the recent growth in emerging economies' R&D activity undermining the traditional position of technological leadership enjoyed by the United States and other advanced industrial economies?

Using US patent data, we examine the innovative explosions in India and China. We trace the dramatic growth of US patents received by inventors residing in India and China across time, technological fields, organizational boundaries, and geographic space. We examine the quality of patents, as evidenced by patent citations, with a focus in this chapter on activity in India.

In doing so, we make two contributions to the literature. First, we find that the rapid growth in US patents awarded to private sector inventors based in India and China is driven, to a great extent, by multinational corporations (MNCs) from advanced industrial economies and are highly dependent on collaborations between local inventors and other inventors in advanced economies.[1] Therefore, India and China's striking innovation surge may represent less of a challenge to conventional models of trade, economic growth, and development than it appears at first glance. The view that the increases in innovation in India and China are undermining the traditional position of technological leadership enjoyed by the United States and other advanced industrial economies might therefore also be exaggerated.

---

1. While Chen, Jang, and Chang (2013) and Jang, Wang, and Chen (2012) have shown the importance of international coinvention in the context of invention in certain emerging markets, we go well beyond this by examining all multinational patents and by comparing the quality of multinational and indigenous patents.

Second, we find evidence of an increasing trend of an international division of R&D labor—or, phrased differently, a vertical distintegration of R&D, with various stages of the R&D process now being conducted in different locations around the world. The general phenonmenon of a vertical disintegration of manufacturing has been studied in the international economics literature (e.g., Yi 2003; Hummels, Ishii, and Yi 2001; Krugman, Cooper, and Srinivasan 1995). We find conceptually similar changes in R&D. As the innovation networks of MNCs span the globe, emerging economies like India and China that possess both a huge scientific and engineering talent pool and large markets have become an important part of these global innovation networks. By undertaking R&D in emerging economies, MNCs can now provide innovative technologies to global markets at a lower cost, and introduce products more suitable for local and other emerging markets.

The rest of the chapter is organized as follows: section 5.2 provides the background of the rise of innovation in India and China and briefly explains how existing theoretical models may explain the rise of innovation in these two economies. Section 5.3 describes our data and presents descriptive features of the rise of innovation in India and China. Section 5.4 presents empirical models and detailed regression results focusing on patent quality as well as quantity. Section 5.5 provides insights from a field study of MNC R&D activity in these emerging markets. Section 5.6 discusses policy implications and presents our conclusions.

5.2 Background

Industrial R&D activity within the borders of mainland China has increased at a very rapid pace over the last fifteen years, and has now reached levels that are quite impressive by the standards of developing economies. It is also one of the favorite destinations for multinational R&D investment. Over the 1997–2007 period, the total amount of US multinational R&D spending increased thirty-three fold in China, from 35 million to 1.17 billion US dollars. The growth of R&D in India has been slower. Its R&D intensity was 0.76 percent of the gross domestic product (GDP) in 2007, essentially unchanged since 2000 (OECD 2012). Nevertheless, the total amount of US multinational R&D spending increased sixteenfold in India, from 22 million to 382 million US dollars over the 1997–2007 period.[2]

Tracking patents granted by the US Patent and Trademark Office (USPTO) to inventors residing in India and China provides another useful way of

2. The number is for majority-owned affiliates of nonbank US parent companies in India or China. A "majority-owned affiliate" is an Indian or Chinese affiliate in which the combined direct and indirect ownership interest of all US parents exceeds 50 percent. (Source: US Department of Commerce, Bureau of Economic Analysis, US Direct Investment Abroad: Financial and Operating Data for U.S. Multinational Companies, http://www.bea.gov/iTable/iTable .cfm?ReqID=2&step=1. Retrieved on August 8, 2012.)

measuring the expansion of R&D within these countries.[3] Anyone seeking to protect intellectual property within the borders of the United States must apply for patent protection from the USPTO. Given the importance of the US economy to the world in general, it is reasonable to regard patents taken out in the United States by inventors residing in India and China as a useful indicator of innovative activity there.[4]

Figure 5.1 shows the annual number of US utility patent grants with at least one inventor residing in India. We can see that US patents granted to Indian inventors grew rapidly. Over the 1996–2010 period, the total number of US patents granted to Indian inventors increased twenty-fivefold. Figure 5.2 presents the annual number of US utility patent grants with at least one inventor residing in China from 1981 to 2010. One can clearly see that the number of US patents granted to Chinese inventors exploded in recent years. Over the 1996–2010 period, the total number of US patents granted to Chinese inventors increased forty-sixfold. A similar explosion can be observed using Chinese domestic patent data (Hu and Jefferson 2009). Over the 2000–2009 period, the total number of invention patents granted by the State Intellectual Property Office of the People's Republic of China (SIPO) increased twentyfold.[5]

Using US patent data, one can further disaggregate patents generated in India and China into ones in which all listed inventors at the time of invention were based in those regions, ones that were created by international teams of inventors, and patents generated by inventors residing in India and China but owned by MNCs. Over 90 percent of US patents granted to American inventors are generated by teams of inventors in which every

3. After we had produced the first draft of this chapter, we discovered that Chen, Jang, and Chang (2013) had used USPTO data to examine R&D cooperation between Chinese and foreign inventors and Jang, Wang, and Chen (2012) used USPTO data to compare this activity in India and China. While there is some overlap between the purely descriptive parts of their papers and our work here, there are also important differences. Our chapter considers both coinvention and multinational sponsorship of indigenous inventor teams, whereas they consider only the former. The econometric approach taken by these two papers differs entirely from ours. In particular, we focus on patent quality as revealed by patent citations, whereas their work does not.

4. US patents have been used to measure inventive output in Britain (Griffith, Harrison, and Van Reenen 2006), Japan (Branstetter and Sakakibara 2002), and Israel (Trajtenberg 2001). At the same time, we recognize that the use of US patents as an indicator of inventive output of another country poses potential problems, and we include a discussion of these later in the chapter.

5. The SIPO grants three types of patents: invention, utility model, and design patents. In principle, applications for invention patents need to pass a substantive examination for novelty and nonobviousness and the utility model and design patents do not. In this sense, a Chinese invention patent is similar to a US utility patent. However, the degree to which Chinese patent examiners hold domestic applicants to the same standards of novelty and nonobviousness as US or European patent examiners is open to question. We will discuss this issue in the latter part of this chapter. Source: The State Intellectual Property Office of PRC website at http://www.sipo.gov.cn/sipo2008/ghfzs/zltj/gnwszzlsqzknb/2009/201001/t20100122_488402.html. Retrieved August 14, 2010.

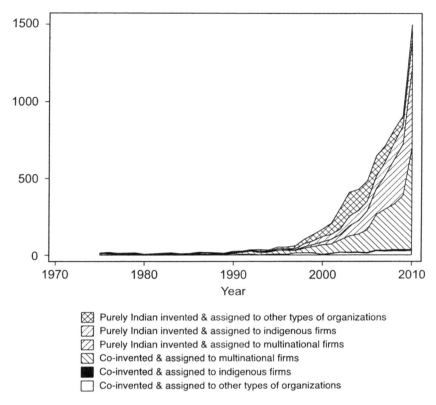

Fig. 5.1 The rise of coinvented and MNC-sponsored USPTO patents in India

inventor is residing in the United States at the time of application. The same is true of US patents granted to Japanese inventors, where over 90 percent of such patents are generated by exclusively Japanese inventor teams.[6] However, this is not true of US patents being generated in India and China. A large and growing fraction of patents with Indian or Chinese inventors result from something we call international coinvention—teams of researchers based in different countries combining their skills and knowledge to generate patented inventions.[7] In addition, a growing fraction of the patents produced by purely Indian or Chinese inventor teams is created under the sponsorship of MNCs. In fact, as illustrated in figure 5.1 and figure 5.2, patents resulting from international coinvention and MNC sponsorship account

6. Danguy (2012) provides a purely descriptive overview of the frequency of international coinvention in an expansive sample of USPTO and EPO patents, and finds it to be relatively rare.
7. To the best of our knowledge, the first use of the term international coinvention was in Branstetter et al. (2008).

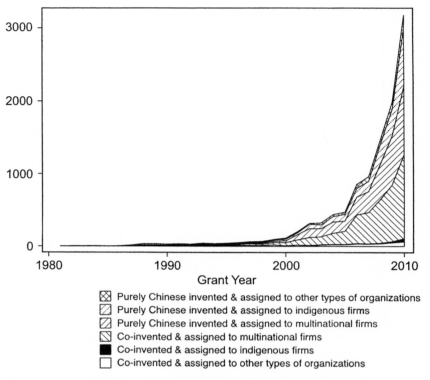

Fig. 5.2 The rise of coinvented and MNC-sponsored USPTO patents in China

for the majority of new US patents granted to Indian or Chinese inventors in recent years.[8]

India and China's patent increases also differ quite substantially from the innovation explosions in Taiwan and South Korea that preceded them. A breakdown of US patent grants to Taiwan-based inventors and South Korea–based inventors is provided in figure 5.3. As can be seen, starting in the late 1980s and proceeding through the 1990s, both Taiwan and South Korea underwent a sharp transition from being almost pure imitators to being increasingly aggressive innovators. The speed of this transition is reminiscent of India and China's more recent invention surges, but the composition of inventor teams is not. The Taiwanese and South Korean patent explosions were generated almost entirely by purely indigenous teams of inventors. The important role of foreign firms in India and China's invention explosions may help explain why they are occurring at an even earlier stage of economic development than did the invention surges in South Korea and Taiwan.

8. This finding was first documented in Branstetter et al. (2008).

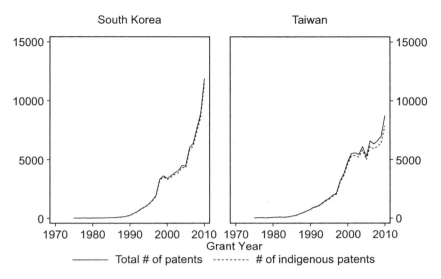

South Korea   Taiwan

Grant Year
————— Total # of patents   ········· # of indigenous patents

Fig. 5.3 Patterns of USPTO patenting from South Korea and Taiwan

Patents granted by the Indian and Chinese national patent offices also bear witness to the importance of foreign firms. In China, foreigners account for more than 50 percent of the total number of invention patents granted by the SIPO over the 1990–2008 period. In 2009, the number of domestic invention patents slightly exceeded foreign invention patents, yet foreign invention patents still had a share of 49 percent.[9] In India, the Office of the Controller General of Patents, Designs & Trade Marks (CGPDTM) granted between 59–84 percent of patents to foreign applicants during the period from 2000–2001 to 2010–2011.[10]

In the same way that USPTO patent data help trace the explosive growth of innovative activity in India and China, they also help put their current levels into perspective. In figure 5.4, we look at patents granted to inventors based in eight different countries from 1996 to 2010, and it is clear that, in spite of the fact that China's inventive output as measured by US patents places it head and shoulders above India and other so-called BRIC (Brazil, Russia, India, and China) economies of Russia and Brazil, China's generation of patents still lags far behind that of the leading advanced industrial economies, and even behind that of newly industrializing economies such as Taiwan and South Korea. Despite being among a population less than

9. Source: The State Intellectual Property Office of PRC website at http://www.sipo.gov.cn/sipo2008/ghfzs/zltj/gnwszzlsqzknb/2009/201001/t20100122_488402.html. Retrieved on August 14, 2010.
10. Source: CGPDTM annual report 2010–11 at http://ipindia.gov.in/main_text1.htm. Retrieved on November 19, 2012. The authors made the calculation.

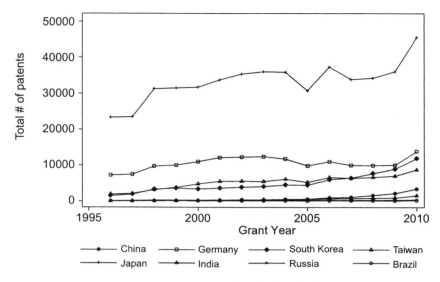

Fig. 5.4 India- and China-based USPTO patenting in comparative perspective

one-tenth that of China, or about one-tenth of India, Japanese inventors received thirteen times as many US patent grants as those based in China in 2010. Taiwan's national population is lower than that of the municipality of Chongqing in the Chinese interior, yet Taiwanese inventors received nearly three times as many patents as mainland Chinese inventors in 2010. India and China's explosive growth in US patents has come from a very low base, and these two countries have a long way to go before they can claim to be a vital part of the global innovation system. However, if China's current international patenting growth rates persist, it will start to rival the patent output of Taiwan and South Korea within a few years. It will clearly take longer for India to reach Taiwan's and South Korea's current levels.

By either assuming or predicting that innovation occurs exclusively in "the north," the product life cycle theory (Vernon 1966) and its current variants (Krugman 1979; Grossman and Helpman 1990, 1991) rules out the possibility of innovation in "the south." This reflects the situation at the time when these theories were established. Research and development in developing countries at the time was sporadic, usually incremental in nature, and lacked real technological breakthroughs.

However, this stylized pattern has begun to change since the mid-1990s. First, multinationals are doing an increasing amount of R&D in emerging economies, notably in India, China, and the leading nations in Eastern Europe (Zhao 2006; Branstetter et al. 2008; Branstetter and Foley 2010). This shift has occurred against a backdrop of rising globalization of R&D, more generally. Between 1999 and 2009, the R&D expenditure of all foreign affiliates of US firms almost doubled (Barefoot and Mataloni 2011). In China, these expenditures more than doubled just between 2004 and 2010, and in India they grew by a factor of ten over the same period.[11] Second, the nature of multinational R&D in emerging economies has changed from a pure adaptation of existing technologies to include some cutting edge R&D on par with that undertaken in developed economies (UNCTAD 2005).

Some work has been done to address these changes. Grossman and Rossi-Hansberg (2008) provide a theoretical model of offshoring that includes skill-intensive tasks. Puga and Trefler (2010) investigated innovation in emerging markets in a theoretical context in which it was treated as mostly incremental. Zhao (2006) suggested that by using closely knit internal technological structures as an alternative mechanism to protect their intellectual property in countries with weak IP legal environments, MNCs are increasingly conducting R&D in countries with less developed intellectual property rights systems, such as India and China. However, systematic study on this topic is still insufficient.

The MNCs' leveraging of their innovation competencies across borders per se is not a new phenomenon (Cantwell 1995; Kogut and Zander 1993), but using coinvention as a vehicle to create novel innovations in emerging economies is. The clear importance of international coinvention in the data on US patents granted to Indian and Chinese inventors may suggest something extremely interesting: the possibility that the R&D process itself can now be sliced into multiple stages, and countries may participate in different stages according to their competitive advantages. This phenomenon is often referred to as "vertical specialization" or "vertical disintegration" in the trade literature (Krugman, Cooper, and Srinivasan 1995; Hummels, Ishiii, and Yi 2001; Yi 2003).

If India and China's emergence in the global innovation system follows this economic canon, coinvention created in India and China is likely to be characterized by a division of labor in the research process. As such, Indian and Chinese researchers may undertake more repetitive, codified, and relatively routine research tasks while researchers in advanced countries may provide more sophisticated, creative, and high-level intellectual input. Combining the two, MNCs can produce more (and more impactful) innovative output with a given amount of R&D expenditure (Romer 1990). As a consequence, an increase in R&D activity induced in China and India through this process might not be a direct substitute for the higher-level R&D inputs from the Western advanced countries, but rather a strong complement to it. However, this notion of complementarity could fade over time. Local Indian and Chinese inventors who initially collaborated with Western inventors through coinvention partnerships could acquire and accumulate high-level skills through this collaboration, and then engage in high-level, original inventive activity

---

11. See Yorgason (2007) and Barefoot (2012).

without the need for input from Western inventors. In this case, coinvention could, over time, lead to greater direct substitution between Western and local invention. But, it is also possible that after acquiring and accumulating high-level skills, these local Indian and Chinese inventors would continue to collaborate with Western inventors (Kogut and Zander 1992; Weitzman 1998; Singh 2008). We will return to this issue in the latter part of this chapter.

5.3 Data Sources and Trends

Our analysis in this chapter will focus primarily on US patent grants as an indicator of inventive output. This is principally because prior research has established that the real economic value of most patents is extremely small (Jaffe and Trajtenberg 2002), but the more valuable patents tend to be patented not just in the home country but in other major markets as well. Because India and China are developing countries with still-developing patent systems, a patent grant in India or China is less likely to represent an important advance over the global state of the art. However, the USPTO will apply the same standards to patent applications originating in India or China that it applies to patent applications originating in California. These US patent grants are far more likely to be reflective of economically valuable new-to-the-world inventions than is an "invention" for which we find Indian or Chinese patent grants but no US patent grants. Furthermore, significant changes in the domestic patent systems in India and China make Indian and Chinese patent data inconsistent over time.

Our data come from several sources. The first is the selected bibliographic information from the US Patents DVD (2009 December) released by the USPTO, which contains bibliographic information for all granted patents from 1969 to 2009.[12] The second is the Disambiguation and Co-Authorship Networks of the US Patent Inventor Database (Lai et al. 2011), which contains bibliographic information for granted patents and citations data for patents granted during the period of 1975–2010.[13] The third is the COMETS database 1.0 (Zucker and Darby 2011), which we used to verify and supplement citation data from the Disambiguation and Co-Authorship Networks of the US Patent Inventor Database. The fourth is the USPTO Patent Full-Text and Image Database (online), as well as the Patent Assignment Database (online), which we used to identify and verify some important information in our data set.

---

12. The information from this data source has been included in the Disambiguation and Co-Authorship Networks of the US Patent Inventor Database (Lai et al. 2011). However, when we began work on this research project, the Disambiguation and Co-Authorship Networks of the US Patent Inventor Database had not come out.

13. In earlier versions of this paper, the citation data were extracted from the NBER Patent Data Project (PDP) citation file (1976–2006) downloaded from Professor Bronwyn Hall's website and the Patent Grant Bibliographic Data/XML Version 4.2 ICE (Text Only) 2007, 2008, and 2009, downloaded from the USPTO website. These have been included in the Disambiguation and Co-Authorship Networks of the US Patent Inventor Database.

We dropped withdrawn patents from our data sets, updated patent classes to current classifications as of the end of 2010, and standardized the assignee codes and names according to the USPTO's assignee harmonization system.[14]

By combining the first three data sets, we identified and characterized 3,983,050 utility patents granted from 1975 to 2010. We then used these patents to track citation relationships and counted the number of citations received (or "forward citations") for each patent.

For the purposes of our research, we separated Hong Kong and Taiwan from mainland China.[15] A total of 12,419 patents are identified as those with at least one inventor residing in China at the time of invention during the period 1981–2010.[16] A total of 7,754 patents are identified as those with at least one inventor residing in India at the time of invention during the period 1975–2010.[17]

The USPTO has classified all patents into the seven types of assignees:

1. Unassigned
2. Assigned to US nongovernment organizations
3. Assigned to non-US nongovernment organizations
4. Assigned to US individuals
5. Assigned to non-US individuals
6. Assigned to the US federal government
7. Assigned to non-US governments

However, we want to distinguish patents granted to a firm entity from those granted to a nonfirm entity. To do so, we manually screened all first assignees' information listed on patents, including original type code, name, address, and so forth, and consulted Dun & Bradstreet's Million Dollar Database,

14. See USPTO, http://www.uspto.gov/patents/process/search/withdrawn.jsp (retrieved January 24, 2012). According to a Cassis2 DVD-ROM, Patents Class: Current Classifications of US Patents Issued 1790 to Present (2010 December). In earlier versions of this chapter, the harmonized assignee codes were extracted from the selected bibliographic information from US Patents DVD (2009 December). We combined them with assignee codes for patents granted in 2010 according to the files downloaded from the USPTO website. http://www.uspto.gov/web/offices/ac/ido/oeip/taf/data/misc/data_cd.doc/assignee_harmonization/. Retrieved July 13, 2012.

15. One issue arises for the years after 1997, when the United Kingdom returned sovereignty over Hong Kong to China. Some inventors residing in Hong Kong continued to list Hong Kong as their inventor country, while others began to list China as their inventor country. (Note: politically, Hong Kong has never been a country, but USPTO designates a separate country code to it for classification purposes.) Before and after 1997, we identify Hong Kong addresses and consider them to be geographically distinct from mainland China. Similar mistakes can be found when a Taiwanese inventor listed Republic of China, the official name of Taiwan, as her home country. A small number of Taiwanese patents have been mistakenly classified with an inventor country code of "CN" (which stands for China) instead of "TW" (which stands for Taiwan) by the USPTO. We corrected these mistakes by looking up an inventor's full address.

16. The first China-based patent was granted to Dynapol, a chemical company in 1981. The patent counts are based on grant years.

17. Similar to what happened to the China-based data, in a few cases, Indonesia and the state of Indiana were mistakenly assigned with an inventor country code of "IN" (which stands for India). We corrected all of these mistakes.

Table 5.1   Top ten firm assignees of India-based USPTO patents

Rank	Assignee name	Nationality	Number	Share (%)
1	General Electric Company	United States	464	8.12
2	IBM	United States	450	7.87
3	Texas Instruments, Inc.	United States	418	7.31
4	Cisco Technology, Inc.	United States	162	2.83
5	Intel Corporation	United States	151	2.64
6	STMicroelectronics Pvt. Ltd.[a]	France & Italy	151	2.64
7	Honeywell International, Inc.	United States	126	2.20
8	Symantec Operating Corporation	United States	116	2.03
9	Ranbaxy Laboratories, Limited	India	102	1.78
10	Microsoft Corporation	United States	96	1.68

[a] STMicroelectronics Pvt. Ltd. is the Indian subsidiary of STMicroelectronics, a French-Italian multinational electronics and semiconductor manufacturer.

LexisNexis Corporate Affiliations, Hoover's Online, and assignees' websites to assign the proper assignee types for all China- and India-related assignees. After this procedure, we find that 78 percent of all 12,419 US utility patents granted to Chinese inventors were assigned to a firm entity, 12 percent to an individual or identified as unassigned, 9 percent to universities and research institutes, and 1 percent to other entities such as governments, hospitals, and so forth. For India-based patents, 74 percent of total 7,754 US utility patents were assigned to a firm entity, 5 percent to an individual or identified as unassigned, 20 percent to universities and research institutes, and 2 percent to other entities. It can be concluded that firms are the main contributors of the recent increase of US patents in India and China.

Who owns these patents? For India, US MNCs own the majority of India-based US patents. Of all 5,716 India-based patents assigned to a firm entity, 70 percent are assigned to US MNCs, 18 percent are assigned to Indian indigenous firms, 3 percent are assigned to Germany, and 3 percent are assigned to France and Italy. These patents are owned by a single firm, STMicroelectronics Pvt. Ltd., the Indian subsidiary of French-Italian multinational electronics and semiconductor manufacturer STMicroelectronics. The remaining 5 percent is distributed among all other countries. For China, at the assignee nationality level, Taiwanese and US MNCs own the majority of Chinese patents, even more than Chinese indigenous enterprises. Of all 9,744 China-based patents assigned to a firm entity, 35 percent are assigned to Taiwanese MNCs or their Chinese subsidiaries, 29 percent are assigned to US MNCs, and 23 percent are assigned to Chinese indigenous firms. Other important nations and areas include Hong Kong, Germany, and Japan, which account for 3 percent, 2 percent, and 2 percent, respectively. The remainder as a whole accounts for 6 percent.

At the firm level, table 5.1 lists the top ten firm assignees of India-generated US patents. Among them, eight are US MNCs and one is a French-Italian

MNC. The only Indian indigenous firm in the list is Ranbaxy, one of the world's top generic pharmaceutical companies. Table 5.2 lists the top ten firm assignees of China-generated US patents. Among them, Hon Hai, a Taiwanese manufacturing firm, also known by its English name Foxconn, leads the list. As the largest manufacturer of electronics and computer components worldwide, Hon Hai conducts intensive R&D in China and has 2,958 US utility patents, or 30 percent of total China-based, firm-owned US patents. Microsoft, with 765 patents, or 8 percent, is a distant second. The third is Huawei, an indigenous Chinese firm that has quickly become one of the leading networking and telecommunication equipment suppliers in the world.

To measure what kinds of invention have been done in India and China, we aggregate all China- and India-based US patents that are owned by a firm entity into the widely used technology categories created by Hall, Jaffe, and Trajtenberg (2001). We will refer to their taxonomy as the HJT categories. The results presented in figure 5.5 show that, with regard to India-based patents, computers and communications is the leading field. India has been well-known for its software industry, so one question worth asking is: To what extent have software patents contributed to India's US patent surge? Among all India-based US patent grants before 2007, about 10 percent are software patents. A large proportion of China-based patents taken out in the United States are in two HJT categories: computers and communications and electrical and electronic. During the same period, the share of China-based software patents is about 5 percent. It can also be seen in figure 5.5 that coinvention plays an important role across all categories in both countries.

By extracting the geographic information on inventors included in patent documents, we found that among the 20,088 inventor addresses that indicate the inventor was living in India, 20,045 addresses can be associated with a particular state in India. Karnataka, Maharashtra, Andhra Pradesh, Uttar Pradesh, and Delhi are the top five states/territories that host most Indian inventors. Together, they account for 76 percent of the frequency distribution of India inventor addresses. These areas are also where technology business incubators (TBIs), science and technology entrepreneurs parks (STEPs), software technology parks of India (STPIs), and universities and research institutions tend to concentrate (Sharma, Nookala, and Sharma 2012).

We found 27,238 inventor addresses indicating that the inventor was located in China. Of these addresses, 27,177 were sufficiently complete that we could associate the address with a particular Chinese province. We find that Chinese inventors are highly clustered in three areas: Beijing Municipality, Guangdong province, and the greater Shanghai regional economy, comprised of Shanghai and the bordering provinces of Jiangsu and Zhejiang. Those areas account for 86 percent of the frequency distribution of

Table 5.2   **Top ten firm assignees of China-based USPTO patents**

Rank	Assignee name	Nationality	Number	Share (%)
1	Hon Hai Precision Ind. Co., Ltd.[a]	Taiwan	2,958	30.36
2	Microsoft Corporation	United States	765	7.85
3	Huawei Technologies Co., Ltd.	China	430	4.41
4	Intel Corporation	United States	197	2.02
5	Inventec Corporation[b]	Taiwan	177	1.82
6	China Petrochemical Corporation (Sinopec)[c]	China	173	1.78
7	Semiconductor Manufacturing International (Shanghai) Corporation	China	139	1.43
8	IBM	United States	129	1.32
9	Sae Magnetics (H. K.), Ltd.[d]	Hong Kong/Japan	128	1.31
10	Metrologic Instruments, Inc.	United States	92	0.94

[a] Figure here represents the sum of patents taken out under Hon Hai (Foxconn) and its China-based subsidiaries.

[b] Figure here represents the sum of patents taken out under Inventec Corporation, Inventec Appliance and Inventec Electronics (Nanjing) Co.

[c] The original data set confused China Petrochemical Corporation (Sinopec), a Chinese company, with China Petrochemical Development Corporation (CPDC), a Taiwanese company. The figure presented here is after correction.

[d] Sae Magnetics (H. K.), Ltd. is a wholly owned subsidiary of TDK, a Japanese multinational electronics manufacturer. However, Sae Magnetics (H. K.), Ltd. itself has manufacturing branches in mainland China. For our research purpose, we will treat it as a Hong Kong firm in our analysis.

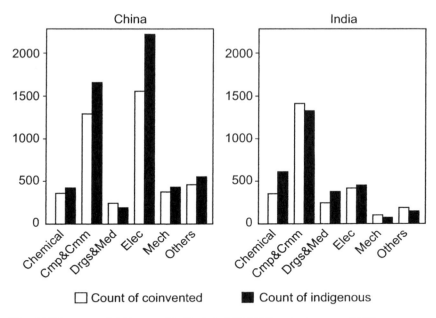

Fig. 5.5 Firm-owned China- and India-based USPTO patents across HJT technology categories

Chinese inventor addresses. These areas are not only the most developed areas in China, but also the places where most multinational R&D centers are located.[18]

All of the above features are based on the analysis of US patent data, which have limitations as indicators of invention in India and China. The most obvious one is that US patent data may exaggerate the roles of US MNCs, since companies usually patent more in their home market than somewhere else. Moreover, although the United States is the largest national economy in the world and grants a large number of patents, patents granted by its patent office may still not be able to capture the whole picture of the rise of innovation in India and China. For these reasons, we have also analyzed European Patent Office (EPO) patent data as a robustness check. The major patterns revealed by US patent data also hold using EPO data, including the importance of coinvention and MNC sponsorship, technological concentration in information technology (IT)-related fields, and geographic

18. As of the end of 2009, 465 multinational R&D centers were established as independent legal entities with approval of the Ministry of Commerce of the People's Republic of China. These centers are mainly concentrated in Shanghai, Beijing, Guangdong, Jiangsu, and Zhejiang. Source: People's Daily online, http://english.peopledaily.com.cn/90001/90778/90861/6921243 .html. Retrieved August 17, 2010.

clustering of Indian and Chinese inventors. It is worth pointing out that even EPO data indicate that US MNCs play a more important role than European MNCs in India-based patenting. This probably reflects the fact that US MNCs are following more aggressive strategies of conducting R&D in India than are MNCs from other places. Figures and tables presenting the results using the EPO patent data are available upon request.

Before moving on to the next section, we need to answer another important question: To what extent can we base our inference about innovation in India and China on the relatively small number of US patents, especially when there is a tidal wave of patents being issued in China itself (Hu and Jefferson 2009)? The numbers of Chinese domestic patents granted by China's State Intellectual Property Office (SIPO) in the most recent years are mind-boggling. In 2011 alone, the SIPO issued nearly one million patents of various kinds. The majority of these grants go to indigenous/domestic applicants. Is our focus on the international patents of Chinese inventors generating a distorted picture of the true innovation going on inside China? To answer this question, we have carefully examined Chinese invention patents, using SIPO microdata on Chinese grants over the 1985–2012 period.

The first thing we want to point out is that the overwhelming majority of SIPO grants are actually utility models or design patents. These are not true patents, in the usual Western sense of the word. Neither requires a substantive examination or a significant technical advance over the existing state of the art. When we focus on China's so-called invention patents, which do require a substantive examination and, in principle, an advance over the existing technical state of the art, we see significantly smaller numbers and significantly less growth. With this more restrictive definition, the total grant number drops by about 80 percent in recent years.

Second, as of the end of 2012, about half of SIPO invention patents are granted to foreign applicants, and among these foreign-owned invention patents, 90 percent of them possess a foreign priority claim. That means more than 40 percent of Chinese invention patents are inventions initially created abroad and then patented in China. Even this may understate the role of foreign inventors. Patents granted to MNC Chinese subsidiaries and joint ventures in China are classified as domestic grants by the SIPO. However, many of these patents are generated using intellectual inputs from outside China, including resources and capabilities located in the multinationals' R&D centers far beyond China's borders. These considerations further limit the degree to which China's impressive headline patent numbers can be taken as evidence of globally significant innovative activity.

Third, we find that domestic applicants allow their patents to expire earlier than foreign applicants by failing to pay maintenance fees over the full legal life of the patent. This result is consistent with Huang (2012), who found similar results for invention patents initially applied for during the 1987–1989 period. This suggests that there is a quality difference between

Chinese patents granted to foreign inventors and Chinese patents granted to domestic inventors.

These concerns are strongly reinforced by the extremely low propensity for Chinese inventors, in the aggregate, to apply for and receive patent protection for their Chinese inventions in patent jurisdictions outside China. For decades, researchers seeking to quantify innovation have used the fraction of domestic patents for which foreign patent protection is sought, and the number of foreign jurisdictions in which patent protection for a given invention is sought, as indicators of patent quality. Even in advanced countries with mature patent systems, the need to file patent applications quickly in order to avoid being foreclosed by rival innovators means that firms often begin the patent application process, at least at home, at a very early stage in the R&D process, before even the inventors themselves have a clear sense of the ultimate economic value of the patent. Under international patenting rules, firms have up to a year after the domestic filing during which they can start the process of seeking patent protection in foreign jurisdictions, with the "priority date" established by their home patent application filing date. We generally observe that firms are more selective in their foreign patenting, applying for the patent protection in major markets outside the home country only when the invention appears sufficiently important to merit the additional time and expense of multiple foreign patent filings.

In the context of these considerations, it is interesting to observe that the top 100 US patent applicants seek to protect nearly 30 percent of their domestic patents in at least one major foreign market, such as Japan or the EPO. In striking contrast, we find that the top 100 Chinese domestic applicants seek patent protection for less than 6 percent of their inventions in the United States, only 4 percent in Europe, and only 1 percent in Japan. If we look at the number of domestic patent grants for which Chinese firms pursue the patent application process all the way to the successful receipt of a foreign patent grant, the numbers are vanishingly small. So far, fewer than 3 percent of Chinese invention patents have been successfully patented in any major market outside China, and fewer than 1 percent have been patented in two or more major markets. Even after decades of rapid growth, China is still a significantly smaller economy than the United States or Western Europe. If Chinese inventors possess new-to-the-world technology, it would seem to be well worth their while to take out patents in these markets that are still larger, in aggregate terms, than China itself. The fact that Chinese inventors forego the opportunity to do so for the overwhelming majority of their domestic patents appears to represent a striking vote of no confidence in the quality of their inventions.

A more detailed discussion of the SIPO data and the unique features of the Chinese domestic patent system is beyond the scope of this chapter, but a number of recent studies call into question the degree to which China's flood of domestic patents really indicates a substantial degree of indigenous innovation.[19] We note here two recent studies of special interest. Lei, Sun, and Wright (2012) note widespread government subsidies for domestic patenting in China, but show that an increase in these subsidies appears to increase the number of patents but not the quantity of innovation. Eberhardt, Helmers, and Yu (2011) study domestic and foreign patenting by Chinese firms, and conclude that the only Chinese firms engaging in real innovation are those also taking out significant numbers of patents outside China.

5.4 Empirical Model and Results

5.4.1 Hypotheses

In section 5.2, we argued that the R&D process itself can be sliced into multiple stages, and countries participate in different stages according to their comparative advantage (Krugman, Cooper, Srinivasan 1995; Hummels, Ishii, and Yi 2001; Yi 2003). Previous research found that invention being generated in developing countries is incremental in nature (Zhao 2006; Puga and Trefler 2010). These findings suggest that Indian and Chinese researchers under the sponsorship of MNCs are likely to undertake low-end tasks, while Western researchers undertake high-end tasks. As such, we might expect that a comparison between Indian or Chinese invention with and without Western intellectual input would suggest that patents with Western intellectual input are of substantially higher quality than those without. The same considerations might suggest that even within the same MNC, patents with Indian or Chinese input might be of lower quality than the patent output of all-Western inventor teams. Interviews with India-based R&D personnel and managers suggest that R&D in India largely follows this pattern.

We will seek to validate that perception by measuring the relative quality of India-based USPTO patents in multiple ways. First, we will compare coinvention generated in India with patents created by purely domestic researcher teams, and compare inventions created by MNCs with those generated by indigenous enterprises. The traditional theory of vertical specialization would suggest our first hypothesis.

HYPOTHESIS 1 (H1). Coinvention and MNC sponsorship are associated with relatively higher patent quality.

We then compare the quality of the patents MNC generated in India (both coinvention and purely domestic patents) with the patents the same MNCs produced in their home countries (with all inventors residing in the

---

19. Hu and Jefferson (2009), who undertook an early quantitative study of the impressive increase in China's domestic patenting, suggested that it was primarily driven by an increase in the propensity to patent rather than an increase in real innovative effort.

MNC's home country). A view based on traditional theory would suggest our second hypothesis.

HYPOTHESIS 2 (H2). Patents produced by MNCs in India are of lower quality than those produced by the same MNCs in their home countries, and even coinvented patents generated in emerging economies are of marginally lower quality.

Besides overall comparisons, we also want to assess the dynamics of patent quality across different patent categories. Conversations with multinational R&D managers suggest that it takes time for talented researchers in emerging economies to become "mature." This implies our next hypothesis.

HYPOTHESIS 3 (H3). The quality gap between patents (including coinvention) produced by MNCs in India and patents produced by the same MNCs in their home countries declines as MNCs gain more experience doing R&D in India.

We have pointed out the possibility that international coinvention could accelerate the advancement of indigenous innovative capability. After a period of time working under the tutelage of multinationals, talented Indian engineers could put their skills and experience to work for indigenous firms that increasingly compete directly with the MNCs. It will therefore be important to see if the measured gaps in patent quality between indigenous firms and multinationals are closing over time.

At this stage, however, it is difficult to capture such a dynamic process. There is no clear turning point in our data set at which we could usefully divide the data into an "early period" with limited catch-up and a later period with more complete convergence of innovative capacity. This stems in part from the fact that some multinationals entered the Indian market early and began building strong R&D operations ten or more years ago, whereas other multinationals have only begun to establish their research capacity much more recently. It may be that early entrants have not only incubated a strong team of local engineers within their labs, but also seeded a number of local spin-off entrants with seasoned R&D personnel. But the innovative performance of these veterans of MNC R&D activity is diluted by an inflow of newly graduated and relatively inexperienced local Indian researchers. Thus, the ideal way to measure convergence would be to compare MNC-employed engineers and engineers employed in indigenous firms who are at the same stage in their inventive careers. This requires the tracking of individual inventors over time. This will be the focus of future research, but we will not attempt such a fine-grained comparison in this chapter. Instead, we will arbitrarily divide the data into three periods to see if there is evidence of a declining gap in relative invention quality between indigenous enterprises and MNCs over time. This leads us to our last hypothesis.

HYPOTHESIS 4 (H4). The gap in patent quality between indigenous enterprises and MNCs is declining over time.

Throughout this section of the chapter, our focus will be on MNC R&D in India. In ongoing research, however, we are conducting a similar empirical investigation of MNC R&D in China, and, in the sections below, we will offer a qualitative comparison of our results for India with those we obtain when running similar regressions on a Chinese data set.

5.4.2 Empirical Model

As already noted, we regard patent citations as an indicator of patent quality. Patent citations serve an important legal function because they delimit the scope of the property rights awarded by the patent. Thus, if patent B cites patent A, it implies that patent A represents a piece of previously existing knowledge upon which patent B builds, and over which B cannot have a claim (Hall, Jaffe, and Trajtenberg 2001). Alcácer and Gittelman (2006) showed that patent citations are an imperfect measure of knowledge spillovers between inventors because examiners add a significant fraction of the citations after the initial patent application. It is obviously problematic to consider these examiner-added citations as reflecting the sources of inspiration of the inventor herself. However, we use citations as an indicator of patent quality rather than a measurement of knowledge spillover. Prior literature has shown that total citations received are highly correlated with the underlying quality of the invention (Trajtenberg 1990; Jaffe, Trajtenberg, and Fogarty 2000; Harhoff et al. 1999; Hall, Jaffe, and Trajtenberg 2005). More valuable invention is more frequently cited by subsequently granted patents. Thus citations received can be used to proxy for the quality of each patent.

Two issues arise when using patent citations as a measure of patent quality: truncation due to time and difference due to technological fields. Prior research has demonstrated that it takes time for patent citations to occur (Hall, Jaffe, and Trajtenberg 2001). The number of citations made to a patent granted just one year ago may be only a small fraction of citations that will occur over the following fifteen years. It is easy to see that patents of different vintages are subject to different degrees of "citation truncation" (Hall, Jaffe, and Trajtenberg 2001) as one cannot simply tell that a patent from 2005 with twenty-five citations is better or worse than a 2008 patent with only ten citations. Similarly, one cannot tell that an electronic device patent granted in 2000 with twenty-five citations is better or worse than a pesticide patent granted in the same year with only five citations. To address the issue of truncation, we will control for patent grant years and use count models with "exposure" (Cameron and Trivedi 1998) in all regressions. To address the issue of technological difference, we control for major technological fields in our empirical analysis.

Our basic model regresses the citations a patent has received on a number of control variables. These variables include a dummy variable indicating whether or not it was a product of international coinvention, a dummy variable indicating whether or not it was produced under multinational sponsorship, and so forth. A significantly positive coefficient on a control variable of interest indicates a higher number of citations received and suggests a higher quality of the patent.

We apply the Poisson quasi-maximum likelihood (PQML) estimation to our regressions for two reasons. First, patent citations are integer counts and have a minimum value of zero. Second, our data are overdispersed and the PQML estimator is consistent under the weaker assumption of the correct conditional mean specification and no restriction on the conditional variance (Wooldridge 1999, 2002; Cameron and Trivedi 2005; Gourieroux, Monfort, and Trognon 1984; Hall, Griliches, and Hausman 1986). While a negative binomial (NB) mode can also deal with the "overdispersion" issue, it assumes that the conditional variance has a gamma distribution. The trade-off between the NB model and the PQML model is obvious: if the gamma assumption about the conditional variance is correct, then the NB estimator will be more efficient, but if the gamma assumption does not hold, then the NB estimator will be biased. Overall, the PQML model is more likely to result in lower significance levels than the NB model. Thus, we tend to regard the PQML model as preferable. However, we also run regressions using the NB model as a robustness check. The NB estimation results are qualitatively consistent with the PQML results presented in this chapter and are available upon request. The PQML estimators can be obtained by estimating an unconditional Poisson model with robust standard errors (Wooldridge 1999; Cameron and Trivedi 2005).

Our dependent variable Y is the number of citations a patent has received, a quantity also referred to in the literature as the count of "forward citations." We count the cumulative number of citations a patent has received as of the end of 2010, when our current citation data set ends. We drop the patents granted in 2010 in order to get at least one year of citation counts for the patents used in regression analysis.

We exclude self-citations in the citation counts. We do this because we are concerned that an inventor working in the Indian R&D subsidiaries of a multinational might have a higher propensity to cite her own or her colleagues' patents than an inventor working in the MNC's home country or somewhere else. This problem is exacerbated by the very rapid growth of India-based US patents in recent years. In addition, Zhao (2006) has suggested that patents created in a developing country and resulting from multinational sponsorship are subject to more self-citations than those created in advanced countries due to MNCs' internal IP protection arrangements. Based on these considerations, we regard the number of citations a patent receives, excluding self-citations, as a better indicator of the "true" quality than those including self-citations.

5.4.3 Cross-Firm, Within-Country Comparisons

To test H1, we run regressions on our India-based patent sample, including all USPTO patents that are granted to Indian inventors by the end of 2009 and assigned to a firm entity. Our regressions take the following form:

(1) $$E(C_i) = \text{PatAge}_i \cdot \exp(\alpha_0 + \alpha_1 \text{Coinv}_i$$
$$+ \alpha_2 \text{MNC}_i + \alpha_3 \text{PatStock}_f$$
$$+ \alpha_4 \text{TeamSize}_i + \alpha_5 \text{Gdelay}_i + H_i + T_i).$$

where C is the total number of non-self-citations an Indian-based patent i receives by the end of 2010. The key coefficients of interest are those on Coinv and MNC, which are dummy variables indicating whether or not a patent is coinvented and whether or not it is assigned to a multinational assignee. The key task here is to compare coinvented and MNC-owned patents generated in India with patents generated by Indian indigenous firms. In addition, we also control for other factors that may influence citations. PatStock denotes the assignee f's three-year patent stock (three-year cumulative sum of US patents) before the date of application. We used PatStock as a proxy for a company's inventive productivity at the time of patent creation. TeamSize is the total number of inventors on the patent. If larger teams are required for more "fundamental" (and potentially more valuable) inventions (Jones 2009), and larger teams are more likely to be international teams, then this could introduce a mechanical positive association between coinvention and quality—the TeamSize variable helps us control for this. Gdelay is the year delay between the patent's application date and grant date. As mentioned earlier, forward citations are truncated in a sense that recently granted patents have less time to garner citations than earlier ones. To correct this, we estimate the PQML mode with "exposure" (Cameron and Trivedi 1998). PatAge is the age of the patent, which serves as the exposure variable and is calculated as the days between the application date and the end of 2010. Thus the natural log of PatAge enters as an offset in the conditional mean. Inclusion of the "exposure" variable controls for the effect of the passage of time on the likelihood of citation if that effect is constant over time. However, this may not be true, so we also include grant year fixed effects T and HJT subcategory fixed effects.

Table 5.3 shows the results. Column (1) includes a coinvention dummy, column (2) includes a MNC assignee dummy, and column (3) includes both dummies. Across all three specifications, the coefficients on the coinvention dummy are positive and significant. The coefficient of 0.239 in column (1) can be interpreted as suggesting that coinvented patents receive 27 percent (exp(0.239)–1) more non-self-forward citations than purely Indian-generated patents. Similarly, column (2) suggests that MNC-sponsored patents—with a multinational assignee—receive 45 percent more citations than ones under the sponsorship of Indian indigenous enterprises, whether they are coin-

Table 5.3   Cross-firm comparison within India (1979–2009)

DV: no. of non-self-citations received as of the end of 2010	Control for coinvention (1)	Control for assignee type (2)	Control for both (3)
Indian coinvention	0.239**		0.171*
	(0.0851)		(0.0806)
Multinational assignee		0.375**	0.264
		(0.144)	(0.138)
Three-year patent stock	0.00760	−0.000294	0.00358
prior to application date (in thousands)	(0.0119)	(0.0121)	(0.0122)
Grant delay in years	0.123***	0.132***	0.127***
	(0.0282)	(0.0285)	(0.0281)
Team size	0.0465***	0.0526***	0.0479***
	(0.00615)	(0.00566)	(0.00603)
Constant	−8.643***	−8.817***	−8.793***
	(0.586)	(0.578)	(0.577)
Grant year dummy	Yes	Yes	Yes
HJT subcat. dummy	Yes	Yes	Yes
Observations	4,280	4,280	4,280
Offset	ln(pat_age)	ln(pat_age)	ln(pat_age)
Log pseudolikelihood	−13,541.4	−13,545.0	−13,512.2
Chi-square	2,037.3	2,087.0	2,051.7
Pro > chi-square	0	0	0

Note: Robust standard errors in parentheses.
***Significant at the 1/10 of 1 percent level.
**Significant at the 1 percent level.
*Significant at the 5 percent level.

vented or not. It turns out that almost all coinvention in India is found in MNC-sponsored patents, so we cannot estimate much of a separate coefficient for the MNC assignee dummy when we also control for a coinvention dummy in column (3). The two dummies are highly collinear. It is also notable that team size has a positive and significant effect on patent quality.

We acknowledge that the biases and issues that beset patent citation data may especially complicate quality comparisons between indigenous patenting and MNC patenting, so we want to proceed with caution. But the data suggest that coinvented and MNC-sponsored patents are more technologically sophisticated and valuable than indigenous patents, as well as more numerous. Running similar regressions on a parallel China-based patent data set yields similar results. With China-based patent data, it is possible to separately estimate multinational sponsorship and international coinvention effects. Both are positive, statistically significant, and collectively point to a multinational patent premium that is roughly the same magnitude as that observed in India.

5.4.4 Cross-Border Comparisons within MNCs

Next, we want to know whether patents produced by MNCs in India are of lower quality than those produced by the same MNCs in their home countries (H2). To do so, we keep only those India-based US patents that are assigned to (owned by) MNCs from 1996 to 2009. We then match them to the patents that are created by inventors in the MNCs' home countries, with the same firm assignee code, three-digit technological class, and grant year. Patents without a match are dropped. We drop patents granted in years before 1996 to ensure that we have a reasonable number of Indian domestic patents for comparison. Undertaking the same matching procedure as described above, we construct a second sample that only includes MNCs with more than thirty India-based patents by the end of 2010. Our specification is as follows:

$$(2) \qquad E(C_i) = \text{PatAge}_i \cdot \exp(\beta_0 + \beta_1\text{Coinv}_i$$
$$+ \beta_2\text{Domestic}_i + \beta_3\text{PatStock}_f + \beta_4\text{TeamSize}_i,$$
$$+ \beta_5\text{Gdelay}_i + F_i + H_i + T_i)$$

where Coinv is a dummy variable indicating whether or not an MNC-sponsored patent is coinvented. Domestic is a dummy variable indicating whether or not an MNC-sponsored patent is generated exclusively by domestic inventor teams in India. Since we compare patents within the boundaries of the MNC, we also include F_i, which denotes assignee (firm) fixed effects. All other variables are defined as in specification (1).

We also want to investigate the dynamics of the quality difference between patents produced by MNCs in India and those produced by the same MNCs in their home countries over time (H3). Using the basic specification as in equation (2), we interact Coinv and Domestic dummies with period dummies that are based on the length of a firm's experience generating USPTO patents through the work of India-based inventors when the patent application was filed. We divide our data into three periods: one to five years of India experience, six to ten years of India experience, and more than ten years of India experience.

Results are presented in tables 5.4 and 5.5. Table 5.4 shows that patents generated by MNCs in India appear to get systematically fewer non-self-citations than those generated at home. In most cases, the differences are statistically significant at the standard levels. With regard to the dynamics of the quality difference, the point estimates for interaction terms in table 5.5 are all negative. Statistically significant negative quality differences fade for both coinvented patents and those with purely Indian inventor teams. Depending on how one looks at it, one can see limited evidence of a relative quality improvement over time for coinventions in India, but the results are still quite weak. These results are consistent with a division of labor between Indian R&D personnel and Western R&D personnel within the boundaries of the MNC in which much of the more fundamental, more frequently cited work is disproportionately likely to be conducted on the Western side.

Table 5.4   **Cross-border comparisons within MNCs (India, 1996–2009)**

DV: no. of non-self-citations received as of the end of 2010	Full sample (1)	Firms with >30 IN patents (2)
Indian coinvention	−0.212***	−0.182*
	(0.0613)	(0.0895)
Purely Indian invention	−0.229*	−0.189
	(0.109)	(0.113)
Three-year patent stock prior to application date (in thousands)	−0.00344	0.00461
	(0.0180)	(0.0232)
Grant delay in years	0.101***	0.115***
	(0.0130)	(0.0138)
Team size	0.0626***	0.0638***
	(0.00490)	(0.00556)
Firm fixed effects	Yes	Yes
Grant year dummy	Yes	Yes
HJT subcat. dummy	Yes	Yes
Observations	40,324	32,633
Offset	ln(pat_age)	ln(pat_age)
Number of firms	234	21
Log pseudolikelihood	−14,2502.3	−116,929.6
Chi-square	8.45260e+11	5.53888e+10
Pro > chi-square	0	0

Note: Robust standard errors clustered by the MNC in parentheses.
***Significant at the 1/10 of 1 percent level.
**Significant at the 1 percent level.
*Significant at the 5 percent level.

Interestingly, similar regression analyses on Chinese data reveal a different pattern. There is no statistically significant difference in quality between China-generated multinational invention and invention generated in the multinational's home country. Efforts to track the evolution of quality differences over time, as in table 5.5, suggest that, in the case of China, there has been rapid relative improvement in the measured quality of the MNC invention conducted in China. Understanding the differences in the Indian results reported here and the results obtained from our Chinese data is the focus of ongoing research.

5.4.5 The Dynamics of the Quality Gap between MNCs and Indigenous Firms

Is the quality difference between MNCs and indigenous firms narrowing over time? (See H4.) We can examine this by dividing the patents used for specification (1) into time periods according to their grant year and interacting our Coinv and MNC dummies with these period dummies. We arbitrarily divide our data into three periods: grant years before 2000, grant years from 2000 to 2004, and grant years from 2005 to 2009.

We specify the regressions as follows:

Table 5.5   Cross-border comparisons within MNCs over time (India, 1996–2009)

DV: no. of non-self-citations received as of the end of 2010	Full sample (1)	Firms with >30 IN patents (2)
Coinvention*1–5 years of India experience	−0.251*** (0.0638)	−0.195* (0.0987)
Coinvention*6–10 years of India experience	−0.348*** (0.0710)	−0.389*** (0.116)
Coinvention*more than 10 years of India experience	−0.0986 (0.118)	−0.114 (0.127)
Purely Indian invention*1–5 years of India experience	−0.295* (0.125)	−0.278 (0.146)
Purely Indian invention*6–10 years of India experience	−0.436*** (0.0631)	−0.383*** (0.0520)
Purely Indian invention*more than 10 years of India experience	−0.133 (0.139)	−0.0995 (0.136)
Three-year patent stock prior to application date (in thousands)	−0.00302 (0.0180)	0.00535 (0.0231)
Grant delay in years	0.102*** (0.0130)	0.116*** (0.0136)
Team size	0.0616*** (0.00502)	0.0631*** (0.00559)
Firm fixed effects	Yes	Yes
Grant year dummy	Yes	Yes
HJT subcat. dummy	Yes	Yes
Observations	40,324	32,633
Offset	ln(pat_age)	ln(pat_age)
Number of firms	234	21
Log pseudolikelihood	−142,469.6	−116,903.5
Chi-square	9.45986e+11	4.33773e+10
Pro > chi-square	0	0

Note: Robust standard errors clustered by the MNC in parentheses.
***Significant at the 1/10 of 1 percent level.
**Significant at the 1 percent level.
*Significant at the 5 percent level.

$$
\begin{aligned}
(3)\quad E(C_i) = {}& \mathrm{PatAge}_i \cdot \exp(\varphi_0 + \varphi_1 \mathrm{Coinv}_i * \mathrm{Gyear}_{<2000} \\
& + \varphi_2 \mathrm{Coinv}_i * \mathrm{Gyear}_{2000-2004} + \varphi_3 \mathrm{Coinv}_i * \mathrm{Gyear}_{2005-2009} \\
& + \varphi_4 \mathrm{MNC}_i * \mathrm{Gyear}_{<2000} + \varphi_5 \mathrm{MNC}_i * \mathrm{Gyear}_{2000-2004} \\
& + \varphi_6 \mathrm{MNC}_i * \mathrm{Gyear}_{2005-2009} + \varphi_7 \mathrm{PatStock}_f + \varphi_8 \mathrm{TeamSize}_i \\
& + \varphi_9 \mathrm{Gdelay}_i + H_i + T_i.
\end{aligned}
$$

Results from this regression specification are given in table 5.6. These results suggest that the quality premia associated with coinvention and with MNC sponsorship do not appear to be fading over time in India. Instead, both remain economically and statistically significant. When we run a paral-

Table 5.6   **Cross-firm comparison within India over time (1979–2009)**

DV: no. of non-self-citations received as of the end of 2010	Control for coinvention (1)	Control for assignee type (2)	Control for both (3)
Coinvention*grant year < 2000	0.327 (0.206)		0.277 (0.212)
Coinvention*grant year 2000–2004	0.109 (0.106)		0.0476 (0.118)
Coinvention*grant year 2005–2009	0.332*** (0.0918)		0.252** (0.0939)
Multinational assignee*grant year < 2000		0.351 (0.250)	0.163 (0.247)
Multinational assignee*grant year 2000–2004		0.265 (0.136)	0.233 (0.154)
Multinational assignee*grant year 2005–2009		0.743*** (0.151)	0.614*** (0.155)
Three-year patent stock prior to application date (in thousands)	0.00766 (0.0119)	−0.000550 (0.0121)	0.00283 (0.0123)
Grant delay in years	0.124*** (0.0287)	0.134*** (0.0288)	0.129*** (0.0287)
Team size	0.0463*** (0.00605)	0.0523*** (0.00564)	0.0475*** (0.00598)
Constant	−8.672*** (0.588)	−8.802*** (0.598)	−8.773*** (0.595)
Grant year dummy	Yes	Yes	Yes
HJT subcat. dummy	Yes	Yes	Yes
Observations	4,280	4,280	4,280
Offset	ln(pat_age)	ln(pat_age)	ln(pat_age)
Log pseudolikelihood	−13,523.2	−13,529.2	−13,480.0
Chi-square	1,973.3	2,319.3	2,209.9
Pro > chi-square	0	0	0

Note: Robust standard errors in parentheses.
***Significant at the 1/10 of 1 percent level.
**Significant at the 1 percent level.
*Significant at the 5 percent level.

lel regression on our Chinese data set, we find evidence suggesting that the quality premium associated with international coinvention appears to fade over time, but the quality premium associated with multinational sponsorship remains strong, both in magnitude and in statistical significance.

5.5 Peering inside Coinvention: Lessons from Interviews of Multinational R&D Personnel

To obtain insights into the mechanisms behind the multinational R&D phenomenon in emerging economies, we took a research trip to China in December 2009 to conduct face-to-face interviews with inventors from multinational R&D centers there. We supplemented these interviews with telephone-based interviews of multinational R&D managers in India, and one member of the research team also participated in on-site interviews in Delhi, Mumbai, Hyderabad, and Bangalore. Our interviews focused on several aspects of multinational R&D activity: How are the international research teams formed? What do the backgrounds of Indian or Chinese participants look like? Where do the main ideas in collaborative work come from? How do team members communicate? Does a division of labor exist within international research teams, and if so, to what extent?

We received strong confirmation from all sources that there is an emerging international division of R&D labor within multinational firms, and that a significant fraction of their India-based and China-based research manpower is being used to contribute to global research projects whose ultimate application will be in global markets, not just the local market. Most interviewees emphasized their commitment to a long-run research presence that could engage the large and growing endowment of engineering human resources in the local labor market in the service of their firm's global R&D agenda.

Second, we also received confirmation of the view that while the endowment of raw talent in China and India is immense and impressive, these talent pools still contain relatively few individuals who have become capable of directing a world-class R&D effort in key areas of technology without many years of exposure to multinational best practice.[20] That being said, talented local engineers can and do become "mature" and effective collaborators in international R&D projects, even taking on leading roles, after a few years of intense experience within a multinational R&D lab. In some organizations, it was explicitly acknowledged that the fundamental intellectual insights and the structuring of the research agenda still came from the foreign side. In other organizations, there was much more local autonomy in terms of setting the research agenda. But even in these cases, expatriate R&D managers and/or local staff with extensive educational and work experience in the United States often maintained a key role in directing the R&D activities of younger staff whose education and experience had been obtained entirely in the local market.

Nevertheless, a simple story of collaboration in which US-based engineers come up with the ideas and give the orders and local engineers carry them out was clearly far too simple to reflect the much more complex patterns of

---

20. We are drawing a sharp distinction here between "indigenous" local personnel who are educated in India or China and spend their entire professional lives there, and "multinational" personnel with Western educations and long-term work experience in the West, who may nevertheless have an Indian or Chinese ethnic background. The statement about managerial capability applies to the former, most definitely not to the latter. Interestingly, these judgments were often rendered by multinational managers who had the same ethnic background as their "local" employees.

interaction we heard described in our interviews. There were certainly cases in which important ideas came in the first instance from the local side, as well as cases in which the projects were conceived, developed, and implemented entirely by the local side, with very little Western input.

Many interviewees placed far more stress on the importance of "(re)engineering products for the local market" as a source of coinvention than we initially expected. In many markets for industrial intermediate goods—and even in some markets for consumer goods—the Chinese market is now substantially larger than the US market or even the European market. India now has more cellular phone subscribers than the United States has citizens. However, India and China are still both poor, developing countries, and the trade-off between cost and functionality is quite different for a local customer—even a local corporate customer—than it is for a Western customer. Therefore, a significant fraction of local engineering personnel were employed in the ongoing process of reengineering Western products for the local market—in ways that were both subtle and profound. This activity was taking place in both China and India, but the tone of the interviews suggested a significantly greater intensity of this effort in China. In the context of this reengineering work, it is not surprising that local engineers often take a leading role. However, the division between "reengineering for the local market" and "contributing to the global R&D agenda" was a fuzzy one and, over time, the same local engineer might be involved in both kinds of undertakings. In fact, interviewees noted that some cost-reducing innovations are often applied to products in other developing markets around the world and sometimes to even Western products and processes. For these reasons, reengineering projects could generate coinvented US patents.

Finally, our interviewees generally confirmed both the communications challenges posed by intercontinental research collaboration and the role of modern telecommunications technologies in meeting these challenges. Videoconferencing and software design tools that allowed a globally distributed team to work with the same virtual prototypes were important mechanisms facilitating research collaboration. The R&D engineers noted that videoconferences with collaborators around the world were now a routine practice in most projects. It was also seen as important for the firms to ensure a steady flow of personnel between the various global R&D centers. Face-to-face communications helped provide a foundation of basic understanding and trust that later Internet-mediated interaction could build on. Most interviewers agreed that, without modern communications tools, this kind of globally distributed R&D effort would be impossible.

5.6 Conclusions and Implications

In this chapter, we have analyzed the patterns found in India- and China-based US patents. In doing so, we found that a majority of India's US patents are owned by foreign MNCs, with US firms playing an especially important role. Similarly, a majority of China's US patents are owned by non-Chinese MNCs, with Taiwanese and US firms playing a significant role. We have shown that China- and India-based US patents are technologically concentrated in IT-related fields. We suspect that the prevalence of software-based design and engineering tools in these domains might have facilitated coinvention and long distance R&D efforts. We explored the geographic distribution of Indian and Chinese inventors and found that the majority of Indian and Chinese inventors are clustered in the most economically advanced regions in both countries, where foreign direct investment (FDI) is also concentrated.

We complemented statistical analyses of the patent data with in-person interviews with researchers in multinational R&D subsidiaries. These interviews confirmed that Indian and Chinese R&D personnel are increasingly seen as an integral part of MNCs' global R&D operations, and they are increasingly contributing to innovations whose ultimate market targets are outside of China and India. However, the patterns of international collaboration within MNCs are more complex than those that arise directly out of traditional views of comparative advantage. Our interviews supported the view that modern advances in telecommunications technologies have been instrumental in facilitating international R&D collaborations.

We have used forward citations from patent documents to compare the quality of India-based patents in multiple ways. Our results support our R&D vertical disintegration argument. Our study suggests that the increase in US patents in India and China are to a great extent driven by MNCs from advanced economies and are highly dependent on collaborations with inventors in those advanced economies. As such, India and China's striking rise in innovation may represent less of a challenge to conventional views of development economics. The view that the rise of innovation in India and China is undermining the traditional position of technological leadership enjoyed by the United States and other advanced industrial economies has been exaggerated.

Nevertheless, the world of R&D is indeed undergoing a major change. The increase in R&D activity in emerging economies such as India and China represents a growing international division of R&D labor. By undertaking R&D in emerging economies, MNCs can now provide innovative technologies to global markets at a lower cost and introduce products more suitable for emerging markets.

All of this leads us to the possibility of a "win-win" outcome for a more integrated global innovation system that can benefit both emerging and advanced economies. By participating in MNCs' R&D networks, emerging economies not only bring in more investment and create more employment, they can also participate in the generation of new technology at an earlier stage in the economic development process, even before they have internally developed all of the necessary categories of capabilities required for the complete R&D process. Their participation can also shift the direction of global R&D in a way that creates more goods and services suited to the income levels and conditions of emerging markets. Jones (2009) suggests diminishing productivity in R&D investment in the traditional innovation centers of the West as the burden of knowledge rises, but this can be offset by adding enough new scientists into a globalized innovation process, generating gains at the global level. By letting their companies do R&D in countries like India and China, advanced economies will also benefit from a faster pace of innovation and a more rapidly expanding stock of knowledge.

References

Alcácer, J., and Michelle Gittelman. 2006. "Patent Citations as a Measure of Knowledge Flows: The Influence of Examiner Citations." *Review of Economics and Statistics* 88 (4): 774–79.

Barefoot, Kevin. 2012. "US Multinational Companies: Operations of US Parents and their Foreign Affiliates in 2010." *Survey of Current Business* November:51–74.

Barefoot, Kevin, and Raymond Mataloni. 2011. "Operations of US Multinational Companies in the United States and Abroad." *Survey of Current Business* November:29–48.

Basant, Rakesh, and Sunil Mani. 2012. "Foreign R&D Centres in India: An Analysis of Their Size, Structure, and Implications." IIM Working Paper no. 2012-01-06, Indian Institute of Management.

Branstetter, Lee, and C. Fritz Foley. 2010. "Facts and Fallacies about US FDI in China." In *China's Growing Role in World Trade*, edited by Robert C. Feenstra and Shang-Jin Wei, 513–39. Chicago: University of Chicago Press.

Branstetter, Lee, Itzhak Goldberg, John Gabriel Goddard, and Smita Kuriakose. 2008. *Globalization and Technology Absorption in Europe and Central Asia.* Washington, DC: World Bank Publications. doi:10.1596/978–0–8213–7583–9.

Branstetter, Lee, and Mariko Sakakibara. 2002. "When Do Research Consortia Work Well and Why? Evidence from Japanese Panel Data." *American Economic Review* 92 (1): 143–59.

Cameron, Adrian Colin, and Pravin K. Trivedi. 1998. *Regression Analysis of Count Data.* Cambridge: Cambridge University Press.

———. 2005. *Microeconometrics: Methods and Application.* Cambridge: Cambridge University Press.

Cantwell, John. 1995. "The Globalisation of Technology: What Remains of the Product Cycle Model?" *Cambridge Journal of Economics* 19 (1): 155–174.

Chen, Jennifer, Show-Ling Jang, and Chiao-Hui Chang. 2013. "The Patterns and Propensity for International Co-Invention: The Case of China." *Scientometrics* 94:481–95.

Danguy, Jerome. 2012. "Globalization of Innovation Production: A Patent-Based Industry Analysis." iCite Working Paper no. 2014-009, International Centre for Innovation, Technology, and Education Studies, Solvay Brussels School of Economics and Management. http://www.solvay.edu/sites/upload/files/WP009-2014.pdf.

Eberhardt, Markus, Christian Helmers, and Zhihong Yu. 2011. "Is the Dragon Learning to Fly? An Analysis of the Chinese Patent Explosion." CSAE Working Paper no. 2011-15, Centre for the Studies of African Economies.

Freeman, Richard B. 2006. "Does Globalization of the Scientific/Engineering Workforce Threaten US Economic Leadership?" In *Innovation Policy and the Economy*, vol. 6, edited by Adam Jaffe, Josh Lerner, and Scott Stern, 123–58. Cambridge, MA: MIT Press.

Gourieroux, C., A. Monfort, and A. Trognon. 1984. "Pseudo Maximum Likelihood Methods: Applications to Poisson Models." *Econometrica* 52 (3) : 701–20. doi: 10.2307/1913472.

Griffith, Rachel, Rupert Harrison, and John Van Reenen. 2006. "How Special is the Special Relationship? Using the Impact of US R&D Spillovers on UK Firms as a Test of Technology Sourcing." *American Economic Review* 96 (5): 1859–75. doi:10.2307/30035000.

Grossman, Gene M., and Elhanan Helpman. 1990. "Comparative Advantage and Long-Run Growth." *American Economic Review* 80 (4): 796–815. doi:10.2307/2006708.

———. 1991. "Quality Ladders in the Theory of Growth." *Review of Economic Studies* 5(1): 43–61. doi:10.2307/2298044.

Grossman, Gene M., and Esteban Rossi-Hansberg. 2008. "Trading Tasks: A Simple Theory of Offshoring." *American Economic Review* 98 (5): 1978–97. doi:10.2307/29730159.

Hall, Bronwyn H., Zvi Griliches, and Jerry A. Hausman. 1986. "Patents and R and D: Is There a Lag?" *International Economic Review* 27 (2): 265–83. doi:10.2307/2526504.

Hall, Bronwyn H., Adam B. Jaffe, and Manuel Trajtenberg. 2001. "The NBER Patent Citation Data File: Lessons, Insights and Methodological Tools." NBER Working Paper no. 8498, Cambridge, MA.

———. 2005. "Market Value and Patent Citations." *RAND Journal of Economics* 36 (1): 16–38. doi:10.2307/1593752.

Harhoff, Dietmar, Francis Narin, F. M. Scherer, and Katrin Vopel. 1999. "Citation Frequency and the Value of Patented Inventions." *Review of Economics and Statistics* 81 (3): 511–15.

Hu, Albert Guangzhou, and Gary H. Jefferson. 2009. "A Great Wall of Patents: What Is Behind China's Recent Patent Explosion?" *Journal of Development Economics* 90 (1): 57–68.

Huang, Can. 2012. "Estimates of the Value of Patent Rights in China." UNU-MERIT Working Paper no. 004, Maastricht Economic and Social Research Institute on Innovation and Technology.

Hummels, David, Jun Ishii, and Kei-Mu Yi. 2001. "The Nature and Growth of Vertical Specialization in World Trade." *Journal of International Economics* 54 (1): 75–96. doi:http://dx.doi.org/10.1016/S0022-1996(00)00093-3.

Jaffe, Adam B., and Manuel Trajtenberg. 2002. *Patents, Citations and Innovations: A Window on the Knowledge Economy*. Cambridge, MA: MIT Press.

Jaffe, Adam B., Manuel Trajtenberg, and Michael S. Fogarty. 2000. "Knowledge Spillovers and Patent Citations: Evidence from a Survey of Inventors." *American Economic Review* 90 (2): 215–18. doi:10.2307/117223.

Jang, Show-Ling, Tzu-Ya Wang, and Jennifer Chen. 2012. "Cross-Border Patenting in Emerging Countries—The Cases of China and India." National Taiwan University Working Paper.

Jones, Benjamin F. 2009. "The Burden of Knowledge and the 'Death of the Renaissance Man': Is Innovation Getting Harder?" *Review of Economic Studies* 76 (1): 283–317. doi:10.1111/j.1467–937X.2008.00531.x.

Kogut, Bruce, and Udo Zander. 1992. "Knowledge of the Firm, Combinative Capabilities, and the Replication of Technology." *Organization Science* 3 (3): 383–97.

———. 1993. "Knowledge of the Firm and the Evolutionary Theory of the Multinational Corporation." *Journal of International Business Studies* 24 (4): 625–45.

Krugman, Paul. 1979. "A Model of Innovation, Technology Transfer, and the World Distribution of Income." *Journal of Political Economy* 87 (2): 253–66. doi: 10.2307/1832086.

Krugman, Paul, Richard N. Cooper, and T. N. Srinivasan. 1995. "Growing World Trade: Causes and Consequences." *Brookings Papers on Economic Activity* 1995 (1): 327–77. doi:10.2307/2534577.

Lai, Ronald, Alexander D'Amour, Amy Yu, Ye Sun, Vetle Torvik, and Lee Fleming. 2011. "Disambiguation and Co-Authorship Networks of the US Patent Inventor Database." Harvard Institute for Quantitative Social Science. https://thedata.harvard.edu/dvn/dv/patent/faces/study/StudyPage.xhtml?studyId=70546&versionNumber=5.

Lei, Zhen, Zhen Sun, and Brian Wright. 2012. "Patent Subsidy and Patent Filing in China." Working Paper, Penn State University and University of California, Berkeley.

National Research Council. 2007. "Rising above the Gathering Storm: Energizing and Employing America for a Brighter Economic Future." *Academy of Science, National Academy Of*. Washington, DC: The National Academies Press.

———. 2012. "OECD Science, Technology and Industry Outlook 2012." OECD Publishing.

Puga, Diego, and Daniel Trefler. 2010. "Wake Up and Smell the Ginseng: International Trade and the Rise of Incremental Innovation in Low-Wage Countries." *Journal of Development Economics* 91 (1): 64–76. doi:10.1016/j.jdeveco.2009.01.011.

Romer, Paul M. 1990. "Endogenous Technological Change." *Journal of Political Economy* 98 (5): S71–S102. doi:10.2307/2937632.

Sharma, P., S. B. S. Nookala, and Anubhav Sharma. 2012. "India's National and Regional Innovation Systems: Challenges, Opportunities and Recommendations for Policy Makers, Industry and Innovation." *Industry and Innovation* 19 (6): 517–37.

Singh, Jasjit. 2008. "Distributed R&D, Cross-Regional Knowledge Integration and Quality of Innovative Output." *Research Policy* 37 (1): 77–96. doi:10.1016/j.respol.2007.09.004.

The Royal Society. 2010. "The Scientific Century: Securing Our Future Prosperity." London: The Royal Society.

Trajtenberg, Manuel. 1990. "A Penny for Your Quotes: Patent Citations and the Value of Innovations." *RAND Journal of Economics* 21 (1): 172–87. doi:10.2307/2555502.

———. 2001. "Innovation in Israel 1968–1997: A Comparative Analysis Using Patent Data." *Research Policy* 30 (3): 363–89. doi:10.1016/S0048-7333(00)00089-5.

United Nations Conference on Trade and Development (UNCTAD). 2005. *World Investment Report 2005: Transnational Corporations and the Internationalization of R&D*. New York: United Nations.

Vernon, Raymond. 1966. "International Investment and International Trade in the Product Cycle." *Quarterly Journal of Economics* 80 (2): 190–207. doi:10.2307/1880689.

Weitzman, Martin L. 1998. "Recombinant Growth." *Quarterly Journal of Economics* 113 (2): 331–60. doi:10.2307/2586906.

Wooldridge, Jeffrey M. 1999. "Distribution-Free Estimation of Some Nonlinear Panel Data Models." *Journal of Econometrics* 90 (1): 77–97. doi:10.1016/S0304-4076(98)00033-5.

————. 2002. *Econometric Analysis of Cross Section and Panel Data.* Cambridge, MA: MIT Press.

Yi, Kei-Mu. 2003. "Can Vertical Specialization Explain the Growth of World Trade?" *Journal of Political Economy* 111 (1): 52–102. doi:10.1086/344805.

Yorgason, Daniel. 2007. "Research and Development Activities of US Multinational Corporations: Preliminary Results from the 2004 Benchmark Survey." *Survey of Current Business* March:22–39.

Zhao, Minyuan. 2006. "Conducting R&D in Countries with Weak Intellectual Property Rights Protection." *Management Science* 52 (8): 1185–99. doi:10.1287/mnsc .1060.0516.

Zucker, Lynne G., and Michael R. Darby. 2011. "COMETS Data Description, Release 1.0." Los Angeles, CA: UCLA Center for International Science, Technology, and Cultural Policy.

Information Technology and the Distribution of Inventive Activity

Chris Forman, Avi Goldfarb, and Shane Greenstein

6.1 Introduction

Vannevar Bush's publication *Science: The Endless Frontier* frames a range of questions about the localization of information, and about how the costs of knowledge transmission, dissemination, and collaboration increase in distance (Jaffe, Trajtenberg, and Henderson 1993; Glaeser, Kerr, and Ponzetto 2010; Delgado, Porter, and Stern 2010; Saxenian 1994; and others). Over the years this conceptualization has motivated a range of research questions about the collocation of inventive activity. The creation and maintenance of geographic clusters of invention, and their links to regional economic growth, have been an important part of innovation policy.

In the years since the publication of that book, several factors have potentially altered the importance of agglomeration for inventive activity. First, globalization and the vertical disintegration of supply chains—in which increasingly many different companies manufacture the components that make up a final product—has increased the premium on invention as a source of regional competitiveness, thereby reinforcing preexisting differences in

Chris Forman is the Brady Family Term Professor at the Scheller College of Business at the Georgia Institute of Technology. Avi Goldfarb is professor of marketing at the Rotman School of Management, University of Toronto, and a research associate of the National Bureau of Economic Research. Shane Greenstein is the Kellogg Chair of Information Technology and professor of management and strategy at the Kellogg School of Management, Northwestern University, and a research associate of the National Bureau of Economic Research.

We thank Adam Jaffe, Ben Jones, Scott Stern, and participants at the preconference and conference for helpful comments and suggestions. We thank Yasin Ozcan for outstanding research assistance. We also thank Harte-Hanks Market Intelligence for supplying data. All opinions and errors are ours alone. For acknowledgments, sources of research support, and disclosure of the authors' material financial relationships, if any, please see http://www.nber .org/chapters/c13032.ack.

the geographic distribution of invention. Second, declines in communications costs—engendered by the widespread diffusion of the Internet—have substantially reduced the cost of certain kinds of communication, leading to changes in the geographic distribution of innovation and invention (Agrawal and Goldfarb 2008; Forman and van Zeebroeck 2012). These two changes push in opposite directions.

In this chapter, we ask whether invention, as measured in patent data, has become more geographically concentrated between the early 1990s and the early twenty-first century. We address this topic in order to explore the overall net effect of the two forces pushing for or against geographic agglomeration of invention. We also explore the potential role of Internet technology in explaining this pattern. Either an increase or decrease in the geographic concentration of invention is possible. By increase in concentration, we mean that the places that served as the location for the majority of the inventions in the past serve as a source for an even greater share in the future. The places rich with inventions become richer. By decrease, we mean the opposite, that the places that are not rich with invention become richer.

This chapter builds on our research agenda examining how the diffusion of the Internet altered the geographic concentration of activity (Forman, Goldfarb, and Greenstein 2002, 2005, 2008, 2012). The approach of this study resembles our approach in Forman, Goldfarb, and Greenstein (2012), which examined how geographic variation in business Internet adoption shaped US wage growth over the late 1990s. This chapter examines a different outcome, and hence, a different question, namely, whether those counties that were leading innovators (as measured by patents) between 1990 and 1995 increased or decreased their relative rate of patenting between 2000 and 2005. Then we explore how Internet adoption correlates with this change, and whether it increases or decreases the rate of concentration in patenting.

We undertake this exercise with the view that economic theory does not give clear guidance to the expected result. There are good reasons to expect the Internet to have increased the geographic concentration of invention or to have decreased it.

On the side of increasing concentration: the literature on the economics of information technology (IT) often finds a localization of the adoption of IT (Forman, Goldfarb, and Greenstein [2008] and Forman and Goldfarb [2006] reviews the literature). The effective use of advanced Internet technology draws on frontier IT skills that are found disproportionately in urban areas, and it builds on existing links between business use of IT, support services, and specialized labor markets in urban areas. Furthermore, while the Internet reduces communication costs for both local and distant communication, most communication and most social contacts are local (Wellman 2001; Hampton and Wellman 2002). Much of the literature on Internet adoption and usage, including much of our own prior work, shows a high geographic concentration of economic activity in the areas where the Internet is most frequently adopted (Blum and Goldfarb 2006; Sinai and Waldfogel 2004; Forman, Goldfarb, and Greenstein 2005; Kolko 2002; Glaeser and Ponzetto 2007; Arora and Forman 2007; Forman, Goldfarb, and Greenstein 2012; Agrawal, Catalini, and Goldfarb 2011; and others).

On the side of decreasing concentration: the Internet is a communications technology, and it can allow people in isolated areas to plug in to the rest of the economy. Communications scholars and others have long argued that the Internet might overcome geographic barriers to economic (and political) activities. Cairncross (1997) and Friedman (2005) provide popular summaries of these ideas, emphasizing the "death of distance" and the "flat world." Moreover, in the specific context of knowledge production and invention, the Internet can reduce collaboration costs and, potentially, the importance of collocation in inventive activity. The empirical literature also has some findings suggesting that the Internet might increase cross-institutional and cross-regional collaboration over time (Jones, Wuchty, and Uzzi 2008; Agrawal and Goldfarb 2008; Ding et al. 2010). The setting most closely resembling the one we study in this chapter (Forman and van Zeebroeck 2012) also shows that Internet adoption leads to increased distant collaboration for patents issued to researchers in a given multiestablishment firm.

Our findings generally favor the view that the Internet worked against the concentration of invention. Studying the growth rate of patenting across counties, we show this in several steps. First, we show that invention became more geographically concentrated over this period, suggesting a general trend toward increasing concentration of invention. Specifically, our raw data suggest that patenting grew 27 percent during this period. For the top quartile of patenting counties from 1990–95, patenting grew 50 percent. For those below the median, patenting did not grow at all. We highlight differences between our setting and findings and a line of research that has found convergence in economic growth rates across countries and geographic regions (e.g., Barro and Sala-i-Martin 1991; Magrini 2004; Delgado, Porter, and Stern 2010, 2012). While differences between our results and this research line may reflect differences in our measure of local economic activity (patents vs. economic output or wage growth), we also show that our findings are driven in particular by substantial increases in the concentration of patenting at the very top of the distribution.

We next demonstrate how county-level growth in patenting is shaped by business Internet adoption and the prior concentration of patents. While the geographic concentration of patenting increased over the time period we study, the Internet appears to have mitigated, rather than exacerbated, that trend. In particular, the overall concentration of invention rose but, among counties that were leading Internet adopters, we see little change in the concentration of invention. Furthermore, our results suggest that this relationship is strongest for long-distance collaboration. Although it is important to recognize that we cannot rule out the possibility that an omitted factor caused both

Internet adoption and growth in patenting in the set of Internet-adopting counties with that were behind in patenting in the early 1990s, our results are more consistent with the Internet reducing the geographic concentration of invention than with the Internet increasing that concentration.

To summarize, our chapter provides evidence about the net effects of opposing factors that have influenced the concentration of inventive activity since the publication of Bush's book. We highlight the effects of the Internet. Recent literature has shown that scientific collaboration across institutions has increased over time and that IT is partly responsible. We contribute the first direct evidence that the diffusion of the Internet is correlated with a reduction in the geographic concentration of inventive activity, suggesting that the diffusion of the Internet has the potential to weaken the long-standing importance of the geographic localization of innovative activity. Our results also raise intriguing questions about whether the Internet's impact on the geographic concentration of invention is distinct from its impact on the geographic concentration of other economic activity, such as wages, business adoption of IT, hospital productivity, and so on. That is, the Internet may be a force for weakening the links between the geography of inventive activity and the geography of other economic activity.

6.2 Data

We use a variety of data sources to examine how adoption of advanced internet among firms will affect local inventive activity. We match data on IT investment from the Harte-Hanks Market Intelligence computer intelligence database with patent data from the United States Patent and Trademark Office (USPTO) between 1990 and 2005. We further combine this with data from the US decennial census. Our sample construction is shaped by key features of our data and the setting. First, we expect a significant lag between the time when IT investments are made and when they influence the creation of new invention. Second, there is significant year-to-year variability in patent output at the county level and particularly at the industry-county level. Third, as with our prior work, we exploit the historical circumstances that led to the deployment of the Internet. Instead of creating a gradual deployment and adoption, circumstances created a rather abrupt change in a short time span, leading to a period "before the Internet diffused" and a period "after the Internet diffused." As a result, in our core analyses our base period and reference period both include six years—that is, we look at the difference in patent output between 1990 and 1995 (before the diffusion of the Internet) and 2000 and 2005 (after its diffusion).[1]

1. We have experimented with alternative specifications for the base and reference years. Our results are robust to these changes, though we do sometimes lose significance for some results in some years.

6.2.1 Patent Data

Our data on local inventive output are measured using patent data from the USPTO. We use application rather than grant date to measure the timing of inventive activity because the application-to-grant delay varies over time, and because the application date is closer to the time when the invention occurred.[2]

To measure the effect that Internet adoption will have on local inventive activity, we match patents to counties using inventor locations.[3] For patents with multiple inventors that reside in multiple counties, we allocate patents to all of the counties where inventors reside. We use county as the unit of observation rather than metropolitan statistical areas (MSA) to facilitate comparison with prior work that has studied the implications of Internet investment on local economic outcomes (Forman, Goldfarb, and Greenstein 2012). Our procedure will accurately assign patent output to the correct county to the extent that inventors work where they reside, but may make some errors in assignment when inventors commute between counties.[4]

In our analyses we use a combination of raw patent counts and five-year citation-weighted patents as our measure of inventive output. As is well known, not all inventions meet the US Patent and Trademark Office (USPTO) criteria for patentability (Jaffe and Trajtenberg 2002). Further, inventors must make an explicit decision to patent an invention rather than relying on some other method to appropriate the value of their invention. There will be incremental inventive activity that is not patented and therefore is not reflected in patent statistics (e.g., Cohen, Nelson, and Walsh 2000). However, so long as the propensity of firms in a location to patent does not vary significantly over time in a way that is correlated with Internet adoption, this should not bias our estimates of the key parameters of interest. It is also well known that patent values are very skewed. Weighting by citations is one way to address this problem; citation-weighted patents have been shown to be correlated with a firm's stock market value above and beyond the information provided by patent counts (Hall, Jaffe, and Trajtenberg 2005).

Our baseline analyses explore whether Internet adoption is associated with changes in the growth of total patents and citation-weighted patents

---

2. See, for example, Griliches (1990).
3. Specifically, we match the city and state of the inventor location to zip codes, and then match the zip codes to counties.
4. We also believe that using inventor locations, which is often the location of their residence, is superior to the alternative of using the location of the assignee, which is the location of a firm or corporate building in the vast majority of patents. The latter does not necessarily correspond with the location of the invention, particularly in corporations that assign all patents to headquarters, irrespective of their origins.

over time. However, we also explore how our results vary by county-industry group. To do this, we utilize the 2011 USPTO concordance between patent classes and North American Industry Classification System (NAICS) manufacturing industries.[5] In these analyses, our unit of analysis is county-industry-year rather than county-year. To facilitate comparisons between our county and county-industry analyses, all of our patents have a primary class that can be mapped to NAICS using the 2011 concordance. Thus, our measures of patent growth will miss some inventive activity that cannot be used downstream in manufacturing.

We perform several additional analyses over different subsets of the patent data. First, we reestimate our models over the set of patents with more than one inventor. We label these as collaborative patents. Second, we define distant collaborative patents as ones in which there exists a pair of inventors for whom the distance between the centroids of the inventors' home counties are greater than fifty miles apart.

We further explore differences based upon the type of institution to which the patent is assigned. We identify educational institutions based upon a search of key phrases in the assignee name field of the patent.[6] We further use the assignee role field in the patents to identify whether the patent is from a US private company or corporation or a US government agency.[7] Last, we examine how our results vary by technological field, using the Hall, Jaffe, and Trajtenberg (2001) technology categories.

A primary question in this chapter is whether Internet investments by firms contribute to changes in the distribution of inventive activity. In particular, our interest is in exploring whether Internet investments have contributed to more or less concentration in outcomes. To facilitate this, we construct measures of the total number of patents in the county between 1990 and 1995 to measure concentration in innovative activity prior to the diffusion of the Internet.

5. For more details on the correspondence, see http://www.uspto.gov/web/offices/ac/ido/oeip/ taf/data/ naics_conc/2011/read_me.txt. To perform the correspondence, we use the primary USPTO class in the patent document. In cases where a given USPTO class is related to several industries, we weight the patent equally across the industries to which it is related.

6. Specifically, we define educational institutions as those that have any of the following phrases in the assignee name (not case sensitive): "university"; "institute of technology"; "college"; "school of medicine"; "school of mines"; "school of engineering"; and some permutations on these phrases. Further, we identified several specific research active institutions for which these key words were not accurate predictors of educational status. As a result, we also added the following phrases: "georgia tech"; "cornell research foundation"; "wisconsin alumni"; "board of regents for education"; "oregon graduate center"; "iowa state research foundation"; and "board of governors for higher education, state of rhode island"

7. We also explored whether our results differed for private firms who were small (below 500 employees) and large using the small entity status field on USPTO data on maintenance fee payments. We found that many of our main results were qualitatively similar for small and larger entities, though the economic magnitudes were somewhat weaker among small firms.

6.2.2 Information Technology Data

As mentioned above, our IT data come from the Harte-Hanks Market Intelligence computer intelligence database (hereafter CI database).[8] The database contains rich establishment- and firm-level data including the number of employees, the number of personal computers and servers, and adoption of Internet applications. Harte-Hanks collects these data to resell to the marketing divisions of technology companies. Interview teams survey establishments throughout the calendar year; our sample contains the most current information as of December 2000.

Harte-Hanks tracks over 300,000 establishments in the United States. We exclude government, military, and nonprofit establishments because the availability of advanced Internet for these establishments and the relationship between advanced Internet adoption and patent output may be different than for private firms. Our sample contains nonfarm business establishments with over 100 employees and includes a total of 86,879 establishments. Prior work has demonstrated that these data are among the best establishment-level data about the use of IT in the United States, and include half of all establishments with 100 or more employees in the United States (Forman, Goldfarb, and Greenstein 2005). While our sample includes only relatively large establishments, this is not a significant problem because very few small establishments adopted advanced Internet technology during this time.

The construction of our measure of advanced Internet is identical to that used in our previous study of the effects of advanced Internet adoption on local wage growth (Forman, Goldfarb, and Greenstein 2012). It includes those facets of Internet technology that became available after 1995 in a variety of different uses and applications. The raw data in the CI database include at least twenty different specific applications, from basic Internet access to software for Internet-enabled enterprise resource planning (ERP) business applications.

Our measure of advanced Internet adoption involves investment in frontier technologies, often with significant adaptation costs. As we have done in our prior work, we use substantial investments in e-commerce or e-business to identify advanced Internet investment. Specifically, we looked for evidence of investment in two or more of the following Internet-based applications: ERP, customer service, education, extranet, publications, purchasing, and technical support. Not all of these applications are directly involved in the production of new inventions, however all support intra- or interestab-

---

8. This section draws heavily from Forman, Goldfarb, and Greenstein (2012). Data from Harte-Hanks Market Intelligence have been used in a variety of previous studies (including our own) studying the adoption of IT (Bresnahan and Greenstein 1997; Forman, Goldfarb, and Greenstein 2005), the productivity of IT investments (Bresnahan, Brynjolfsson, and Hitt 2002; Brynjolfsson and Hitt 2003; Bloom, Sadun, and Van Reenen 2012), and the effects of IT investments on local wage growth (Forman, Goldfarb, and Greenstein 2012).

lishment communication and coordination, and often involve significant changes to business processes. Our measure of advanced Internet investment should be viewed as a proxy for a firm's propensity to invest in frontier IT that facilitates communication and collaboration, rather than a direct measure of IT investments that are used as part of the production process in science. As a result, it is possible this will generate some attenuation bias in our estimates.[9]

We aggregate our establishment-level indicators of advanced Internet investment to the county to obtain location-level measures of the extent of advanced Internet investment. Because the distribution of establishments over industries may be different in our sample of firms from that of the population, as we have done in prior work we weight the number of establishments in our database using the number of establishments by two-digit NAICS industry in the Census Bureau's 1999 County Business Patterns data.

This measure has several attractive properties.[10] For one, industry-level measures of this variable correlate with Bureau of Economic Analysis measures of industry-level differences in IT investments. The measure also highlights significant regional differences in advanced Internet use (Forman, Goldfarb, and Greenstein 2005). Advanced Internet adoption is high in locations that include Internet-intensive and IT-intensive industries, such as the San Francisco Bay Area, Seattle, Denver, and Houston. In such regions, advanced Internet adoption is high even for establishments that are not producing in traditionally IT-intensive industries.

As noted above, variance in our IT measure will come from differences in adoption rates among large nonfarm business establishments at the county level. Because we do not directly measure the IT investment behavior of public and educational institutions, our analyses of the effects of IT investment on patenting behavior in these institutions must be treated with some caution.

6.2.3 Controls

We combine these IT and patent data with additional county-level information from a variety of sources. First, we use information from the 1990 US Census on population, median income, and percentage of population with a university education, high school education, below the poverty line, African American, and above sixty-four years old. We further use the 2000 US Census to control for changes in factors such as population and change in percent African American, university education, high school education, and over sixty-four years old. We obtain county-level information on additional factors that will influence the propensity of a county to innovate such

9. Unfortunately, the CI database collects little information on applications that directly facilitate knowledge sharing or knowledge management. See Forman and van Zeebroeck (2012) for further details.

10. Here we summarize some highlights. For further details, see Forman, Goldfarb, and Greenstein (2012).

as enrollment in Carnegie tier 1 research universities in 1990, the fraction of students enrolled in engineering programs, and the 1990 percentage of the county's workforce in professional occupations.[11] To control for differences in growth rates based on the scale of economic activity, we also include controls for employment, establishments, and weekly wages in the county from 1999 County Business Patterns data.

Table 6.1 provides the descriptive statistics. While our census data include the population and demographic data of over 3,100 counties, as in our prior work we drop several hundred counties for which we have no IT data. Generally, these are very low population counties with few firms and patents. There are 2,734 counties for which we have IT data. There are also some counties that we drop from our analysis because there are no patents in either the 1990–1995 or 2000–2005 period, though results are robust to assuming that these are zero growth counties. If there are no patents in both periods, we set growth in patenting to zero. Across our different dependent variables we have between 2,519 and 2,854 observations. As a result, we have between 2,235 and 2,833 observations in our combined IT and patent data set.

The top part of table 6.1 shows the average percent change in our dependent variable across different categories. The average percent change is decreasing for some variables. Because these variables are the average of the percent changes across counties, this does not mean that total patenting in the United States for that category is decreasing. Some counties in our data have a large percent change but, due to their small size, do not have a large impact on the total amount of patenting.

6.3 Empirical Strategy and Results

Our empirical analysis proceeds in four steps. First, we establish the relationship between patent levels in the 1990–1995 period and growth in patenting between 1990–1995 and 2000–2005. We show an increased concentration in patenting. Second, we show that there is no significant relationship between advanced Internet adoption by firms and growth in patenting. Third, we show that the relationship between prior patent levels and growth in patenting is weaker for counties with high levels of Internet adoption. Fourth, we demonstrate that the effect of Internet on weakening the trend to increased geographic concentration of patenting is driven by changes in distant collaborative patents and private firms.

6.3.1 Increased Concentration of Patenting

Figure 6.1 shows a Lorenz curve for patenting by county comparing 1990–1995 to 2000–2005. The size of the area under the forty-five degree

---

11. Downes and Greenstein (2007) showed that these three help explain the availability of Internet service providers.

Table 6.1   Summary statistics (county-level data)

Variable	Mean	Std. dev.	Min.	Max.	No. obs.
Dependent variables					
Growth in patenting	0.2655	0.7319	-2.4849	3.6376	2,807
Growth in citation-weighted patenting	-0.1126	0.9946	-4.804	4.9200	2,714
Growth in collaborative patenting	-0.1814	0.7166	-2.9444	3.4553	2,792
Growth in citation-weighted collaborative patenting	-0.5320	1.0105	-4.8040	4.3407	2,705
Growth in distant collaborative patenting	-0.5559	0.7558	-3.7612	2.8332	2,840
Growth in citation-weighted distant collaborative patenting	-0.7729	1.0616	-5.7004	3.6109	2,793
Growth in nondistant, collaborative patenting	0.4442	0.6072	-2.5649	3.4553	2,631
Growth in citation-weighted, nondistant collaborative patenting	0.3449	0.8391	-3.4965	4.3758	2,519
Growth in noncollaborative patenting	0.4843	0.3386	-1.6094	2.4849	2,709
Growth in citation-weighted, noncollaborative patenting	0.4820	0.5041	-2.9957	3.3202	2,598
Growth in patenting by educational institutions	0.1082	0.4243	-2.1972	2.7726	2,631
Growth in citation-weighted patenting by educ. institutions	-0.0236	0.5276	-4.0775	4.2341	2,719
Growth in patenting by private firms	0.3789	0.7717	-2.4849	3.7136	2,644
Growth in citation-weighted patenting by private firms	0.0175	1.0018	-4.2485	4.7095	2,604
Growth in patenting by government institutions	-0.0117	0.3170	-3.1499	2.3979	2,793
Growth in citation-weighted patenting by govt. institutions	-0.0681	0.4791	-5.3083	3.4340	2,854
Core covariates					
Advanced Internet	0.0888	0.1329	0	1	2,734
Patenting 1990–1995	7786.7	46,544	0	1808028	3,131
Citation-weighted patenting 1990–1995	24,998	184,073	0	7905438	3,131

Controls

Log employment	8.8320	1.6851	2.77259	15.051	3,125
Log estabs.	6.4598	1.4594	1.94591	12.742	3,125
Log weekly wages	6.2084	0.2267	5.32301	7.335	3,125
Log pop.	10.137	1.3681	4.67283	15.997	3,122
Percent black	0.0859	0.1434	0	0.8624	3,122
Percent university education	0.1356	0.0659	0.03692	0.5366	3,122
Percent high school education	0.6979	0.1039	0.31682	0.9620	3,122
Percent below poverty line	0.1667	0.0794	0	0.6312	3,122
Median HH income	23978	6599.6	8595	59284	3,122
Carnegie tier 1 enrollment	0.0073	0.0651	0	2.6154	3,124
Fraction in engineering	0.0009	0.0055	0	0.1125	3,124
Fraction professional	0.3522	0.0659	0.16019	0.6744	3,124
Percent > 64 years old	0.1486	0.0442	0.0087	0.3409	3,122
Change population	0.0959	0.1337	-0.5506	1.0683	3,122
Change % black	0.0014	0.0175	-0.0994	0.2724	3,122
Change % univ. educ.	-0.0265	0.0210	-0.1461	0.0748	3,122
Change % high school educ.	-0.1873	0.0523	-0.3237	-0.0260	3,122
Change % > 64 years old	-0.0011	0.0145	-0.0919	0.0851	3,122

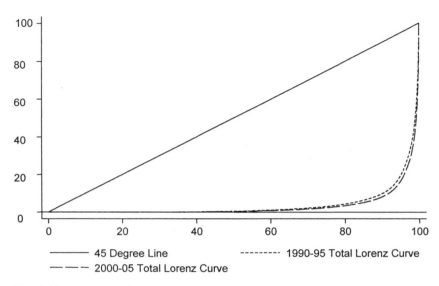

Fig. 6.1 Lorenz curves for concentration of patenting, 1990–1995 versus 2000–2005

line measures the degree of inequality across counties in their patenting behavior. As the curve moves away from the forty-five degree line, it suggests that the geographic concentration of patenting rises in general. Thus, the curve suggests that patenting was somewhat more geographically concentrated in the 2000–2005 period than in the 1990–1995 period.

Table 6.2 shows that the increase in concentration is influenced by a substantial increase in concentration at the very top of the distribution, with the top 0.1 percent of counties (i.e., the top three counties) showing a particularly large increase in the share of patents. That finding suggests that we should make inferences with some caution, as this finding depends on the performance in a very small number of locations. We can have more confidence in the inference since, as noted above, other evidence points in a similar direction. Overall patenting grew 27 percent during this time period. This suggests that patenting increased in the top 30th percentile of the distribution of patenting. It stayed roughly the same between the 30th–70th percentiles, and fell for the remainder of counties.

In table 6.3, we show the related result that those counties that had a large number of patents in the 1990–1995 period had a relatively large increase in their level of patenting. In particular, column (1) contains the following regression:

(1) $\text{Log}(\text{Patents}_{i0005}) - \text{Log}(\text{Patents}_{i9095}) = \alpha + \gamma X_i + \beta_1 \text{Patents}_{i9095} + \varepsilon_i,$

where Patents_{i9095} and Patents_{i0005} are the number of cumulative patents in county i from 1990–1995 and 2000–2005, X_i is a vector of controls including

Table 6.2   Concentration of overall patenting by decile and over time

	Share patenting 1990–1995 (%)	Share patenting 2000–2005 (%)
Top 0.1% of counties	8.51	12.16
Top 1% of counties	36.23	42.34
Top 5% of counties	73.32	76.39
0–10th percentile of counties	85.55	87.88
10th–20th percentile of counties	8.14	7.02
20th–30th percentile of counties	3.59	3.02
30th–40th percentile of counties	1.06	0.79
40th–50th percentile of counties	0.76	0.64
50th–60th percentile of counties	0.43	0.33
60th–70th percentile of counties	0.26	0.19
70th–80th percentile of counties	0.15	0.09
80th–90th percentile of counties	0.06	0.04
90th–100th percentile of counties	0	0

county-level business and demographic data (as listed in table 6.1), and ε_i is a normal i.i.d. error. The positive and significant coefficient in the first row shows that those counties with higher levels of patenting from 1990–1995 had higher rates of patent growth.

The remaining columns of the table show robustness to various alternative specifications. Column (2) weights the patents by citations over five years. Columns (3) and (4) use only collaborative patents to define the dependent variable.[12]

Columns (5) through (8) show robustness to switching the unit of observation to the industry-county. This enables the analysis to account for differences across industries where agglomeration takes place. The industry-level data is challenging to work with as there are many zeros. Therefore, the simple logged difference growth equation cannot be used as it will lead to many missing observations. In addition, the data are highly skewed, with a long positive tail and a fatter-than-normal negative tail in the difference. Instead of the logged difference, we use an ordered probit, splitting the dependent variable into nine groups: $(\infty, -5)$, $[-5, -2)$, $[-2, -1)$, $[-1, 0)$, 0, $(0, 1]$, $(1, 2]$, $(2, 5]$, $(5, \infty)$. The results show that this alternative specification does not yield qualitatively different results: those counties that were leading in patenting from 1990–1995 had relatively rapid growth in patenting.

The controls also yield some interesting, though perhaps unsurprising, correlations. The level of education, and changes in the level of education,

---

12. We maintain total patents 1990–95 as the key covariate as we believe the key measure is the rate of overall patenting in the preperiod. That said, results are robust to using collaborative patents as the key covariate.

Table 6.3 Patenting grew fastest in counties that patented more in 1990

	County-level data				County industry-level data			
	All patents		Collaborative patents		All patents		Collaborative patents	
	Patents (1)	Citation-weighted patents (2)	Patents (3)	Citation-weighted patents (4)	Patents (5)	Citation-weighted patents (6)	Patents (7)	Citation-weighted patents (8)
Patenting 1990–1995 (000s)	0.00064 (0.00021)***	0.00091 (0.00040)**	0.00021 (0.00013)*	0.00020 (0.00011)*	10.5638 (2.1771)***	0.3966 (0.0901)***	5.6169 (1.0899)***	0.1929 (0.0389)***
Log emp.	−0.0182 (0.0536)	−0.1345 (0.0521)***	−0.2357 (0.0732)***	−0.1127 (0.0730)	0.0459 (0.0281)	0.0648 (0.0226)***	0.0439 (0.0264)*	0.0711 (0.0239)***
Log estabs.	−0.0724 (0.0695)	−0.1576 (0.0676)**	−0.0083 (0.0889)	0.0665 (0.0923)	−0.0644 (0.0358)*	−0.0196 (0.0299)	−0.0924 (0.0358)***	−0.0436 (0.0328)
Log weekly wages	0.1504 (0.1062)	0.3106 (0.1023)***	0.2475 (0.1309)*	0.1349 (0.1333)	0.0272 (0.0552)	−0.0525 (0.0431)	0.0552 (0.0513)	−0.0668 (0.0467)
Log pop.	0.1118 (0.0528)**	0.2342 (0.0543)***	0.1219 (0.0718)*	−0.1095 (0.0718)	0.1065 (0.0275)***	−0.0850 (0.0217)***	0.0846 (0.0260)***	−0.1292 (0.0233)***
Percent black	−0.0194 (0.1269)	−0.0819 (0.1152)	−0.1195 (0.1574)	0.1466 (0.1682)	−0.1213 (0.0462)***	0.0204 (0.0392)	−0.1416 (0.0470)***	0.0165 (0.0404)
Percent university education	2.9157 (0.6190)***	3.7058 (0.6191)***	3.4233 (0.8475)***	0.9879 (0.8314)	1.1924 (0.3207)***	−0.1891 (0.2774)	1.2201 (0.3177)***	−0.1427 (0.3008)
Percent high school education	1.8525 (0.4937)***	0.5213 (0.4775)	0.3910 (0.6372)	0.6376 (0.6723)	0.4444 (0.2166)**	0.4339 (0.1844)**	0.4234 (0.2071)**	0.5101 (0.1892)***
Percent below poverty line	−0.2586 (0.4012)	−0.5166 (0.4008)	−0.0220 (0.5239)	0.9880 (0.5431)*	−0.4910 (0.2290)**	−0.3886 (0.1897)**	−0.5436 (0.2277)**	−0.4562 (0.1892)**
Median HH income (000s)	−0.0038 (0.0045)	0.0052 (0.0050)	0.0977 (0.0061)	0.0089 (0.0062)	−0.0010 (0.0044)	−0.0010 (0.0034)***	−0.0055 (0.0043)	−0.0134 (0.0035)***

	(1)	(2)	(3)	(4)	(5)	(6)	(7)	(8)
Carnegie tier 1 enrollment	-0.2523	-0.1234	-0.0473	-0.0056	0.0040	0.0068	0.0059	0.0128
	(0.2225)	(0.2636)	(0.2320)	(0.2473)	(0.19017)	(0.0938)	(0.1824)	(0.1040)
Fraction in engineering	3.4605	4.1533	1.9239	2.2435	6.1849	1.2883	5.8570	0.6267
	(1.7789)*	(2.0558)**	(2.9139)	(2.5279)	(1.5281)***	(1.5782)	(1.5666)***	(1.9420)
Fraction professional	-1.1209	-0.0977	0.0366	0.6745	0.6215	0.0974	0.7066	-0.0646
	(0.3978)***	(0.3751)	(0.5148)	(0.5234)	(0.1813)***	(0.1584)	(0.1780)***	(0.1666)
Percent > 64 years old	0.4407	0.0530	-0.5205	0.5431	-0.4569	-0.6054	-0.7806	-1.0565
	(0.4674)	(0.4743)	(0.6453)	(0.6480)	(0.2635)*	(0.2378)**	(0.2563)***	(0.2189)***
Change population	1.0615	0.7838	0.3383	0.2912	0.6106	0.4974	0.6854	0.5021
	(0.1244)***	(0.1262)***	(0.1770)*	(0.1772)	(0.0662)***	(0.0591)***	(0.0663)***	(0.0593)***
Change % black	-1.7615	-1.3385	-0.8581	-0.3806	-0.3419	-0.5315	-0.3086	-0.5910
	(0.7871)**	(0.7412)*	(0.9339)	(1.1084)	(0.4043)	(0.3192)*	(0.3864)	(0.3423)*
Change % univ. educ.	6.1112	7.0145	7.0145	4.5995	4.5567	2.2008	4.2466	1.9685
	(1.2503)***	(1.2419)***	(1.6676)***	(1.6587)***	(0.5824)***	(0.5230)***	(0.5965)***	(0.5358)***
Change % high school educ.	2.7721	1.6390	1.9857	0.7946	0.5027	0.2296	0.5669	0.4838
	(0.9188)***	(0.8710)*	(1.1579)*	(1.2148)	(0.3481)	(0.3049)	(0.3428)*	(0.3161)
Change % > 64 years old	-2.5195	-4.2944	-4.1308	-3.9618	-1.4758	-0.1890	-1.4842	-0.4454
	(1.2037)**	(1.1384)***	(1.4394)***	(1.5700)**	(0.5091)***	(0.4513)	(0.4906)***	(0.4550)
Constant	-1.8417	-2.6327	-1.3764	-0.7248	N/A	N/A	N/A	N/A
	(0.7104)***	(0.6845)***	(0.8983)	(0.8975)				
Observations	2,793	2,700	2,809	2,750	80,892	80,892	80,892	80,892
R-squared	0.10	0.04	0.10	0.06	N/A	N/A	N/A	N/A
Log likelihood	N/A	N/A	N/A	N/A	-121,013	-117,261	-115144	-112,603

Note: Columns (1)–(4) are ordinary least squares regressions with county as the unit of observation. Columns (5)–(8) are ordered probit regressions with county industry as the unit of observation and include industry fixed effects. Robust standard errors in parentheses.

***Significant at the 1 percent level.

**Significant at the 5 percent level.

*Significant at the 10 percent level.

are strongly and positively correlated with growth in patenting. In addition, the fraction of the local students in engineering is highly correlated with growth in patenting. An increased population is associated with increased growth in patenting while an increased elderly population is associated with decreased growth in patenting.

To summarize, these results suggest that regions where patenting had previously been concentrated experienced the greatest increase in patent growth between the early 1990s and early in the first decade of the twenty-first century. That is interesting because these findings differ qualitatively from findings on the literature on regional growth and convergence (e.g., Barro and Sala-i-Martin 1991; Magrini 2004; Delgado, Porter, and Stern 2010, 2012), which have documented evidence of convergence in aggregate growth rates across countries and regions in a range of settings. There may be several reasons for this difference in findings. Our focus is on growth in patenting rather than growth in economic output or wages. Further and related, our results are particularly influenced by increases in the concentration of inventive activity at the very top of the patenting distribution, a result that may have no analog for other measures of economic activity, such as wage growth. In addition, recent work has found that the presence of clusters of related industries may have a significant impact on growth in employment and patenting (Delgado, Porter, and Stern 2010, 2012). These clusters may have been particularly influential in influencing the top tail of the distribution of inventive activity over our sample. We stress these different effects, because it highlights the open question motivating our study. There are different mechanisms at work, and they push in different directions, and it is important to know whether they operate to the same degree and direction on all economic activity.

6.3.2 Business Adoption of the Internet and Growth in Patenting

Before assessing whether the Internet might enhance or reduce the rate of concentration in patenting, it is important to establish the baseline relationship between Internet adoption and growth in patenting. Table 6.4 shows that there is no significant correlation between Internet adoption and growth in patenting overall and a weakly significant correlation for collaborative patents. Column (1) shows the results of the following regression:

$$(2)\ \mathrm{Log(Patents}_{i,0005}) - \mathrm{Log(Patents}_{i9095}) = \alpha + \gamma X_i + \beta_2 \mathrm{AdvancedInternet}_i + \varepsilon_i,$$

where AdvancedInternet$_i$ measures the extent of advanced Internet investment by businesses in county i in 2000. Columns (2) through (8) mirror the columns in table 6.3, and while the coefficients are positive, there is no significance in any of the noncollaborative patent specifications. In this table, and in all remaining tables, we do not report the coefficients on the controls because they are not the focus on the analysis and the signs and significance are similar to those found in table 6.3.

Table 6.4 Internet adoption is not significantly correlated with growth in patenting

| | County-level data | | | | County industry-level data | | | |
| | All patents | | Collaborative patents | | All patents | | Collaborative patents | |
	Patents (1)	Citation-weighted patents (2)	Patents (3)	Citation-weighted patents (4)	Patents (5)	Citation-weighted patents (6)	Patents (7)	Citation-weighted patents (8)
Advanced Internet	0.1722 (0.1371)	0.2649 (0.1328)**	0.1637 (0.2178)	0.3260 (0.2108)	0.0565 (0.0378)	0.0508 (0.0344)	0.0726 (0.0409)*	0.0702 (0.0364)*
Observations	2,540	2,510	2,441	2,409	72,576	72,576	72,576	72,576
R-squared	0.10	0.12	0.05	0.05	N/A	N/A	N/A	N/A
Log Likelihood	N/A	N/A	N/A	N/A	-115,059	-110,747	-109,970	-106,638

Note: Columns (1)–(4) are ordinary least squares regressions with county as the unit of observation. Columns (5)–(8) are ordered probit regressions with county industry as the unit of observation and include industry fixed effects. All regressions include the same set of controls as table 6.3. Robust standard errors in parentheses.

***Significant at the 1 percent level.

**Significant at the 5 percent level.

*Significant at the 10 percent level.

6.3.3 Business Adoption of the Internet
and the Concentration of Patenting

Table 6.5 examines whether Internet adoption increases or reduces the rate of concentration in patenting. Column (1) shows the results of the following regression:

$$(3) \quad \text{Log(Patents}_{i0005}) - \text{Log(Patents}_{i9095}) = \alpha + \gamma X_i + \beta_1 \text{Patents}_{i9095}$$
$$+ \beta_2 \text{AdvancedInternet}_i$$
$$+ \beta_3 \text{Patents}_{i9095} \text{AdvancedInternet}_i + \varepsilon_i.$$

The core coefficient of interest is β_3, the interaction between preperiod patenting and Internet adoption. The result suggests that Internet adoption is correlated with a reduction in the growth in concentration of patenting (as measured by the correlation between growth in patenting and patenting in the preperiod). The quantitative importance is not apparent from the coefficient, so we separately calculate the implied marginal effect. It suggests that an increase in advanced Internet by one standard deviation reduces the increase in concentration by 57 percent, which is quite substantial. In other words, among counties that were leaders in Internet adoption, the rate of patent growth between the early 1990s and early in the twenty-first century is only weakly correlated with the level of patenting in the 1990 to 1995 period.

Put another way, for a county in the 25th percentile of Internet adoption, moving from the 25th percentile in patenting to the 90th percentile in patenting in the early 1990s yields an implied increase in the growth of patenting of 5.4 percent. For a county in the 75th percentile of Internet adoption, the same move yields an implied increase in patenting of 2.3 percent. For a county in the 90th percentile of Internet adoption, the same move yields an implied increase in patenting of just 0.4 percent.[13] Thus, Internet adoption is correlated with a reduction in this divergence: high Internet adopting locations that were not leaders in patenting did not fall behind.

As in tables 6.2 and 6.3, the alternative specifications in columns (2) through (8) are broadly consistent with column (1). The qualitative results are similar if patents are weighted by five-year citation rates, if only collaborative patents are used, and if the unit of observation is the county-industry.

One potential concern with this analysis is that $\text{AdvancedInternet}_i$ and Patents_{i9095} are highly correlated and therefore the interaction term captures an unusual part of the distribution. Figure 6.2 addresses this concern. It presents a scatter plot of $\text{AdvancedInternet}_i$ on the horizontal axis and Patents_{i9095} on the vertical axis. Figure 6.2 shows that, while $\text{AdvancedInternet}_i$ and Patents_{i9095} are indeed highly correlated, there is plenty of variation.

13. The increase estimated from the regression is substantially smaller than might be suggested by the descriptive statistics presented in the introduction because the regressions include controls for county-level demographics that are highly correlated with growth in patenting, such as education and population growth.

Table 6.5   **Internet adoption mutes the correlation between prior patents and growth in patenting**

	County-level data				County industry-level data			
	All patents		Collaborative patents		All patents		Collaborative patents	
	Patents (1)	Citation-weighted patents (2)	Patents (3)	Citation-weighted patents (4)	Patents (5)	Citation-weighted patents (6)	Patents (7)	Citation-weighted patents (8)
Advanced Internet	0.1852	0.2883	0.1907	0.3466	0.1273	0.0726	0.1457	0.0947
	(0.1376)	(0.1331)**	(0.2190)	(0.2117)	(0.0373)***	(0.0354)**	(0.0392)***	(0.0375)**
Patenting 1990–1995 (000s)	0.0037	0.0058	0.0021	0.0018	16.410	1.6063	20.5669	1.6495
	(0.00081)***	(0.0011)***	(0.0005)***	(0.00044)***	(2.871)***	(0.2629)***	(3.4910)***	(0.2575)***
Patenting 1990–1995 (000s) x advanced Internet	−0.0160	−0.0271	−0.0101	−0.0086	−70.815	−7.4655	−74.7057	−7.7062
	(0.0039)***	(0.0051)***	(0.0023)***	(0.0021)***	(14.826)***	(1.3657)***	(17.1913)***	(1.3039)***
Observations	2,540	2,448	2,509	2,409	72,576	72,576	72,576	72,576
R-squared	0.10	0.04	0.07	0.05	N/A	N/A	N/A	N/A
Log Likelihood	N/A	N/A	N/A	N/A	−113,555	−110,289	−108,544	−106,469

Note: Columns (1)–(4) are ordinary least squares regressions with county as the unit of observation. Columns (5)–(8) are ordered probit regressions with county industry as the unit of observation and include industry fixed effects. All regressions include the same set of controls as table 6.2. Robust standard errors in parentheses.

***Significant at the 1 percent level.

**Significant at the 5 percent level.

*Significant at the 10 percent level.

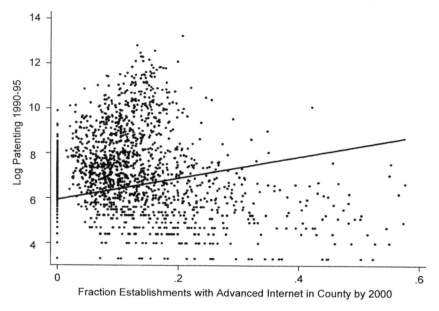

Fig. 6.2 Internet adoption and patenting (1990–1995)

There are many locations with high levels of AdvancedInternet$_i$ and low levels of Patents$_{i9095}$ and there are many with low levels of AdvancedInternet$_i$ and high levels of Patents$_{i9095}$.

Broadly, table 6.5 is suggestive that Internet overcomes isolation in invention, though we need to be cautious as it also could be an omitted variable driving both increased invention and increased Internet. Next, we provide some suggestive evidence that the Internet facilitated communication by inventors, providing some support for a causal interpretation of table 6.5.

6.3.4 Collaboration, Firm Type, and Local Growth in Patenting

Table 6.6 reproduces the first four columns of table 6.5, but with alternative dependent variables. Instead of measuring patents and collaborative patents, column (1) looks at the growth in the number of distant collaborators, as defined in section 6.2.1. Column (2) looks at the growth in the number of collaborative patents by county in which none of the collaborators are distant from each other. Column (3) looks at noncollaborative patents. Columns (4) through (6) show the same analysis, but with citation-weighted patents.

In our previous results we documented that advanced Internet adoption was associated with decreasing concentration in innovative activity. One possible explanation for this result is that advanced Internet adoption made innovative activity in less innovative places relatively more attractive through

Table 6.6 Comparing patents with distant collaborators, patents with nondistant collaborators, and noncollaborative patents

	Patents			Citation-weighted patents		
	Patents with distant collaborators (1)	Patents with nondistant collaborators (2)	Noncollaborative patents (3)	Patents with distant collaborators (4)	Patents with nondistant collaborators (5)	Noncollaborative patents (6)
Advanced Internet	0.0827	0.1697	−0.0737	0.0370	0.1741	0.0330
	(0.1262)	(0.0766)**	(0.0560)	(0.1816)	(0.1165)	(0.0969)
Patenting 1990–1995 (000s)	0.0047	0.0016	−0.0016	0.0021	0.0003	−0.0004
	(0.00096)***	(0.0008)**	(0.0004)***	(0.00052)***	(0.0002)	(0.0001)***
Patenting 1990–1995 (000s) x advanced Internet	−0.0216	−0.0076	0.0084	−0.0099	−0.0014	0.0020
	(0.0046)***	(0.0037)**	(0.0020)***	(0.0025)***	(0.0011)	(0.0006)***
Observations	2,498	2,370	2,469	2,445	2,339	2,334
R-squared	0.13	0.09	0.04	0.14	0.17	0.05

Note: Ordinary least squares regressions with county as the unit of observation. Distant is defined as more than fifty miles apart. Regressions include the same controls as in table 6.3. Robust standard errors in parentheses.

***Significant at the 1 percent level.

**Significant at the 5 percent level.

*Significant at the 10 percent level.

a decline in the costs of collaboration. Another possibility is that the Internet increased the productivity of innovative activity in less innovative regions relative to more innovative ones by, for example, more easily accessing labor, consultants, or ideas developed elsewhere. While we are unable to identify between these hypotheses, we view the results of table 6.6 as suggestive that advanced Internet adoption reduced the extent of geographic concentration for inventions developed through distant collaborations more than other types of inventions.

In particular, the Internet is primarily a communications technology that reduces the cost of both distant and local communication, but the impact of patenting by firms is largest for distant collaborations (Forman and van Zeebroeck 2012). As in table 6.5, columns (1) and (4) (row 3) of table 6.6 show that, for counties with low rates of advanced Internet adoption, leading counties in the preperiod increased distant collaborations much faster than other counties. For counties with high rates of advanced Internet adoption, leading counties in the preperiod did not increase distant collaborations much faster.

In contrast, for nondistant collaborations (columns [2] and [5], row 3) and for noncollaborative patents (columns [3] and [6], row 3) we see no difference between counties with high and low rates of advanced Internet adoption, leading counties in the preperiod, and the increase in patenting. Thus, the correlation in table 5 between patenting in the preperiod, advanced Internet, and patent growth does not hold for noncollaborative patents and short-distance collaborative patents, even though it holds for long-distance collaborative patents.

Because the role of the Internet is likely to facilitate distant collaboration, and because prior work suggested that the Internet increased distant patenting between firms (Forman and van Zeebroeck 2012), this suggests that the results of table 6.5 may suggest a causal relationship rather than only a spurious relationship measuring counties that were becoming more innovative overall (and therefore becoming more innovative in terms of both patenting and internet adoption).

Table 6.7 separates patents assigned to US-based private firms, patents assigned to educational institutions, and patents assigned to governments. Consistent with the suggested mechanism, and consistent with the fact that our data on advanced Internet represents US-based private firms and not educational institutions or government, our results are strongest for US-based private firms.

We have conducted a number of additional robustness checks on our main results. While not shown here to save space, qualitative results are robust to several alternative specifications including slightly different years, dropping controls, assigning a value of 1 to counties with zero patents in a given period to avoid dropping missing values, and to using alternative threshold choices for the ordered probit in the results at the industry-county level.

Table 6.7   Results by type of patenting institution

	Private firms				Educational institutions				Government institutions			
	Patents (1)	Citation-weighted patents (2)	Patents (3)	Citation-weighted patents (4)	Patents (5)	Citation-weighted patents (6)	Patents (7)	Citation-weighted patents (8)	Patents (9)	Citation-weighted patents (10)	Patents (11)	Citation-weighted patents (12)
Patenting 1990–1995 (000s)	0.00026 (0.00013)**	0.00012 (0.00006)**	0.0016 (0.00080)**	0.0013 (0.00035)***	−0.00013 (0.00013)	0.00005 (0.00005)	0.0011 (0.00096)	0.00083 (0.00043)*	−0.00074 (0.00019)***	−0.00025 (0.00011)**	−0.0016 (0.0014)	−0.0012 (0.00045)***
Advanced Internet			0.4061 (0.1612)**	0.3349 (0.2280)			−0.0261 (0.0306)	0.0110 (0.0385)			−0.0488 (0.0232)**	−0.0537 (0.0300)*
patenting 1990–1995 (000s) x advanced Internet			−0.0066 (0.0039)*	−0.0062 (0.0017)***			−0.0069 (0.0052)	−0.0041 (0.0021)*			0.0048 (0.0069)	0.0053 (0.0022)**
Observations	2,631	2,591	2,356	2,302	2,610	2,699	2,235	2,325	2,773	2,833	2,399	2,457
R-squared	0.10	0.03	0.10	0.03	0.21	0.01	0.21	0.01	0.03	0.14	0.03	0.15

Note: Ordinary least squares regressions with county as the unit of observation. Regressions include the same controls as in table 6.3. Robust standard errors in parentheses.

***Significant at the 1 percent level.

**Significant at the 5 percent level.

*Significant at the 10 percent level.

6.3.5 Differences in Concentration of Patenting
across Technology Categories

Table 6.8 shows the results by technology category. We use the six broad technology categories defined by Hall, Jaffe, and Trajtenberg (2001). We find that the results are broadly robust across categories with the exception of computers and communication, which does not display increase concentration in patenting activity over time. While this finding merits additional investigation, it is interesting to note the recent findings by Ozcan and Greenstein (2013) of decreasing concentration of inventive activity among firms in this technology category. We see increased geographic concentration in patenting across all other technological categories, including chemical, drugs and medical; electrical and electronic; and mechanical. We see that the interaction of Internet is associated with reduced geographic concentration for these categories, too. This is particularly interesting in light of the findings for electrical and electronic industries, the area closest to computing and communications. This suggests no simple explanation will suffice, not one that stresses simple differences between hardware and software or upstream and downstream industries. This is another important question for future work.

Conclusion

We have explored the geographic concentration of invention. We first find evidence that suggests that the geographic concentration of patenting increased from 1990–1995 to 2000–2005. Overall patenting grew 27 percent, but patenting in the top quartile of counties grew 50 percent. While this result seems to contrast with work in the convergence literature, we emphasize the use of different methods and the importance of the very top of the patenting distribution in our findings. Then we showed that advanced Internet adoption by businesses works against the general increase in the geographic concentration of patenting, leading to different experiences across the regions of the United States. We find that the correlation is strong for distant collaborations and disappears for nearby collaborations and for noncollaborative patents, which suggests that the Internet's availability and growth drove at least part of the overall reduction in the growth in concentration of invention.

As noted above, our analysis helps us understand the net impact of two fundamental changes in the years since the publication of Vannevar Bush's *Science: The Endless Frontier*: (a) globalization and its implications for innovation and invention as a source of regional competitiveness, and (b) the impact of the Internet and associated reductions in communication and coordination costs. Our results suggest that while the net effect of these changes on the concentration of innovation is positive, Internet technology has played a role in mitigating this effect.

Our analysis contains a number of limitations that limit the generalizability of our findings. First, we study one type of invention and patenting

Table 6.8   Robustness to HJT categories

Category	Chemical		Computers and communications		Drugs and medical		Electrical and electronic		Mechanical		Others	
	(1)	(2)	(3)	(4)	(5)	(6)	(7)	(8)	(9)	(10)	(11)	(12)
Advanced Internet		0.0232		−0.1428		−0.5575		−0.3143		0.2155		0.1577
		(0.0254)		(0.3589)		(0.3445)		(0.2579)		(0.1829)		(0.1650)
patenting 1990–1995 (000s)	0.1462	0.3630	−0.0058	−0.0565	0.1373	0.3618	0.0727	0.5562	0.3049	0.9579	0.2757	1.1160
	(0.0611)**	(0.2320)	(0.0106)	(0.0822)	(0.0362)***	(0.1246)***	(0.0398)*	(0.1244)***	(0.0952)***	(0.3821)**	(0.0665)***	(0.3256)***
Patenting 1990–1995 (000s) x advanced Internet		−1.2146		0.2635		−1.2419		−2.5902		−4.2682		−5.3300
		(1.4579)		(0.3876)		(0.6356)*		(0.5999)***		(2.2643)*		(1.9028)***
Observations	1,414	1,390	1,065	1,055	1,073	1,063	1,303	1,290	1,970	1,921	2,183	2,114
R-squared	0.06	0.07	0.14	0.14	0.06	0.06	0.09	0.10	0.06	0.06	0.06	0.07

Note: Ordinary least squares regressions with county as the unit of observation. Regressions include the same controls as in table 6.3. Robust standard errors in parentheses.

***Significant at the 1 percent level.

**Significant at the 5 percent level.

*Significant at the 10 percent level.

in a particular time period. The Internet might have increased patenting but not invention, for example, by simplifying the process of applying for a patent through Internet lawyers rather than causing any increase in invention per se. Hence, our results beg questions about whether other measures of invention—for example, nonpatented inventions, new product development, entrepreneurial founding in technologically intensive markets—follow a similar pattern.

In addition, and as mentioned, our findings are consistent with two different explanations. First, it could be the causal explanation, perhaps by allowing relatively isolated inventors to collaborate with inventors located elsewhere. Second, it could be driven by an omitted variable that caused both increased patenting and Internet adoption. For example, for counties that were not leaders in patenting in the early 1990s, Internet adoption might be a symptom rather than a cause of increased attention to invention and a growth in the rate of Internet adoption by firms. While the results on distant collaboration versus noncollaborative patents are suggestive, they are not definitive. Hence, our findings beg questions about how to instrument for Internet adoption to identify truly exogenous variation across the United States.

Notwithstanding these limitations, our results here, combined with prior work on the impact of the Internet on the concentration of economic activity, suggest that the impact can depend on the particular activity and context being studied. It seems to lead to increased concentration in wages (Forman, Goldfarb, and Greenstein 2012) and hospital efficiency (Dranove et al. 2013), but a decreased concentration in retailing (Choi and Bell 2011), and, as suggested above, in patenting and invention. Those findings also raise intriguing questions about whether the Internet's impact on the geographic concentration of invention is distinct from its impact on the geographic concentration of other economic activity, such as wages, business adoption of IT, hospital productivity, and so on. If that is the case, then the Internet could act as a broad force for weakening the links between the geography of inventive activity and spatial patterns of downstream use of it. We speculate that such a broad trend, if sustained for a long time period, would manifest in numerous measurable economic activities. Hence, our findings also motivate questions comparing changes in the geographic concentration of different parts of the value chain over the very long run.

References

Agrawal, Ajay, Christian Catalini, and Avi Goldfarb. 2011. "The Geography of Crowdfunding." NBER Working Paper no. 16820, Cambridge, MA. http://www .nber.org/papers/w16820.

Agrawal, Ajay, and Avi Goldfarb. 2008. "Restructuring Research: Communication Costs and the Democratization of University Innovation." *American Economic Review* 98 (4): 1578–90.

Arora, Ashish, and Chris Forman. 2007. "Proximity and Information Technology Outsourcing: How Local are IT Services Markets?" *Journal of Management Information Systems* 24 (2): 73–102.

Barro, Robert J., and Xavier Sala-i-Martin. 1991. "Convergence across States and Regions." *Brookings Papers on Economic Activity* 1:107–82.

Bloom, Nicholas, Raffaella Sadun, and John Van Reenen. 2012. "Americans Do IT Better: US Multinationals and the Productivity Miracle." *American Economic Review* 102 (1): 167–201.

Blum, Bernardo, and Avi Goldfarb. 2006. "Does the Internet Defy the Law of Gravity?" *Journal of International Economics* 70 (2): 384–405.

Bresnahan, Timothy, Erik Brynjolfsson, and Lorin Hitt. 2002. "Information Technology, Work Organization, and the Demand for Skilled Labor: Firm-Level Evidence." *Quarterly Journal of Economics* 117 (1): 339–76.

Bresnahan, Timothy, and Shane Greenstein. 1997. "Technical Progress and Co-Invention in Computing and in the Use of Computers." *Brookings Papers on Economics Activity: Microeconomics* (January):1–78.

Brynjolfsson, Erik, and Lorin Hitt. 2003. "Computing Productivity: Firm-Level Evidence." *Review of Economics and Statistics* 85 (4): 793–808.

Bush, Vannevar. 1945. *Science: The Endless Frontier.* Washington, DC: National Science Foundation.

Cairncross, Frances. 1997. *The Death of Distance.* Cambridge, MA: Harvard University Press.

Choi, Jeonghye, and David Bell. 2011. "Preference Minorities and the Internet." *Journal of Marketing Research* 58:670–82.

Cohen, Wesley M., Richard R. Nelson, and John P. Walsh. 2000. "Protecting Their Intellectual Assets: Appropriability Conditions and Why US Firms Patent (Or Not)." NBER Working Paper no. 7552, Cambridge, MA.

Delgado, Mercedes, Michael Porter, and Scott Stern. 2010. "Clusters and Entrepreneurship." *Journal of Economic Geography* 10 (4): 495–518.

———. 2012. "Clusters, Convergence, and Economic Performance." NBER Working Paper no. 18250, Cambridge, MA.

Ding, Waverly, Sharon Levin, Paula Stephan, and Anne Winkler. 2010. "The Impact of Information Technology on Academic Scientists' Productivity and Collaboration Patterns." *Management Science* 56 (9): 1439–61.

Downes, Tom, and Shane Greenstein. 2007. "Understanding Why Universal Service Obligations May Be Unnecessary: The Private Development of Local Internet Access Markets." *Journal of Urban Economics* 62 (1): 2–26.

Dranove, David, Chris Forman, Avi Goldfarb, and Shane Greenstein. 2013. "The Trillion Dollar Conundrum: Complementarities and Health Information Technology." NBER Working Paper no. 18281, Cambridge, MA.

Forman, Chris, and Avi Goldfarb. 2006. "Diffusion of Information and Communication Technologies to Businesses." In *Handbook of Information Systems, Volume 1: Economics and Information Systems*, edited by Terrence Hendershott, 1–52. Amsterdam: Elsevier.

Forman, Chris, Avi Goldfarb, and Shane Greenstein. 2002. "Digital Dispersion: An Industrial and Geographic Census of Commercial Internet Use." NBER Working Paper no. 9287, Cambridge, MA.

———. 2005. "How Did Location Affect the Adoption of the Commercial Internet? Global Village vs. Urban Density." *Journal of Urban Economics* 58 (3): 389–420.

———. 2008. "Understanding the Inputs into Innovation: Do Cities Substitute for Internal Firm Resources?" *Journal of Economics and Management Strategy* 17 (2): 295–316.

———. 2012. "The Internet and Local Wages: A Puzzle." *American Economic Review* 102 (1): 556–75.

Forman, Chris, and Nicholas van Zeebroeck. 2012. "From Wires to Partners: How the Internet Has Fostered R&D Collaborations within Firms." *Management Science* 58 (8): 1549–68.

Friedman, Thomas L. 2005. *The World is Flat: A Brief History of the Twenty-First Century*. New York: Farrar, Straus, and Giroux.

Glaeser, Edward L., William R. Kerr, and Giacomo A. M. Ponzetto. 2010. "Clusters of Entrepreneurship." *Journal of Urban Economics* 67 (1): 150–68.

Glaeser, Edward L., and Giacomo A. M. Ponzetto. 2007. "Did the Death of Distance Hurt Detroit and Help New York?" NBER Working Paper no. 13710, Cambridge, MA.

Griliches, Zvi. 1990. "Patent Statistics as Economic Indicators: A Survey." *Journal of Economic Literature* 28 (4): 1661–707.

Hall, Bronwyn H., Adam Jaffe, and Manuel Trajtenberg. 2001. "The NBER Patent Citations Data File: Lessons, Insights, and Methodological Tools." NBER Working Paper no. 8498, Cambridge, MA.

———. 2005. "Market Value and Patent Citations." *RAND Journal of Economics* 36 (1): 16–38.

Hampton, Keith, and Barry Wellman. 2002. "Neighboring in Netville: How the Internet Supports Community and Social Capital in a Wired Suburb." *City and Community* 2 (3): 277–311.

Jaffe, Adam, and Manuel Trajtenberg. 2002. *Patents, Citations, and Innovations: A Window on the Knowledge Economy*. Cambridge, MA: MIT Press.

Jaffe, Adam B., Manuel Trajtenberg, and Rebecca Henderson. 1993. "Geographic Localization of Knowledge Spillovers as Evidenced by Patent Citations." *Quarterly Journal of Economics* 108 (3): 577–98.

Jones, Benjamin F., Stefan Wuchty, and Brian Uzzi. 2008. "Multi-University Research Teams: Shifting Impact, Geography, and Stratification in Science." *Science* 322 (21): 1259–62.

Kolko, Jed. 2002. "Silicon Mountains, Silicon Molehills: Geographic Concentration and Convergence of Internet Industries in the US." *Information Economics and Policy* 14 (2): 211–32.

Magrini, Stefano. 2004. "Regional (Di)Convergence." In *Handbook of Regional and Urban Economics, Volume 4: Cities and Geography*, edited by J. Vernon Henderson and Jacques-François Thisse, 2741–96. Amsterdam: Elsevier.

Ozcan, Yasin, and Shane Greenstein. 2013. "The (de)Concentration of Sources of Inventive Ideas: Evidence from ICT Equipment." Working Paper, Northwestern University.

Saxenian, Annalee. 1994. *Regional Advantage: Culture and Competition in Silicon Valley and Route 128*. Cambridge, MA: Harvard University Press.

Sinai, Todd, and Joel Waldfogel. 2004. "Geography and the Internet: Is the Internet a Substitute or a Complement for Cities?" *Journal of Urban Economics* 56 (1): 1–24.

Wellman, Barry. 2001. "Computer Networks as Social Networks." *Science* 293: 2031–34.

III

Entrepreneurship and Market-Based Innovation

7
Innovation and Entrepreneurship in Renewable Energy

Ramana Nanda, Ken Younge, and Lee Fleming

7.1 Introduction

At the time of Vannevar Bush's writing of *The Endless Frontier*, energy supply was not a major policy concern. To the extent that anyone thought about research related to energy, it was based on the belief that nuclear power would soon be "too cheap to meter." Today, nearly 90 percent of the world's energy is still produced from coal, oil, and natural gas.[1] We find

Ramana Nanda is associate professor of business administration at Harvard Business School and a faculty research fellow of the National Bureau of Economic Research. Ken Younge is assistant professor in strategic management at the Krannert School of Management at Purdue University. Lee Fleming is professor of industrial engineering and operations research at the University of California, Berkeley.

This chapter was prepared for an NBER conference volume, *The Changing Frontier: Rethinking Science and Innovation Policy*. We thank the editors of the volume, Adam Jaffe and Ben Jones, and our discussant, Steven Kaplan, for very helpful comments. In addition, we are grateful to Lee Branstetter, Guido Buenstorf, Ronnie Chaterji, Jason Davis, Chuck Eesley, Mazhar Islam, Matt Marx, David Popp, and the participants at the NBER preconference, the Petersen Institute for International Economics, and Stanford Social Science and Technology Seminar for helpful comments. Dan DiPaolo and Guan-Cheng Li provided excellent research assistance. For support we thank the Division of Faculty Research and Funding at Harvard Business School, the Kauffman Foundation's Junior Faculty Fellowship, the Fung Institute for Engineering Leadership at UC Berkeley, and the National Science Foundation (1064182). Parts of this chapter draw extensively on Ghosh and Nanda (2014). For acknowledgments, sources of research support, and disclosure of the authors' material financial relationships, if any, please see http://www.nber.org/chapters/c13048.ack.

1. Data from the BP Statistical Review of World Energy (2012) shows that 87 percent of the energy was produced from "conventional energy," namely coal, oil, and natural gas. On the other hand, solar, wind, biomass, hydro, and other renewables accounted for a mere 8 percent of global energy produced in 2010. The BP study only reports data on commercially traded fuels, including renewable energy that is commercially traded. The International Energy Agency (IEA) estimates a slightly higher share of renewables based on estimates of the use of wood chips, peat, and other biomass used in developing countries that is not commercially traded. Even so, their estimate of renewables including hydroelectricity is 13 percent compared to the 8 percent estimated by BP.

ourselves with nuclear power facing huge challenges and the development of other noncarbon energy sources a high priority, thus making the role of innovation in renewable energy a first-tier policy concern.

Indeed, the global demand for energy is projected to almost triple over the next several decades. Estimates suggest that a growing world population, combined with rising living standards, will lead global energy consumption to reach about 350,000 terawatt hours (TWh) in 2050 from the 2010 level of 130,000 TWh. To put this increase in perspective, it will require the equivalent of setting up 750 large coal-burning power plants *per year* for forty years in order to meet the increased demand for energy in the coming decades.

In addition to the challenges of meeting the growing energy needs of the world's population with conventional sources of energy, the implications of continued dependence on fossil fuels are believed to be particularly stark for climate change. The shale gas revolution in the United States in recent years has implied a reduced dependence on coal. Nevertheless, the benchmark of trying to achieve "zero emissions" has led the United States and several European countries to focus more intensely on promoting innovation in renewable energy technologies in recent years. While there is no clear winning alternative at present, there is also a growing belief that progress will come from radical innovations that will allow us to make the jump from the status quo, whether it is in renewable energy or other more conventional sources of energy production. In this chapter, we examine the technological and organizational sources of such innovation in renewable energy, with a particular focus on the possible role of venture capital-backed entrepreneurship.

Ghosh and Nanda (2014, 1) point out that "venture capital has been a key source of finance for commercializing radical innovations in the United States, particularly over the last three decades (Kortum and Lerner 2000; Gompers and Lerner 2002; Samila and Sorenson 2011). The emergence of new industries such as semiconductors, biotechnology and the Internet, as well as the introduction of several innovations across a spectrum of sectors such as healthcare, IT and new materials, have been driven in large part by the availability of venture capital for new startups. A key attribute of venture-backed innovation in the US has been the ability of private capital markets to finance a wide variety of approaches in a specific area, as opposed to choosing a specific winner." Since it is hard to know, ex ante, which technological trajectory will be successful ex post, in order to make rapid technological progress, we are likely to need to proceed by conducting numerous "economic experiments" in the energy sector (Rosenberg 1994; Stern 2005; Kerr, Nanda, and Rhodes-Kropf 2014). This makes venture capital (VC) an ideal candidate to play a role in financing radical innovation in renewable energy technologies.

In fact, venture capital financing for renewable energy start-ups rose dramatically in the middle of the first decade of the twenty-first century after being consistently low in the previous decades. Between 2006 and 2008, several billion dollars were channeled into start-ups focused on clean technologies, and in particular solar and biofuels-related start-ups. In the last few years, however, venture capital investment in renewable energy technologies has plummeted, falling as a share of overall VC investment and even within clean tech, shifting away from renewable energy production to investments in energy efficiency, software, and storage.

We investigate the role of venture capital in renewable energy innovation by comparing the patenting activity of VC-backed start-ups with other types of organizations engaged in renewable energy innovation. We not only examine patenting rates, but also the characteristics of the patents being filed by the different types of organizations. Understanding these factors will help determine the extent to which falling VC investment in renewable energy should be seen as a cause for concern as opposed to being easily substitutable by innovation by others such as large incumbent firms.

We address these questions by using patent data from the US Patent and Trademark Office (USPTO) over the thirty-year period from 1980 through 2009. We find that large incumbent firms have dominated patenting in renewable energy for several decades. For example, the top twenty firms accounted for 50 percent of the renewable energy patents and the top fifty firms account for nearly 70 percent of such patents filed at the USPTO in the early 1990s. Innovation became more widespread in the first decade of the twenty-first century when patenting by VC-backed firms grew, but the top twenty firms still accounted for over 40 percent of the patenting activity in 2010. Despite accounting for the largest share of patents, however, we find incumbents are more likely to file patents that are either completely uncited or are self-cited, suggesting a greater focus on incremental or process innovation. Furthermore, they are less likely to have extremely influential patents, that we define as being in the top ten percentiles of forward citations in a given technology area and given year. Finally, we create a measure of novelty using textual analysis of the patent documents that does not depend on citations. This independent measure also suggests that on average, incumbent firms have been engaged in less novel patenting than venture capital-backed start-ups, even more so in the period when VC funding for start-ups increased dramatically. Given the more influential and novel patenting associated with VC-backed start-ups, our results suggest that the sharp fall in financing available for such firms could have implications for the nature of innovation we may see going forward in this sector.

The rest of the chapter is structured as follows. In section 7.2, we outline the data used for our analysis. Section 7.3 provides a detailed description of our main results on the differences in innovation across incumbent and venture capital-backed firms. In section 7.4, we discuss the challenges faced by venture capital investors in sustaining the financing of renewable energy start-ups, and section 7.5 concludes.

7.2 Data

7.2.1 Sample Selection Criteria

Our focus in this chapter is on patenting in sectors related to renewable energy production, namely solar, wind, biofuels, hydroelectric power, and geothermal technologies. Before moving to a description of the data, however, we first outline the criteria for selecting our sample.

Our approach was to define a set of technologies that would first, allow us to build a comprehensive and well-delineated data set of patenting activity within the chosen technology, and second, enable us to compare the characteristics of innovation between venture capital-backed start-ups and other firms engaged in innovation.

This led us to leave out some technologies that are often associated with clean energy production, but are not renewable energy. For example, although natural gas has a lower carbon footprint than oil and coal, it is difficult to break out innovations related to energy production in this area, as opposed to other businesses pursued by oil and gas companies. On the other hand, we have also left out other "clean-tech" sectors that receive VC finance but are not energy production. For example, venture capital has been involved in financing a number of innovations in software related to smart grid and energy efficiency. These innovations are extremely difficult to isolate in a systematic manner from other software patents that start-ups could be working on (e.g., a GPS software that helps route trucks in a manner that conserves fuel is hard to distinguish from other GPS patents, even when manually classifying patents). Our focus, therefore, is on renewable energy production technologies that have been patented at the USPTO.[2] Although our scope is narrower than either "energy production" or "clean tech," our hope is that our trade-off buys us greater confidence in defining a clear and consistent set of technologies within which we can characterize both the trends in patenting over time, and the differences in the nature of patenting across the various organizational forms.

7.2.2 Data Used to Create the Sample

We created our sample using three steps. First, we worked with a private research firm, IP Checkups, to define a set of renewable energy patents at the US Patent and Trademark Office (USPTO) in each of the energy production sectors of solar, wind, biofuels, hydro, and geothermal. IP Checkups has particular expertise in clean energy, including a database of clean technology patents filed at the US and foreign patent offices (we consider only patents

---

2. The focus of our chapter is therefore different and complementary to Popp, Hascic, and Medhi (2011), who look at worldwide patenting of renewable energy technologies. Our emphasis is on organizational differences in the type of patenting with a focus on the USPTO. As we discuss further below, however, our patent sample for the United States seems to correspond well with Popp, Hascic, and Medhi's data on the United States.

filed with the USPTO). They provided us with a sample of 17,090 renewable energy patents whose application dates were between January 1980 and December 2009 across the five subsectors listed above.[3]

Second, we developed a procedure to validate and extend the sample from IP Checkups in order to ensure that the sample was comprehensive. Specifically, we used the patents from IP Checkups as a training set, and applied the LIBLINEAR machine classifier algorithm (Fan et al. 2008) to search through every patent title and abstract in the universe of approved utility patents at the USPTO with application dates between January 1980 and December 2009. The machine classifier algorithm aimed to identify other patents (based on their titles and abstract) that looked similar to those in the training set provided by IP Checkups. The assumption behind this approach is that IP Checkups may have missed patents at random, but would not have a systematic bias in the types of patents they did not provide us. In this case, the algorithm would be able to search efficiently among the over 4.3 million patents in the universe of patents for others with similar titles and abstracts that may have been overlooked by IP Checkups. The classifier returned an additional 31,712 patents for consideration.

Finally, we contracted with IP Checkups to have a PhD expert in clean technologies manually review each of the candidate patents identified by the machine classifier and select appropriate ones for inclusion into the final sample. An additional 5,779 patents were selected for inclusion, resulting in a final sample size of 22,869 patents.

We believe that this three-step process outlined above has produced a comprehensive sample of patents looking specifically at renewable energy. Given the systematic and replicable approach used by the machine-learning sample, we believe this method will allow subsequent researchers to easily update the sample, as well as apply similar techniques to identify patents in other sectors that share the property with renewable energy not easily demarcated by specific technology classes at the USPTO. Our approach is therefore complementary to that used by Popp, Hascic, and Medhi (2011), who look at global innovation in renewable energy and also focus on specific patent classes. Although the time period used by our study is somewhat different (our analysis starts in 1980 and extends until 2009), the trends in patenting rates over time are similar.[4]

Having thus identified our five primary categories of clean-tech patents by technology type, we further categorized each patent into one of four organizational types: academia and government, VC-backed start-ups, non-VC-backed firms, and unassigned. Unassigned patents were those with no assignee provided in the patent application. These have typically been assumed to be

---

3. Although the USPTO data goes as far as 2012, we truncate the sample at the end of 2009 to allow for our analysis of forward citations.
4. Our data set contains a larger number of patents than the Popp, Hascic, and Medhi (2011) database does for the United States. This is likely due to the fact that our search procedure extended beyond the primary set of patent classes used by these technologies.

independent inventors, but may also be corporate patents with just a missing assignee field. As we show in the following section, unassigned patents seem significantly different in terms of their characteristics. While we do report some analyses that include unassigned patents, the majority of our analyses focus on comparisons between VC-backed start-ups, non-VC-backed firms, and inventors in academic institutions or government labs. We classified firms as venture-capital backed if the assignee name and location corresponded with firms in either the Cleantech i3 or the Bloomberg New Energy Finance database of venture capital-backed financings.[5] To classify assignees as university or government, we used a text-matching process followed by manual review to identify academic institutions (assignees with words such as "university," "universitaet," "ecole," "regents," etc.) and governmental organizations (assignees with words such as "Department of Energy," "United States Army," "Lawrence Livermore," "Bundesrepublik," etc.).

Our residual category, therefore, is the category of assignees that are not VC backed and not from academic institutions or the government. The residual category can therefore be thought of as incumbent firms (keeping in mind the qualifications described above). As far as possible, we manually matched subsidiaries to the parent company's name, so that, for example, all known subsidiaries of General Electric were classified as GE. While this categorization is imperfect, cases where we missed matching a subsidiary to a parent company will tend to bias us toward finding less concentration in patenting, and our findings should be seen as a lower bound to the true level of concentration across organizations involved in renewable energy patenting.

7.3 Results

7.3.1 Patenting Rates in Renewable Energy

We begin by providing an overview of the patenting landscape in renewable energy technologies. Figures 7.1A and 7.1B report the absolute and relative amount of renewable energy patenting at the USPTO, broken down by technology. They show that renewable energy patents fell over the 1980s, both in absolute and relative terms. While the patenting rate increased slightly in the 1990s, it rose considerably in the first decade of the twenty-first century, increasing at a disproportionate rate relative to overall patenting activity at the USPTO. In fact, both the number of patents filed per year and the share of patents filed in the USPTO approximately doubled over the ten-year period from 2000–2009. They also show that the increase was due to solar, biofuels, and wind patenting in particular, while hydro and geothermal patents

5. Both databases have more comprehensive coverage of venture capital financings in clean energy than Thompson Venture Economics and Dow Jones Venture Source, the two databases typically used for studies on venture capital-backed start-ups.

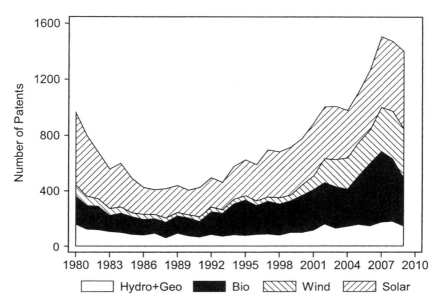

Fig. 7.1A Count of renewable energy patents at USPTO by technology, 1980–2009

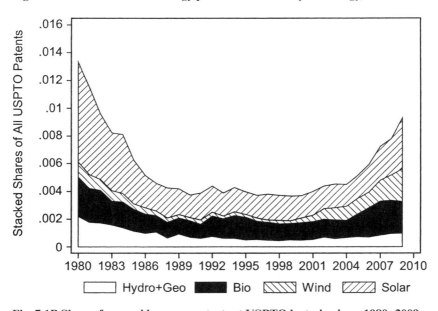

Fig. 7.1B Share of renewable energy patents at USPTO by technology, 1980–2009

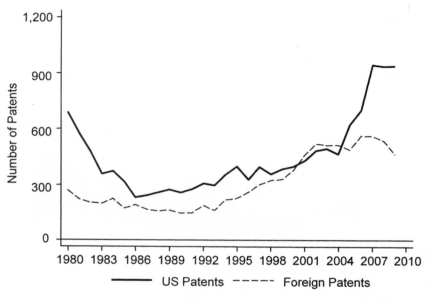

Fig. 7.2 USPTO patents granted to foreign and US-based inventors, 1980–2009

remained about constant over the period. Figure 7.2 shows that the increase in patenting was much greater among US-based inventors than those based outside the United States. In fact, there was a sharp break in the trend of patenting by US-based investors relative to foreign inventors around 2004.

Tables 7.1A and 7.1B provide a breakdown of the total number of patents used in our sample, broken down by organizational form and technology area. Table 7.1A reports the breakdown for the entire sample, while table 7.1B reports the results for inventors who are based in the United States.[6] Solar and biofuels are the two largest categories and account for about 75 percent of the patents in our sample. Incumbent firms account for nearly two-thirds of the patents in the data set and about 55 percent of the patents filed by US-based inventors.

Table 7.2 provides more detail by listing the most active US-based assignees patenting in renewable energy in recent years and the number of patents

6. Since our sample looks only at patents at the USPTO, "foreign inventors" are those who live outside the United States and have chosen to patent their inventions in the United States. Of course, there are likely to be significant numbers of renewable energy inventions by foreign inventors that are not patented at the USPTO. For example, a number of patents related to solar in Germany are not patented in the United States. However, given that the United States is such an important market, our prior is that important patents would in fact be patented in the United States in addition to other countries. Anecdotal evidence suggests that this is indeed the case. Nevertheless, the structure of our sample does not allow us to make substantive conclusions about US versus foreign patents, or speak to differing trends in patenting between US and foreign inventors in renewable energy over time.

Table 7.1   **Patenting rates in renewable energy, by technology and organization type**

	Venture-backed start-ups	Incumbent firms	Academia and government	Unassigned	Total	Percent
A. All renewable energy patents at USPTO (1980–2009)						
Solar	473	5,937	732	2,502	9,644	42
Wind	169	1,679	70	1,129	3,047	13
Biofuels	177	4,995	884	778	6,834	30
Hydroelectric	78	1,132	107	1,058	2,375	10
Geothermal	52	597	54	266	969	4
Total	949	14,340	1,847	5,733	22,869	100
B. US-based inventors only						
Solar	402	2,797	482	1,884	5,565	41
Wind	71	689	39	693	1,492	11
Biofuels	143	2,987	659	513	4,302	32
Hydroelectric	41	643	68	757	1,509	11
Geothermal	29	431	42	219	721	5
Total	686	7,547	1,290	4,066	13,589	100

Notes: This table reports the breakdown of 22,869 renewable energy patents at the USPTO that were granted between 1980 and 2009. Panel A provides a breakdown for the entire sample and panel B provides a breakdown for US-based inventors. Venture-backed start-ups refer to patents where the assignee was matched to a firm that received venture capital finance (identified using data from Cleantech i3 and Bloomberg New Energy Finance). Patents granted to academic institutions or government labs were identified using a text-matching algorithm followed by manual review. Incumbent firms refer to the residual category of assignees who were not classified as either VC-backed or from academia/government. Unassigned patents are those not affiliated with any organization and are typically seen as independent inventors.

associated with these. Specifically, it focuses on the assignees with at least five patents between 2005 and 2009 in each of the technologies. As can be seen from table 7.2, large energy and energy-equipment incumbents account for the disproportionate share of the overall patenting. Firms such as GE, DuPont, Chevron, ExxonMobil, and Applied Materials are among the most active firms patenting in renewable energy. However, a number of VC-backed firms are also on this list. For example, SoloPower, Konarka Technologies, Stion, Nanosolar, Solyndra, MiaSolé, Twin Creeks Technologies, and Solaria are all VC-backed firms, so that eight of the top twenty assignees with US-based inventors patenting in solar between 2005 and 2009 were VC-backed start-ups. Similarly, Amyris, KiOR, and Ceres in biofuels; Clipper Windpower and FloDesign Wind Turbines in wind; and Ocean Power Technologies and Verdant Power in hydro are all venture capital-backed firms.[7]

7. Appendix A provides a more detailed list of the top assignees from VC-backed start-ups, incumbents, and academia/government, including both US and foreign inventors patenting at the USPTO and over the period 2000–2009. Given that the list includes assignees with many foreign inventors, other familiar names such as Vestas, Sanyo, Sharp, Gamesa, and Schott AG are now also among the leading assignees involved in renewable energy innovation.

Table 7.2 Assignees with five or more patents to US-based inventors

Rank	Solar	Patent count	Biofuels	Patent count	Wind	Patent count	Hydro	Patent count	Geothermal	Patent count
1	Applied Materials, Inc.	56	Stine Seed Company	212	GE	204	GE	16	Kelix Heat Transfer Systems	7
2	SunPower Corporation	35	DuPont	78	Genedics Clean Energy, LLC	9	Ocean Power Technologies	12	Earth to Air Systems, LLC	5
3	SoloPower	33	Merschman Seeds	76	Clipper Windpower Technology	8	Alticor Corporate Enterprises	11	GE	5
4	GE	31	Novo Group	30	Northern Power Systems, Inc.	6	Lockheed Martin	6		
5	Boeing	25	UOP	30	FloDesign Wind Turbine Corp.	5	Verdant Power	6		
6	Konarka Technologies	24	Chevron	29	Frontier Wind, LLC	5				
7	Stion Corporation	24	Monsanto	29	RenScience IP Holdings, Inc.	5				
8	IBM	20	Syngenta AG	25						
9	DuPont	18	MS Technologies, LLC	22						
10	EMCORE Solar Power, Inc.	17	Dow	21						
11	Nanosolar, Inc.	17	Michigan State University	17						
12	Guardian Industries Corp.	16	ADM	13						
13	Xerox	13	University of Illinois	13						
14	Solyndra, LLC	12	University of Wisconsin	13						
15	TE Connectivity	12	Amyris	12						
16	Twin Creeks Technologies	12	Royal DSM	12						
17	Lockheed Martin	11	NewMarket Corporation	11						
18	MiaSolé	11	US Department of Agriculture	11						
19	Solaria Corporation	9	GE	10						
20	UTC	9	University of California	10						

	Company	Count
21	Energy Innovations, Inc.	8
22	GM	8
23	Solexel, Inc.	8
24	SolFocus, Inc.	8
25	University of California	8
26	World Factory, Inc.	8
27	Foxconn	7
28	Iostar Corporation	7
29	North Carolina State University	7
30	University of Central Florida	7
31	Xantrex Technology, Inc.	7
32	Chevron	6
33	Genedics Clean Energy, LLC	6
34	Qualcomm, Inc.	6
35	NASA	6
36	United Solar Ovonic, LLC	6
37	Varian Semiconductor	6
38	Ampt, LLC	5
39	Advanced Energy Industries, Inc.	5
40	Alcatel-Lucent	5
41	Apple Inc.	5
42	Architectural Glass and Aluminum	5
43	Batelle/MRIGlobal	5
44	Plextronics, Inc.	5
45	Primestar Solar, Inc.	5

Company	Count
Battelle Memorial Institute	9
Ceres, Inc.	8
University of Southern California	7
BASF	6
ConocoPhillips Company	6
ExxonMobil	6
ZeaChem, Inc.	6
Agritope/Aventis	5
Battelle	5
Catalytic Distillation Technologies	5
Coskata, Inc.	5
Luca Technologies, Inc.	5
LenLo Chem, Inc.	5
Pennsylvania State University	5
Rice University	5
Syntroleum Corporation	5
Xyleco, Inc.	5

Figure 7.3 helps address the apparent discrepancy that stems from comparing table 7.1 (where VC-backed start-ups have a small share of patents over the entire period) to table 7.2 (where VC-backed start-ups are prominent in the last five years). It shows how VC-backed start-ups increased their proportional share of patenting by US inventors the most over this period, increasing the share of patenting from under 5 percent in 2000 to almost 20 percent of the patents filed in 2009. Table 7.2 and figure 7.3 highlight how VC-backed start-ups have grown to become much more important contributors to innovation in renewable energy in the last few years.

Despite the sharp increase in patenting by VC-backed start-ups, however, patenting in renewable energy still remains concentrated in a relatively small number of firms. Figure 7.4 documents the share of total patents filed by US inventors working at either incumbents or venture capital-backed firms that are attributed to the ten, twenty, and fifty most actively patenting firms in each year. As can be seen in figure 7.4, the top twenty firms accounted for about half of all the renewable energy patents filed by firms in the early late 1980s and early 1990s. Although the concentration has fallen from that peak, it is still over 40 percent in 2009.

7.3.2 Characteristics of Patenting by Incumbent versus VC-Backed Firms

We next compare the characteristics of the patents filed by the different types of organizations. Our first step is to examine the citations to the patents that they file. Since citations tend to have a highly skewed distribution, we report the results from count models. Table 7.3 reports the results from negative binomial regressions, where the dependent variable is the count of citations received for each patent. Although we include technology and year fixed effects to account for fixed differences in patenting propensities across technologies and to account for cohort differences in the number of citations, we nevertheless also account for the fact that patents in 1980 would have received more citations than those in 1995 by looking at the cumulative citations received by patents five years from the year of application. Our measure of citations excludes self-citations, so we examine the influence of the patents on other assignees.

Panel A of table 7.3 reports results on both US and foreign inventors, while panel B restricts the sample to US-based inventors. Columns (1), (2), and (3) of both panels report the results for all technologies together, while columns (4), (5), and (6) split out the three most prevalent technologies—solar, biofuels, and wind. We use academic and government patents as our reference group, as they are likely to have remained the most stable over the entire period.

Table 7.3 shows some interesting patterns. First, as noted above and consistent with prior findings (Singh and Fleming 2010), unassigned patents seem to be far less influential than patents with assignees, both in the full sample and for US-based inventors. When interpreted as incidence rate ratios, panel A, column (1) implies that unassigned patents are associated with a 75 percent lower citation rate than academic and government patents. Second, patents

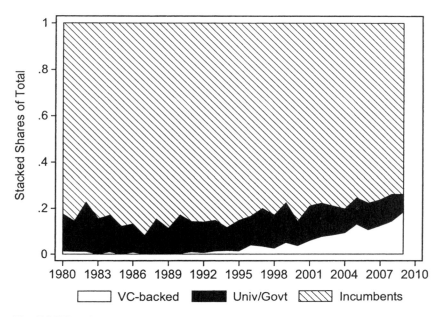

Fig. 7.3 US assignees of renewable energy patents by organization type, 1980–2009

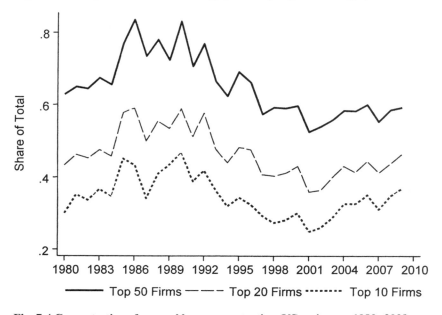

Fig. 7.4 Concentration of renewable energy patenting, US assignees, 1980–2009

Table 7.3 **Citations to patents**

	(1)	(2)	(3)	(4)	(5)	(6)
			A. Full sample			
(A) venture capital-backed start-up	0.606***	0.639***	0.672***	0.948***	0.675**	−0.016
	(0.165)	(0.158)	(0.177)	(0.238)	(0.267)	(0.323)
(B) incumbent firms	0.118	0.113	0.099	0.134	0.007	−0.295
	(0.073)	(0.075)	(0.108)	(0.126)	(0.162)	(0.187)
(C) unassigned	−1.365***					
	(0.176)					
P-value on chi2 test for difference between (A) and (B)	0.002***	<0.001***	<0.001***	<0.01**	<0.01**	0.310
Patent application year fixed effects	Y	Y	Y	Y	Y	Y
Technology fixed effects	Y	Y	Y			
Observations	22,869	17,136	11,611	4,568	4,201	1,669
			B. US-based inventors only			
(A) venture capital-backed start-up	0.601***	0.641***	0.670***	0.781***	0.718**	−0.107
	(0.181)	(0.173)	(0.199)	(0.246)	(0.284)	(0.341)
(B) incumbent firms	0.163*	0.164*	0.143	0.145	0.065	−0.391
	(0.087)	(0.089)	(0.132)	(0.134)	(0.189)	(0.304)
(C) unassigned	−1.472***					
	(0.175)					
P-value on chi2 test for difference between (A) and (B)	0.011**	0.003***	0.002***	<0.01**	<0.01**	0.130
Patent application year fixed effects	Y	Y	Y	Y	Y	Y
Technology fixed effects	Y	Y	Y			
Observations	13,589	9,523	6,155	2,227	2,568	643

Notes: This table reports the results from negative binomial regressions where the dependent variable is the count of cumulative citations received by each patent five years from the patent's application. Column (1) reports results for the full sample. Columns (2) and (3) exclude patents with no assignees. Column (3) looks only at the latter half of the thirty-year period, from 1995–2009. Columns (4), (5), and (6) are the equivalent to column (3) but done separately for solar, biofuels, and wind, respectively. Panel B replicates the regressions in panel A, but for the subsample of US-based inventors only. All regressions include fixed effects for the patent's grant year and regressions (1), (2), and (3) also include a technology fixed effect (for solar, wind, biofuels, hydro, and geothermal). Parentheses report robust standard errors, clustered by assignee.

***Significant at the 1 percent level.

**Significant at the 5 percent level.

*Significant at the 10 percent level.

filed by incumbent firms are slightly more influential than academic patents, but only marginally so. The economic magnitude is small and it is imprecisely estimated. Incumbents are associated with a citation rate that is 1.1 times that of university and government patents. On the other hand, patents filed by VC-backed firms are much more likely to receive subsequent citations. The economic magnitudes are large. The coefficients imply that VC-backed start-ups are associated with a citation rate that is 1.9 times that of university and government patents. In addition, a chi2 test for the difference in the coefficient between citations to VC-backed firms and incumbents shows that the differences are statistically significant. Columns (4), (5), and (6) explore the extent to which these differences come from certain technologies versus others. They highlight that the differences we see in columns (1), (2), and (3) are strongest for solar and biofuels—the two technologies that account for 75 percent of the overall patents in our data. Venture capital-backed start-ups patenting in wind technologies receive more citations than incumbents, but these differences are not statistically significant in panel A and only close to marginal significance in panel B.[8] This seems consistent with the fact that the largest amount of venture capital finance was devoted to solar and biofuels (a fact discussed in great detail in section 7.4).

The difference in the overall level of citations between VC-backed start-ups and incumbent firms could come from two different fronts. First, it is possible that VC-backed firms have fewer marginal or uncited patents, so that the difference stems from the left tail of the citation distribution being better. Second, it is possible that VC-backed firms are more likely to have highly cited patents, so that even if the left tail of the distribution is no better, the intensive margin of citations is higher, including a thicker right tail. To probe these possible explanations, we examine both the share of patents with at least one citation and the share of patents that are highly cited.

Table 7.4 reports the results from ordinary least squares (OLS) regressions where the dependent variable takes a value of one if the patent received at least one citation. Again, unassigned patents are far less likely to receive a single citation. The coefficients imply a 35–38 percentage point lower chance of being cited relative to academic patents on a baseline of a 50 percent citation probability. Both VC-backed start-ups and incumbents have patents that are more likely to receive citations than patents by inventors in university and government labs. This, of course, could be due to the basic nature of academic and government research and development (R&D). When comparing VCs and incumbents, however, we find that VCs have an 11–14 percentage point higher likelihood of being cited relative to academic labs, compared to a 5–7 percentage point higher probability for incumbents. These differences are statistically significant, suggesting that on average,

8. While we do not separately report the effects by technology in subsequent tables, we find exactly this pattern for the other measures that we examine in tables 7.4–7.7.

Table 7.4   **Share of patents with at least one citation**

	Full sample			US-based inventors only		
	(1)	(2)	(3)	(4)	(5)	(6)
(A) venture capital-backed start-up	0.105***	0.130***	0.136***	0.077*	0.114***	0.123***
	(0.034)	(0.027)	(0.029)	(0.040)	(0.030)	(0.032)
(B) incumbent firms	0.055***	0.053***	0.058***	0.061***	0.063***	0.070***
	(0.016)	(0.014)	(0.017)	(0.019)	(0.018)	(0.022)
(C) unassigned	−0.353***			−0.381***		
	(0.018)			(0.020)		
P-value on Wald test for difference between (A) and (B)	0.126	0.002***	0.003***	0.643	0.059*	0.058*
Patent application year fixed effects	Y	Y	Y	Y	Y	Y
Technology fixed effects	Y	Y	Y	Y	Y	Y
Observations	22,869	17,136	11,611	13,589	9,523	6,155

Notes: This table reports the results from OLS regressions where the dependent variable takes a value of one if the patent received at least one citation and zero otherwise. Results are robust to running logit regressions. Column (1) reports results for the full sample. Columns (2) and (3) exclude patents with no assignees. Column (3) looks only at the latter half of the thirty-year period, from 1995–2009. Columns (4), (5), and (6) are the equivalent to columns (1), (2), and (3), respectively, but for the subsample of US-based inventors only. All regressions include fixed effects for the patent's grant year as well as a technology fixed effect (for solar, wind, biofuels, hydro, and geothermal). Parentheses report robust standard errors, clustered by assignee.

***Significant at the 1 percent level.

**Significant at the 5 percent level.

*Significant at the 10 percent level.

VC-backed patents are less likely to be marginal and more likely to influence future R&D.

Table 7.5 reports the results from OLS regressions where the dependent variable is equal to one if the patent was highly cited. Specifically, we define a patent as being highly cited if the citations for that patent are in the top 10 percent of five-year forward citations for patents in that technology and year. Table 7.5 shows that unassigned patents are much less likely to have a highly cited patent. Since the baseline probability is by definition about 10 percent, the coefficients on unassigned patents in columns (1) and (4) of table 7.5 point out that the chance of such a patent being highly influential is essentially zero. On the other hand, VC-backed firms are almost twice as likely as academic patents to be highly cited. Incumbent firms have no statistically significant difference in highly cited patents in the overall sample, and a slightly higher chance among US-based inventors. However, importantly, the difference in the chance of being highly cited between VC-backed firms and incumbents is both statistically and economically significant.

Thus far our analysis has suggested that renewable energy innovation by incumbent firms tends to be less influential. Innovation by incumbents is less likely to be cited at all and when it is, it is less likely to be highly cited. These results are consistent with the literature that has documented that incumbent firms have different goals, search processes, competencies, and opportunity costs that lead them toward more incremental innovation (Tushman and Anderson 1986; Henderson and Clark 1990; Tripsas and Gavetti 2000; Rosenkopf and Nerkar 2001; Akcigit and Kerr 2011), although these papers have not directly compared innovation by incumbents with that by VC-backed start-ups.

To probe our results further, we turn next to directly examine the extent to which incumbents pursue more incremental innovation, by examining the degree to which they cite their own prior work relative to other types of organizations. Following Sorensen and Stuart (2000), we hypothesize that if firms are citing their own patents at a disproportionate rate, then they may be engaged in more "exploitation" rather than "exploration" (March 1991). We therefore study the extent to which inventors in the different organizational settings tend to cite themselves.

Table 7.6 reports the results from negative binomial regressions where the dependent variable is the count of the self-citations a focal patent makes, where a self-citation is defined as citing a patent from the same assignee. The regressions control for the total number of citations the patent made, and technology and patent application year fixed effects. As can be seen from table 7.6, VC-backed firms are no more likely to cite themselves than academic labs. Although the coefficient is in fact negative, it is imprecisely estimated. On the other hand, the coefficient on incumbent firms implies that they are 50 percent more likely to cite themselves compared to academic labs. Again, the difference between VC-backed firms and incumbents is statistically significant, suggesting that part of the reason that incumbents have less

Table 7.5   **Share of patents that are highly cited**

	Full sample			US-based inventors only		
	(1)	(2)	(3)	(4)	(5)	(6)
(A) venture capital-backed start-up	0.075***	0.075***	0.083***	0.086***	0.086***	0.101***
	(0.026)	(0.026)	(0.027)	(0.028)	(0.028)	(0.028)
(B) incumbent firms	0.018	0.018	0.015	0.037**	0.037**	0.031*
	(0.013)	(0.013)	(0.014)	(0.016)	(0.016)	(0.016)
(C) unassigned	−0.125***			−0.114***		
	(0.015)			(0.015)		
P-value on Wald test for difference between (A) and (B)	0.018**	0.018**	0.008***	0.054*	0.054*	0.009***
Patent application year fixed effects	Y	Y	Y	Y	Y	Y
Technology fixed effects	Y	Y	Y	Y	Y	Y
Observations	22,869	17,136	11,611	13,589	9,523	6,155

Notes: This table reports the results from OLS regressions where the dependent variable takes the value of one if the patent was above the 90th percentile in terms of citations received, and zero otherwise. Results are robust to running logit regressions. Percentiles are calculated relative to citations received by other patents in the same technology and application year and are based on cumulative citations received by each patent five years from the patent's grant. Column (1) reports results for the full sample. Columns (2) and (3) exclude patents with no assignees. Column (3) looks only at the latter half of the thirty-year period, from 1995–2009. Columns (4), (5), and (6) are the equivalent to columns (1), (2), and (3), respectively, but for the subsample of US-based inventors only. Since percentiles are calculated with a technology-year cell, all regressions implicitly include fixed effects for the patent's application year as well as a technology fixed effect (for solar, wind, biofuels, hydro, and geothermal). Parentheses report robust standard errors, clustered by assignee.

***Significant at the 1 percent level.

**Significant at the 5 percent level.

*Significant at the 10 percent level.

Table 7.6   Degree of self-citation

	Full sample			US-based inventors only		
	(1)	(2)	(3)	(4)	(5)	(6)
(A) venture capital-backed start-up	−0.340	−0.338	−0.213	−0.135	−0.130	−0.146
	(0.266)	(0.266)	(0.281)	(0.296)	(0.296)	(0.308)
(B) incumbent firms	0.405**	0.405**	0.466**	0.379**	0.380**	0.307
	(0.188)	(0.188)	(0.232)	(0.181)	(0.181)	(0.219)
(C) unassigned	−5.259***			−5.283***		
	(1.045)			(1.082)		
P-value on chi2 test for difference between (A) and (B)	0.007***	0.007***	0.013**	0.056*	0.057*	0.087*
Patent application year fixed effects	Y	Y	Y	Y	Y	Y
Technology fixed effects	Y	Y	Y	Y	Y	Y
Observations	22,869	17,136	11,611	13,589	9,523	6,155

Notes: This table reports the results from negative binomial regressions where the dependent variable is the number of backward citations that are self-citation. All regressions control for the total number of backward citations, so the coefficients reflect the share of prior art being cited that is self-citation. They can therefore be interpreted as the degree to which the assignee is engaged in incremental or exploitative innovation. Column (1) reports results for the full sample. Columns (2) and (3) exclude patents with no assignees. Column (3) looks only at the latter half of the thirty-year period, from 1995–2009. Columns (4), (5), and (6) are the equivalent to columns (1), (2), and (3), respectively, but for the subsample of US-based inventors only. All regressions include fixed effects for the patent's application year as well as a technology fixed effect (for solar, wind, biofuels, hydro, and geothermal). Parentheses report robust standard errors, clustered by assignee.

***Significant at the 1 percent level.
**Significant at the 5 percent level.
*Significant at the 10 percent level.

influential innovations is that they are engaged in more incremental R&D than VC-backed start-ups.

One possible reason for not being cited at all and for citing one's own work could also be that firms are engaged in extremely novel innovations that have not yet yielded citations. This could be particularly true in nascent technologies such as renewable energy. In addition, since patenting activity is concentrated in a few incumbent firms, it is possible that some of the higher self-citation is purely due to the fact that the prior art to be cited is more likely to be that of incumbents or that VC-backed firms do not have many prior patents to cite.

In order to address these concerns, we use a new measure of novelty that is not based on citation measure. Instead, we draw on a textual analysis of patent applications to look at the similarity of patent claims and descriptions for patents in a given technology area. Intuitively, our definition is such that patents with greater textual similarity to neighboring patents are considered to be less novel. Our measure of novelty should be particularly useful in the context of science-based patenting, where technical terms are more unique and therefore more likely to signal differences in the characteristics of innovation, and for more recent time periods, where initial forward citations may be a noisy predictor of ultimate outcomes. The measure also avoids problems with citation-based measures, where citation patterns can suffer from selection biases. A more detailed description of the measure is outlined in appendix B (see also Ullman and Rajaraman 2011, 92–93).

As can be seen from table 7.7, our novelty measure is quite consistent with the other citation-based measures of patenting. First, it highlights that in addition to unassigned patents not receiving many citations, they are also less novel than patents being developed in academic and government labs. The regressions highlight that the novelty of the patents for VC-backed start-ups is no different from that of academic labs. However, incumbent firms have a significantly lower level of novelty. Although the difference between the novelty of patenting by incumbent and VC-backed firms is not significant for the overall sample, it is close to being significant at the 10 percent level for US-based inventors, particularly in the latter part of our sample.

Our results therefore suggest that incumbent firms have been engaged in less novel and exploratory innovation than VCs, in particular in the United States. However, it is also important to weigh these differences in the quality of innovation against the patenting rates discussed before. First, as shown in table 7.1 and figure 7.3, the vast majority of the patents in renewable energy still come from incumbent firms. Despite the fact that a larger share of these are completely uncited and are less likely to be influential, incumbents still account for the largest share of innovation in aggregate. Second, it is important to remember that process improvements and innovations may be particularly important in the energy sector, where large-scale implementation can help reduce cost and make new technologies more competitive and get closer to "grid parity."

Table 7.7 Characteristics of innovation: Patent novelty

	Full sample			US-based inventors only		
	(1)	(2)	(3)	(4)	(5)	(6)
(A) venture capital-backed start-up	-0.020	-0.017	-0.022*	0.003	0.011	-0.000
	(0.014)	(0.015)	(0.012)	(0.018)	(0.022)	(0.017)
(B) incumbent firms	-0.039***	-0.039***	-0.052***	-0.050***	-0.049***	-0.068**
	(0.012)	(0.012)	(0.017)	(0.019)	(0.018)	(0.029)
(C) unassigned	-0.009**			-0.005		
	(0.004)			(0.005)		
P-value on Wald test for difference between (A) and (B)	0.380	0.352	0.222	0.121	0.117	0.10*
Patent application year fixed effects	Y	Y	Y	Y	Y	Y
Technology fixed effects	Y	Y	Y	Y	Y	Y
Observations	22,869	17,136	11,611	13,589	9,523	6,155

Notes: This table reports the results from OLS regressions where the dependent variable is the novelty of the patent claims, calculated as described in appendix B. Column (1) reports results for the full sample. Columns (2) and (3) exclude patents with no assignees. Column (3) looks only at the latter half of the thirty-year period, from 1995–2009. Columns (4), (5), and (6) are the equivalent to columns (1), (2), and (3), respectively, but for the subsample of US-based inventors only. Since the novelty measure is calculated for each focal patent at a given point in time, relative to patents from the three previous years in the same technology area, all regressions implicitly include fixed effects for time and technology. Parentheses report robust standard errors, clustered by assignee.

***Significant at the 1 percent level.

**Significant at the 5 percent level.

*Significant at the 10 percent level.

7.4 Venture Capital Financing of Renewable Energy Start-Ups[9]

Thus far we have documented that VC-backed start-ups have increased their share of patenting most substantially over the past decade and that these start-ups seem to be associated with more radical and novel innovation than that by incumbent firms. We next document that the timing of growth in renewable energy patenting by VC-backed firms is closely associated with venture capital dollars flowing into renewable energy startups. In 2002, only forty-three clean energy start-ups received VC funding in the United States, raising a combined total of $230 million. In 2008, over 200 clean energy start-ups raised $4.1 billion in venture capital in the United States.[10] Figure 7.5A shows points to the fact that VC investment in renewable energy was greatest in solar and biofuels. Figure 7.5B shows that investment in renewable energy almost doubled as a proportion of first financings. When taking later-stage investments into account, the effects are even more pronounced—clean energy investments accounted for about 15 percent of the total dollars invested by VCs in the United States in 2008, of which a majority went to renewable energy technologies. Figures 7.5A, 7.5B, and 7.5C also show that in the last few years venture capital investment in renewable energy technologies has fallen sharply, in absolute terms, as a share of overall VC investment and even within clean tech, shifting away from renewable energy production technologies.

Although our work cannot distinguish whether VCs lead start-ups to engage in more radical innovation or are just able to select more radical innovations than the incumbents tend to fund, it does highlight that venture capital financing seems to be associated with more novel and high impact innovation in renewable energy, particularly late in the first decade of the twenty-first century (Conti, Thursby, and Thursby 2012).[11] This seems important given the need for the widespread experimentation required to make progress in providing low cost, clean energy that will support development without incurring massive costs in terms of climate change. To the extent that the shift in venture capital finance away from such technologies is due to structural factors, it suggests that this will have a noticeable impact on the type of innovation being undertaken in renewable energy.[12]

---

9. This section draws extensively on Ghosh and Nanda (2014), which goes into greater detail on the financing model of venture capital and why it is poorly suited to financing renewable energy start-ups.

10. Source: Ernst and Young, National Venture Capital Association Press Releases, as reported in Ghosh and Nanda (2014).

11. Note that simply looking at the timing of the patents and the investment will not help untangle the causality as VCs will often invest in firms that have promising technologies in the anticipation that they will patent.

12. We should note that this could be equally true either through the treatment or the selection effect of venture capital investment. Even if venture capital was associated with the greater level of innovation due to its role in "picking radical technologies" rather than leading firms to become more innovative, a lack of willingness to finance renewable energy technologies could still impact innovation and commercialization in this sector as it would lead promising technologies to go unfunded.

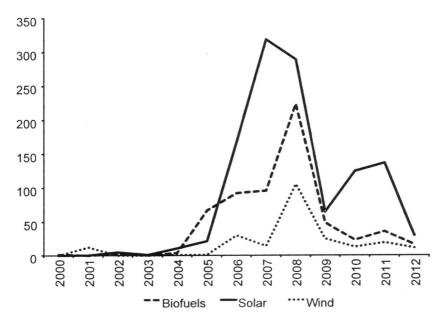

Fig. 7.5A Series A financing for US-based start-ups in solar, wind, and biofuels, by sector (United States only, millions of dollars)

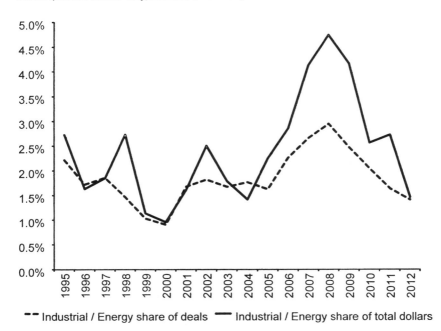

Fig. 7.5B Industrial/energy share of total VC investments (first-series financings only)

Fig. 7.5C Share of series A clean-tech financings by VCs going to solar, wind, and biofuels start-ups

Needless to say, a number of factors are likely responsible for the rapid decline in VC financing for renewable energy start-ups. The economic collapse in 2009 had a chilling effect on all venture capital investment, including clean energy. In addition, improvements in hydrofracking technology that opened up large reserves of natural gas lowered the cost of natural gas considerably and changed the economics of renewable energy technologies in terms of them being close to "grid parity." Nevertheless, our discussions with venture capital investors suggest that there are in fact structural factors, over and above these historical developments, that have led investors to become unwilling to experiment with renewable energy production technologies. In this section, we outline these structural factors that VCs seem to be facing, making sustained funding of entrepreneurship in renewable energy difficult.

7.4.1 Capital Intensity and Time Frame of Energy Production

Two facts about VC investments make staging very attractive. First, the ex post distributions of VC returns tend to be extremely skewed. Hall and Woodward (2010) and Sahlman (1990, 2010) document that about 60 percent of VC investments are likely to go bankrupt and the vast majority of returns are typically generated from about 10 percent of the investments that do extremely well. Second, Kerr, Nanda, and Rhodes-Kropf (2014) document how hard it is for VCs to predict which start-ups are likely to be extremely successful and which will fail at the time of first investment. The VCs therefore invest in stages, in effect buying a series of real options, where the information gained from an initial investment either justifies further financing or the exercise of the VC's abandonment option to shut down the investment (Gompers 1995; Bergemann and Hege 2005; Bergemann, Hege, and Peng 2008; Guler 2007). This helps them to invest as little as possible in start-ups that end up failing and put a larger share of their money in start-ups that ultimately succeed.

Hence, properties of start-ups that maximize the option value of their investments make their portfolio more valuable. For example, investments that are capital efficient (cost of buying the option is less), where step ups in value when positive information is revealed are large relative to the investment (more discriminating "experiments" being run with the money that is invested), and where the information about the viability of a project is revealed in a short period of time are all properties that make investments more attractive for VCs.

Sectors such as IT and software, that have relatively low levels of capital investment, and where initial uncertainty about the viability of the technology is revealed quickly, are therefore ideal sectors for VCs. On the other hand, the unit economics of energy production technologies need to be demonstrated at scale, because even if they work in a lab, it is hard to predict how they will work at scale. This implies that demonstration and first commercial plants face technology (in addition to engineering) risk and hence are too risky to be financed through debt finance. The fact that "risk capital" is required even at the later stages of a renewable energy start-up implies that VCs who back such start-ups therefore need to finance the companies through extremely long and capital intensive investments. The resolution of uncertainty takes much longer, as start-ups often need to build demonstration and first commercial plants before it is clear that the technology is truly viable.

The funds required to prove commercial viability for energy production technologies can reach several hundred million dollars over a five- to ten-year period, compared to the tens of millions that VCs are typically used to investing in any given start-up.[13] This level of investment is not feasible from a typical venture capital fund without severely compromising the diversification of the venture firm's portfolio. For example, investing only eight to fifteen million dollars in a project that is twice as capital intensive halves the dollar return if the start-up is successful (or it requires that the start-up to be twice as valuable at exit. This is typically not the case, as elaborated on below). On the other hand, investing a sufficient amount to retain a large share in a successful exit requires making far fewer investments across the portfolio and hence

13. For example, Solyndra, a company that manufactured photovoltaic systems using thin-film technology, raised $970 million in equity finance in addition to a $535 million loan guarantee from the Department of Energy, prior to its planned IPO in mid-2010. This amount of capital to prove commercial viability is an order of magnitude greater than the $40–$50 million that VCs are typically used to investing in each company to get them to a successful exit.

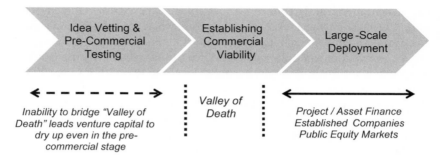

Fig. 7.6 Funding gaps and the "valley of death"
Source: Ghosh and Nanda (2014).

makes the portfolio much more risky. Such investments are thus typically too capital intensive for VCs, given the size and structures of most VC funds today.

The inability to raise either debt or venture capital at the demonstration and first commercial stage has led this stage of the start-up's life to be known as the "valley of death" (see figure 7.6). The fact that investors are now acutely aware of this funding gap before the firm gets to cash flow positive leads to an unraveling of the entire financing chain. That is, since investors forecast that even promising start-ups may have a hard time getting financing when they reach the stage of needing to build a demonstration plant, the benefits of sinking capital in a start-up at the stage before may not be worthwhile. This logic, that VCs refer to as financing risk, works through backward induction to the first investor. Thus, a forecast of limited future funding may lead promising projects to not be funded, even if when fully funded, they would be viable and NPV positive investments (e.g., see Nanda and Rhodes-Kropf 2012).

7.4.2 Exit Opportunities and Financing Risk

The VCs do invest in some industries such as biotechnology, semiconductors, and IT/networks that also share the attributes of huge financing requirements that are outside the scope of a start-up. However, in these instances, VCs bank on an established exit mechanism to hand over their early stage investments before they hit the valleys of death. For example, in the biotechnology industry, the VC model evolved over fifteen to twenty years in such a way that pharmaceutical companies stepped in to buy promising start-ups at a point even before commercial viability had been proven. This is a key part of the innovation ecosystem as it bridges the potential valley of death and thereby facilitates precommercial VC investments in biotechnology. The propensity of pharmaceutical companies to buy promising start-ups also facilitates their initial public offerings (IPOs) at precommercial stages, because public investors believe there is sufficient competition among pharmaceutical firms for biotechnology start-ups with innovative solutions and that they will be acquired well before they hit the valley of death. Cisco, Lucent, HP, and Juniper networks play an equivalent role in the IT/networking industry.

Thus far, however, energy-producing firms and utilities that supply electricity to customers have been far from active in acquiring promising clean energy start-ups. This bottleneck in the scaling-up process has a knock-on effect on the ability for VCs to fund precommercial technologies in this space as well. If early stage venture investors face the risk that they may be unable to raise follow-on funding or to achieve an exit, even for start-ups with otherwise good (but as yet unproven) technologies, they run the danger of sinking increasing amounts of dollars for longer periods of time to keep the start-up alive. With incumbent firms unwilling to buy these start-ups at precommercial stages, the time to exit for the typical start-up is much longer than the three- to five-year horizon that VCs typically target (the time to build power plants and factories is inherently longer than a software sales cycle and can even take longer than the life of a VC fund). Moreover, each time the start-up needs to return to the capital market for ever larger amounts of financing makes them vulnerable to the state of the capital markets for an extended period of time. As shown in figure 7.6, this leads venture capitalists to withdraw from sectors where they could have helped with the precommercial funding, but where they are not certain that they will be able to either fund the project through the first commercial plant, or they are not sure if they can exit their investment at that stage (Nanda and Rhodes-Kropf 2013). In fact, the history of capital-intensive industries such as biotechnology, communications networking, and semiconductors suggests that until the incumbents start buying start-ups, the innovation pipeline does not truly take off.

The biotechnology industry took several years to develop a financing ecosystem that allowed VCs to back start-ups and large companies to buy and scale them. Indeed, average VC returns to biotech investments are low, and even today the challenges associated with this model have led most VCs to exit this sector, leaving only a few specialist investors to focus on the industry and leading others to propose new funding models to drive innovation in drug development (Fernandez, Stein, and Lo 2012).

While in some ways, the renewable energy industry resembles the early days of the biotech industry (thereby giving hope that the ecosystem will develop in time), there are reasons to believe that achieving the same ecosystem may be harder in renewable energy. In the case of the biotechnology industry, a clear exit mechanism was facilitated by a vibrant market for ideas (Gans, Hsu, and Stern 2002) and the fact that the US Food and Drug Administration (FDA) developed well-understood and transparent metrics for success at each stage. Because the set of buyers was uniform and the criteria for a successful exit at each stage had been developed and well understood, VCs could work backward and set their own investment milestones. In this way, the downstream exit process had important consequences for the direction of upstream innovation. The extent to which large energy companies will play an equivalent role in the innovation pipeline for clean energy is not yet clear. There are some signs that this may be changing, with the most promising developments being the rise of a number of corporate venture capital funds among the large energy companies (Nanda and Rothenberg 2011) as well as the growth of some extremely large and dedicated venture capital investors focused on the renewable energy sector.

7.4.3 Global Commodities and Policy Risk

A final important difference between the renewable energy production and the typical VC-backed start-up is that energy is a commodity. Success in energy comes from being a low-cost provider rather than having an innovation that can be priced high due to the willingness of end users to pay (as is the case for biotechnology). While incumbents in other industries compete with each other to acquire start-ups in order to meet end-user demand, the end-user in the energy market cannot distinguish electrons produced from coal, the sun, or the wind, unless the government prices the cost of carbon appropriately. In the absence of appropriate price signals or incentives to invest in renewables, incumbents are therefore not pressed to acquire start-ups in this space. In the case of biofuels, the inputs to their production process are also commodities. Energy producers therefore face commodity risk for both raw materials and end products. Since these markets can exhibit substantial price volatility, it makes running and managing these companies more difficult. For example, second- and third-generation biofuel start-ups producing ethanol or biocrude at $80–$90 per barrel were competitive in 2007 prior to the global recession when conventional oil prices topped $100 a barrel, but most went bust when oil prices plummeted in the subsequent recession.

The challenges of backing a global commodity producer are compounded by the fact that energy and clean energy are sectors with large involvement by governments across the world. Given that clean energy technologies have not yet achieved grid parity, government policy is also critical in determining the prices of inputs and finished products. Some governments choose to either tax carbon content in conventional fuels or to buy clean energy at a premium. Others choose to subsidize clean energy companies through direct grants and subsidies or through tax breaks. Regardless of the policy, it implies that the extent to which a given start-up's product is likely to be profitable depends greatly on whether it is included in the subsidy or credit, the extent to which carbon is taxed, or the price premium at which the government buys the commodity.

Policy changes and uncertainty are thus major factors hindering the potential investment by private sector players across the clean energy investment landscape (Bloom 2009). This is particularly true when the periodicity of the regulatory cycle is smaller than the investment cycle required for demonstrating commercial viability. In such an event, no one is willing to invest in the first commercial plant if they do not know what the regulatory environment is going to be by the time success has been demonstrated (based on the rules of the prior regulatory regime).

7.5 Conclusion

Innovation in renewable energy has grown in recent years, in part due to the sharp rise in venture capital finance for renewable energy start-ups in the early years of the twenty-first century. However, the availability of venture capital finance for renewable energy has fallen dramatically in recent years. In this chapter, we ask whether we should worry that the decline and shift in VC will slow the rate and alter the direction of innovation in renewable energy.

Our results suggest that start-ups backed by venture capital file patents that are more likely to have at least one citation are more likely to be highly cited, have fewer self-citations, and are more likely to be novel than patents filed by incumbent firms. Although the lag in the patent grants do not allow us to directly observe how the falling levels of VC finance relate to the innovations by VC-backed start-ups, our results suggest VC financing is associated with a greater degree of economic experimentation and therefore, their shift away from financing renewable energy start-ups could impact the rate and trajectory of innovation in these industries.

Our chapter has also aimed to shed light on some of the structural factors that have made sustained experimentation by VCs hard, with a particular emphasis on the difficulty of exiting their investments to incumbent firms that have the expertise and capital to finance the scale up of such technologies. Larger/longer funds may be one alternative, since such funds could get past the uncertainty and to a stable place for exit, but measures designed to incentivize incumbents up the financing chain to bridge the valley of death also seem like possible solutions.

Although the US government has played a role in supporting clean technology innovation in the United States, the vast majority of this has been on the "supply side," through the direct support of individual firms (Roberts, Lassiter, and Nanda 2010). In addition to policies that would put a price on carbon, our discussion suggests that facilitating a more vibrant exit environment for start-ups at precommercial stages has the potential to stimulate greater private-sector funding of these start-ups, thereby increasing the degree of innovation and entrepreneurship in renewable energy.

We should note that our analysis is not meant to suggest that the innovations undertaken by incumbent firms are unimportant, or that the focus of VC on other aspects of clean tech is not valuable. Rather, our objective is to highlight the fact that the shifting focus of venture capital is likely to have an impact on both the rate and the characteristics of renewable energy innovation in the coming years. To the extent that there is still a need for experimentation with new technologies and a desire to commercialize radical innovations in renewable energy, our work highlights that there are structural factors that make sustained experimentation by VCs difficult in renewable energy. Although it is still early in the life cycle of this industry, our discussion has outlined some specific factors that may facilitate the deployment of large amounts of risk capital that are necessary to finance renewable energy innovations.

Appendix A

Table 7A.1   Assignees with the most patents between 2000–2009, including US and foreign inventors

Rank	Incumbents	Patent count	VC-backed firms	Patent count	Academia and government	Patent count
1	GE	541	Konarka Technologies, Inc.	52	Industrial Technology Research Institute	38
2	DuPont	235	SoloPower, Inc.	33	University of California	36
3	Stine Seed Company	213	LM Glasfiber A/S	30	University of Wisconsin	27
4	Vestas	136	Repower Systems AG	30	Michigan State University	22
5	Canon	131	Nanosolar, Inc.	29	US Navy	21
6	Siemens	92	Stion Corporation	24	US Department of Agriculture	19
7	Boeing	88	Clipper Windpower Technology, Inc.	18	North Carolina State University	18
8	Mitsubishi Group	88	PowerLight Corporation	14	Fraunhofer Society	17
9	Novo Group	83	MiaSolé	13	University of Illinois	17
10	Merschman Seeds	81	Solyndra, LLC	12	Institut Francais du Petrole	16
11	Sanyo	79	Twin Creeks Technologies, Inc.	12	NASA	15
12	BASF	76	Xantrex Technology, Inc.	11	University of Central Florida	15
13	Sharp	68	Energy Innovations, Inc.	10	University of Florida	14
14	ExxonMobil	65	Hansen Transmissions International NV	9	Princeton University	13
15	DKB Group	64	Ocean Power Technologies, Inc.	9	Massachusetts Institute of Technology (MIT)	12
16	Applied Materials, Inc.	61	Solaria Corporation	9	National Institute of Advanced Industrial Science and Technology	12
17	Chevron	60	Ceres, Inc.	8	Battelle Memorial Institute	11
18	Monsanto	56	Nanosys, Inc.	8	US Army	10
19	Samsung	52	Solexel, Inc.	8	Iowa State University	9
20	Nordex Energy GmbH	48	Solfocus, Inc.	8	California Institute of Technology	8
21	SunPower Corporation	45	ZeaChem, Inc.	8	Korea Advanced Institute of Science and Technology	8

Rank	Company				
22	Kaneka Corporation	44	Metabolix, Inc.	7	Pennsylvania State University	8
23	Lockheed Martin	42	Oryxe Energy International, Inc.	7	University of Southern California	8
24	Dow	39	SunPower Corporation Systems	7	National Research Council of Canada	7
25	Sumitomo	39	Converteam Ltd.	6	National Taiwan University	7
26	Gamesa	38	Enlink Geoenergy Services, Inc.	6	US Department of Energy	7
27	RAG Foundation	37	Tigo Energy, Inc.	6	University of Arizona	7
28	Schott AG	37	Verdant Power	6	Clemson University	6
29	CSIR	34	Coskata, Inc.	5	Queen's University at Kingston	6
30	BP	33	FloDesign Wind Turbine Corp.	5	Rice University	6
31	Royal DSM	32	Kior, Inc.	5	Swiss Federal Institute of Technology (EPFL)	6
32	UTC	32	LUCA Technologies, Inc.	5	University of Colorado	6
33	IBM	31	Marine Current Turbines Limited	5	University of Michigan	6
34	Royal Dutch Shell	31	Plextronics, Inc.	5	University of Toledo	6
35	Sony Corporation	31	Primestar Solar, Inc.	5	Atomic Energy Council	5

Appendix B

New Measure of Novelty

We have developed a new measure of novelty that is not based on citation measures. Instead, we draw on a textual analysis of patent applications to look at the similarity of patent claims and descriptions for patents in a given technology area. Intuitively, our definition is such that patents with greater textual similarity to neighboring patents are considered to be less novel. Our measure of novelty should be particularly useful in the context of science-based patenting, where technical terms are more unique and therefore more likely to signal differences in the characteristics of innovation, and for more recent time periods, where initial forward citations may be a noisy predictor of ultimate outcomes. The general outline for the calculation of the measure is as follows: First, the calculation algorithm reviews every patent claim and description in the sample to build a list of all terms used; the list of terms constitutes a high-dimensional positive space wherein each term represents a dimension into that space. Second, the algorithm positions each patent in the vector space by assigning it a set of coordinates where the magnitude of each dimension is calculated as the "term frequency inverse document frequency" (TF-IDF) of each term in the patent. Intuitively, TF-IDF gives a greater weight to a dimension when a term occurs more frequently in the patent, and gives a lesser weight to a dimension if the word is frequently observed in other patents as well. Third, the algorithm calculates the "similarity" between every possible combination of two patents, by calculating the cosine of the angle formed between their vectors. The measurement of similarity is bounded [0,1], with a measurement of 1 representing a perfect similarity between two patents.

Having thus arrived at a pair-wise list of similarity comparisons between every possible combination of patents in the sample, the algorithm then calculates a measurement of novelty for each focal patent by examining the distribution of similarities relative to a comparison set of patents. The comparison set is drawn from the prior three years and from the same technology area as the focal patent (e.g., "solar"). To assess the novelty of a patent—a concept connoting few neighbors in the technology landscape—we take the 5th percentile of the rank-ordered distribution of similarities tied to the comparison set. For ease of interpretation, we reverse the novelty measure by subtracting it from 1, arriving at a measurement for novelty that is bounded [0,1], where 1 represents a patent that is entirely dissimilar from all other patents. When needed, we average patent-level measures of novelty up to the firm or category level.

References

Akcigit, Ufuk, and William Kerr. 2011. "Growth through Heterogeneous Innovations." Working Paper, Harvard Business School, Harvard University.

Bergemann, Dirk, and Ulrich Hege. 2005. "The Financing of Innovation: Learning and Stopping." *RAND Journal of Economics* 36:719–52.

Bergemann, Dirk, Ulrich Hege, and Liang Peng. 2008. "Venture Capital and Sequential Investments." Cowles Foundation Discussion Paper no. 1682, Cowles Foundation for Research in Economics.

Bloom, Nicholas. 2009. "The Impact of Uncertainty Shocks." *Econometrica* 77: 623–85.

Bloomberg New Energy Finance. 2010. "Crossing the Valley of Death: Solutions to the Next Generation Clean Energy Project Financing Gap." White Papers, June. http://about.bnef.com/white-papers/crossing-the-valley-of-death-solutions-to-the-next-generation-clean-energy-project-financing-gap/.

BP. 2012. Statistical Review of World Energy. http://www.bp.com/content/dam/bp/pdf/Statistical-Review-2012/statistical_review_of_world_energy_2012.pdf.

Conti, Annamaria, Jerry Thursby, and Marie Thursby. 2013. "Patents as Signals for Startup Financing?" Journal of Industrial Economics 22 (2): 592–622.

Fan, Rong-En, Kai-Wei Chang, Cho-Jui Hsieh, Xiang-Rui Wang, and Chih-Jen Lin. 2008. "LIBLINEAR: A Library for Large Linear Classification." *Journal of Machine Learning Research* 9:1871–74.

Fernandez, Jose-Maria, Roger M. Stein, and Andrew W. Lo. 2012. "Commercializing Biomedical Research through Securitization Techniques." *Nature Biotechnology* 30:964–75.

Gans, Joshua, David Hsu, and Scott Stern. 2002. "When Does Start-Up Innovation Spur the Gale of Creative Destruction?" *RAND Journal of Economics* 33:571–86.

Ghosh, Shikhar, and Ramana Nanda. 2014. "Venture Capital Investment in the Clean Energy Sector." Harvard Business School Entrepreneurial Management Working Paper no. 11-020, Harvard University.

Gompers, Paul. 1995. "Optimal Investment, Monitoring, and the Staging of Venture Capital." *Journal of Finance* 50:1461–89.

Gompers, Paul, and Josh Lerner. 2002. *The Venture Capital Cycle*. Cambridge, MA: MIT Press.

Guler, Isin. 2007. "Throwing Good Money after Bad? A Multi-Level Study of Sequential Decision Making in the Venture Capital Industry." *Administrative Science Quarterly* 52:248–85.

Hall, Robert, and Susan Woodward. 2010. "The Burden of the Nondiversifiable Risk of Entrepreneurship." *American Economic Review* 100 (3): 1163–94.

Henderson, Rebecca, and Kim Clark. 1990. "Architectural Innovation—The Reconfiguration of Existing Product Technologies and the Failure of Established Firms." *Administrative Science Quarterly* 35 (1): 9–30.

Kerr, William, Ramana Nanda, and Matthew Rhodes-Kropf. 2014. "Entrepreneurship as Experimentation." *Journal of Economic Perspectives* 3:25–48.

Kortum, Samuel, and Josh Lerner. 2000. "Assessing the Contribution of Venture Capital to Innovation." *RAND Journal of Economics* 31 (4): 674–92.

March, James. 1991. "Exploration and Exploitation in Organizational Learning." *Organizational Science* 2:71–87.

Nanda, Ramana, and Matthew Rhodes-Kropf. 2012. "Financing Risk and Innovation." Harvard Business School Working Paper no. 11-013, Harvard University.

————. 2013. "Investment Cycles and Startup Innovation." *Journal of Financial Economics* 110(2): 403–18.

Nanda, Ramana, and Juliet Rothenberg. 2011. "A Quiet Revolution in Clean Energy Finance." Harvard Business Review (blog). http://blogs.hbr.org/2011/10/quiet-revolution-clean-energy-finance/.

Popp, David, Ivan Hascic, and Neelakshi Medhi. 2011. "Technology and the Diffusion of Renewable Energy." *Energy Economics* 33:648–62.

Roberts, Michael, Joseph Lassiter, and Ramana Nanda. 2010. "US Department of Energy & Recovery Act Funding: Bridging the 'Valley of Death'." Harvard Business School Case 810-144, Harvard University.

Rosenberg, Nathan. 1994. "Economic Experiments." In *Inside the Black Box*, edited by Nathan Rosenberg. Cambridge: Cambridge University Press.

Rosenkopf, Lori, and Atul Nerkar. 2001. "Beyond Local Search: Boundary-Spanning, Exploration, and Impact in the Optical Disk Industry." *Strategic Management Journal* 22(4): 287–306.

Sahlman, W. 1990. "The Structure and Governance of Venture-Capital Organizations." *Journal of Financial Economics* 27:473–521.

————. 2010. "Risk and Reward in Venture Capital." Harvard Business School Background Note no. 811-036, Harvard University.

Samila, Sampsa, and Olav Sorenson. 2011. "Venture Capital, Entrepreneurship and Economic Growth." *Review of Economics and Statistics* 93:338–49.

Singh, Jasjit, and Lee Fleming. 2010. "Lone Inventors as Sources of Breakthroughs: Myth or Reality?" *Management Science* 56 (1): 41–56.

Sorensen, Jesper, and Toby E. Stuart. 2000. "Aging, Obsolescence and Organizational Innovation." *Administrative Science Quarterly* 45:81–112.

Stern, Scott. 2005. "Economic Experiments: The Role of Entrepreneurship in Economic Prosperity." In *Understanding Entrepreneurship: A Research and Policy Report*, edited by Carl J. Schramm. Kansas City, MO: Ewing Marion Kauffman Foundation.

Tripsas, Mary, and Giovanni Gavetti. 2000. "Capabilities, Cognition and Inertia: Evidence from Digital Imaging." *Strategic Management Journal* 21:1147–61.

Tushman, Michael L., and Philip Anderson. 1986. "Technological Discontinuities and Organizational Environments." *Administrative Science Quarterly* 31 (3): 439–65.

Ullman, Jeff, and Anand Rajaraman. 2011. *Mining of Massive Datasets*. New York: Cambridge University Press.

8

Economic Value Creation in Mobile Applications

Timothy F. Bresnahan, Jason P. Davis, and Pai-Ling Yin

8.1 Introduction

No discussion of the great changes in the innovation processes of the twenty-first century would be complete without an examination of one of the newest growth poles: mobile applications. Mobile applications are software programs that run on a new class of mobile devices, smartphones, and tablets, which are typically connected to cell phone networks. The rapid growth of mobile devices has been accompanied by an equally rapid growth in app development, in substantial part because platform providers Apple and Google have lowered the costs of development and distribution of mobile applications.

Timothy F. Bresnahan is the Landau Professor in Technology and the Economy and a professor of economics at Stanford University, and a member of the board of directors of the National Bureau of Economic Research. Jason P. Davis is associate professor of entrepreneurship and family enterprise at INSEAD. Pai-Ling Yin is a social science research scholar at the Stanford Institute for Economic Policy Research.

This research project is based on data collection and analysis over a wide range of data sources. We are very grateful to a number of research assistants who have worked on those data sets, gathered industry information, and joined us in industry interviews. These include Markus Baldauf, Sean Batir, Robert Burns, Jane Chen, Sherry Fu, Osama El-Gabalawy, Carlos Garay, Jorge Guzman, Alireza Forouzan Ebrahimi, Tim Jaconette, Nayaranta Jain, Julia Kho, Sigtryggur Kjarttansson, Xing Li, Derek Lief, Sean Mandell, Laura Miron, Jaron Moore, Yulia Muzyrya, Abhishek Nagaraj, Joe Orsini, Francis Plaza, Hatim Rahman, Sam Seyfollahi, Melissa Sussman-Martinez, Masoud Tavazoei, Sylvan Tsai, Julis Vazquez, Joon Yoo, and Parker Zhao. We are also very grateful to the many industry participants who have shared their time and expertise with us. We appreciate the valuable comments of Josh Lerner, Xibao Li, Ben Jones, Scott Stern, and Adam Jaffe. Pai-Ling Yin and Jason Davis benefited from the Karl Chang (1965) Innovation Fund and Edward B. Roberts (1957) Fund. This chapter represents data and conclusions as of its submission to the volume editors on October 13, 2013. For acknowledgments, sources of research support, and disclosure of the authors' material financial relationships, if any, please see http://www.nber.org/chapters/c13044.ack.

Like any new industry with significant promise, mobile apps have also engendered a long list of conjectures about where economic value might lie. Today, the industry is in the experimental phase of its life cycle. Like any information and communications technology (ICT) industry, much of the uncertainty and experimentation is about commercial rather than technical innovation. What kinds of apps will consumers use? Will consumers pay for them, or will they be advertising supported? Which existing industries will they "disrupt"?

Our goal in this chapter is to examine the supply of apps. We take up three topics that are suggested by the industry structure and by the early conjectures by industry participants:

1. Platform innovation: How is complementary innovation coordinated to create a new industry? How are scientific and technical opportunity linked to demand?

2. Industry evolution: How will experimentation in technologies, markets (and other institutions), and commercialization lead to changes over time?

3. Value creation: What could the largest new industry of our century contribute to economic growth?

Our investigation of platform innovation is laid out in section 8.2. Any innovation platform involves the sharing of general purpose components across applications; this sharing lowers entry costs for diverse, innovative applications. Our setting exhibits the most users of any application platform, the most apps, and the fastest growth thus far in ICT settings. The positive feedback loop appears to be working.

We choose to focus our study on the app developers and examine the rest of the industry participants from their perspective. A robust literature explores the technological history of mobile communication devices, platforms, and users, but the app developers (responsible for private and social value creation) are largely unstudied. Our empirical work has led us to a topic that cuts through all three of these areas: identification of the important bottlenecks to entrepreneurship, experimentation, and value creation that have arisen despite the considerable growth opportunity. The explosion in entrepreneurship has created an explosion in competition and an overwhelming choice set for the consumer. The problem arises in the first market institution required of a new mass market platform industry: the market that matches users and apps. We describe these institutions in section 8.3 and examine their implications throughout. We describe new data sets used in our analysis, some of which we have constructed, in section 8.4.

An application platform does not directly create economic value; instead, it lowers the costs of applications, enabling applications in a wide variety of sectors. We discuss this wide variety of applications sectors for mobile apps in section 8.5. Even though almost all of the most popular apps are not in market competition with one another, the size distribution of app demand is highly concentrated. This is true whether we look at downloads or usage. While in principle this might arise because only a few highly popular app categories or apps have been discovered, we show in section 8.6 that there is a high rate of turnover (churn) in app success. We interpret the combination of concentration and churn as reflecting not only the underlying distribution of app attractiveness, but also competition across all apps categories for consumers' attention.

In section 8.7, we consider the implications of the "top lists" for industry development. We document a new, growing, and important category of apps, "corporate" apps, which form part of a consumer product and service firm's offerings to its customers. We show that these new, "nondisruptive" apps are considerably advantaged over entrepreneurial apps given the current state of the industry's market institutions.

In addition to experiments with a large variety of zones of application, developers are also experimenting with a wide variety of "monetization" strategies for apps, including paid apps, advertising-supported apps, "freemium" apps, and several more. We have gathered unique data on developers' commercialization strategies. These are analyzed in section 8.8.

To access most of their customers for a mass market mobile app, developers must write for the two largest platforms. A substantial body of discussion in the industry suggests that developers should write for iOS (Apple) first and then Android (Google) second. In section 8.9, we examine developers' platform choice behavior in two senses: platform preference (iOS vs. Android) and multihoming (both). Finally, we then consider alternative equilibrium scenarios for the industry in section 8.10.

The broad economic picture of value creation that we are finding in the mobile ecosystem is reminiscent of earlier information and communications technology (ICT) platform industries. Raw technical progress (faster and smaller computers like mobile devices, faster communications, etc.) has a higher rate of change than applications innovation to create economic value. This has been noted in corporate computing, personal computing, business data communications, and the commercial Internet.[1] Like mobile, these earlier platforms had rapid invention in purely technical components and successful exploitation of social scale economies, but slower, though very valuable, innovation in application. Uncertainty about the value proposition for a new technology leads to exploration—not only at the scientific stage, but at the commercialization stage.[2] This is typical of general purpose

1. See Bresnahan and Greenstein (1999) for computing applications and Bar (2001) for communications industries. The widespread use of the Internet is taken up in Greenstein (2001).

2. The importance of a wide variety of *technologies* experimentally chosen by different firms early in the industry life cycle has been emphasized in a large literature, epitomized by Klepper (2002). We emphasize the parallel importance of a wide variety of commercialization experiments. In both cases, later market selection is critical. We also depart from the industry life cycle's standard modeling approach, which links the underlying uncertainty/variety/selection to horizontal industry structure variables, market shares, entry and exit, and so on. We emphasized different observables largely because we examine experimentation by firms that are not in direct competition with one another, so it is not clear what horizontal industry structure to examine.

technologies (GPTs), where the industry does not necessarily know all the uses or even the main uses at the beginning. The slowness of that commercialization and value creation process in ICT platforms has been one of the leading determinants of the aggregate growth rate of the rich economies in recent decades.[3]

Our investigation of platform choice behavior by developers reveals a number of aspects of the barriers to application success in the current state of the industry. Entry barriers appear to be significantly lower for corporate apps than for apps from entrepreneurs. This arises not because of any discernible lack of technical or even marketing capabilities on the part of entrepreneurs, but rather because of the problems of matching users to apps. The current matching problem between developers and their customers slows the rate of innovation in mobile application, just as the low technical entry barriers raise it. The current matching problem also affects the direction of technical change, raising the fixed costs of entrepreneurial apps much more than those of corporate apps. "Disruption"—or even ordinary, high-value entrepreneurship—is disadvantaged versus continuity strategies of existing firms.

8.2 Innovation in Platform-Based Industries

A new platform-based industry always has elements of a general purpose technology (GPT). Increasing returns to scale arise at an industry-wide level because some of the components used in different applications are common. By the same token, coordination issues arise among inventors of applications and/or of the general components.[4] A new platform-based industry can sometimes recombine general purpose components that are already in existence. This, too, creates economies by avoiding reinvention of the existing components and creates coordination problems by bringing existing suppliers into the coordination loop. However the general components are supplied, a new platform-based industry presents applications developers with a partial solution. It is typically up to applications developers to discover or invent valuable uses of the general technology, establish the markets needed for valuable uses, and engage in other social value-creation activities.

In our discussion of the platforms for mobile app development, we begin with the most successful areas so far, the recombination of existing technolo-

3. There is a large literature on this, summarized in Sichel (1997).
4. For the analysis of GPTs, see Bresnahan and Trajtenberg (1995) and Helpman and Trajtenberg (1998). The GPT literature emphasizes the market (uncoordinated) determination of the complementary rates of technical progress in a general purpose technology and applications. The two-sided market or platform-pricing literature (see Rysman 2009) emphasizes the use of a pricing mechanism, sometimes nonlinear, to coordinate supply and pricing and thereby internalize network externalities.

gies and the creation of new general purpose technologies that dramatically lower the costs of inventing a new mobile application.

The first stage of innovation for the mobile app industry began with the recombination of existing technologies to provide an infrastructure for app development and consumer utility. The rapid improvements in the portability and power of mobile devices combined with the rapid improvements in the networking capacity of the mobile telephone and Wi-Fi networks created a huge opportunity for mobile devices to provide much more utility than simply as a communication device. The established consumer familiarity with networked computing also paved the way for rapid consumer adoption of mobile applications. Users already understood the concept of accessing remote sites and downloading software.[5] The increased bandwidth further augmented the ability of the smartphone to become a powerful user interface by allowing much of the storage and processing power to be relegated to the cloud. This allowed the mobile device to become both more portable and more powerful at the same time. The recombination of these existing technologies and the improving savviness of consumers permitted the invention of new GPT components, in particular, the iPhone, iOS, and (expanded) iTunes store, with parallels on the Android side.

The impact of all of this recombination and new invention was threefold. First, an application developer could create a new mobile app spending only a tiny fraction of the overall research and development (R&D) cost of providing it. Much of the R&D cost, including invention of mobile devices, mobile telephony transmission, the commercialized Internet, the cloud, and so forth, and the investment cost in infrastructure, including diffusion of mobile devices to users, had already been sunk and spread over thousands of applications. None of these common R&D or infrastructure investment costs were marginal to a particular app. For mobile applications developers, the fixed costs of offering a working system to users were dramatically lowered. Second, the potential economic return to new invention and investment in the platforms and in the preexisting complementary technologies that were recombined became higher. Third, applications developers were given free reign (and little guidance) to discover these opportunities. We shall return to this theme below. For now, we continue with the positive side of positive feedback.

Network effects imply a second external economy, the "indirect network effects" of attracting both a significant body of demanders and a significant supply of applications so that the platform gets over the hump into viability. In the case of mobile apps, this was achieved by the supply of a modest

---

5. Though there were technical antecedents to mobile applications (games and ringtones on Symbian phones, e-mail and messaging on Blackberries, online stores run by carriers, etc.), none of them launched the mobile app industry due to the lack of the necessary critical mass in use and developers necessary to generate positive feedback loops.

number of influential apps, which, taken together, were sufficient to attract a significant body of demanders.[6] These included a media store, especially for music (iTunes), an app for accessing the Web (mobile browser), an interactive map, and, after a brief interval, some games. Together with some ease-of-use improvements over existing phones, and some economies of purse and pocket space (phone and music in one device), the supply of these "killer apps" led to an initial expansion in user demand for smartphones that could run apps. This created an enormous market for new apps. The combined impact of the existing mobile telephone system, which created the platform products themselves, and killer apps has generated huge growth in mobile device demand. Today, more new smartphones (sixth year of diffusion) are sold worldwide than personal computers (PCs) (twenty-eighth year). For some kinds of consumers in some economies—for example, young adults in South Korea—the diffusion of smartphones has gone further than the diffusion of television. App developers have access to a very large body of demanders, as do advertisers seeking to run ads in apps.

The first thing that is notable in this story of obtaining positive feedback around the two successful mobile platforms is that it was achieved with remarkably little coordination. There was no widespread contract with developers or users—the mass of developers and users were simply offered an arms-length opportunity.

The rapid emergence of many demanders, together with the very low barriers to entry created by the platform providers, has led to a rapid and very substantial expansion in the number of overall apps. Figure 8.1 shows the dramatic growth in both iOS and Android apps. As is easy to see, the iOS growth starts earlier, as the iPhone was widely marketed before any Android phones.

Android apps, measured by number, have caught up to iOS apps. One must be careful, however, about drawing any economic inference from a count of apps: the majority of apps are marginal and have not been downloaded or used by customers. We will revisit the question of the size of the supply of Android versus iOS apps later.

This industry is still in its early stages. The big driver for huge growth in this industry was the establishment of a platform that drastically lowered the costs of entry into mobile application development: it provided both the R&D and the distribution system (iTunes/Google Play stores) at much lower cost than would typically be faced by an entrepreneur. This enabled

---

6. This is not the only way to get over the hump into viability. A large demander, such as the Defense Department, can attract sufficient suppliers to start the positive feedback loop, or a platform sponsor can coordinate the joint attraction of many demanders and suppliers. As with many other commercial computing and communications platforms, however, the leap of mobile platforms over the hump was achieved by the attractiveness of a few "killer apps," that is, apps attractive enough to give users a motivation to buy the product.

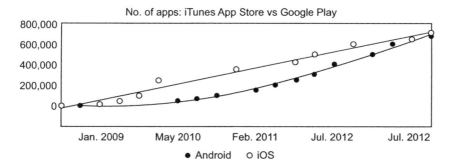

Fig. 8.1 Apps offered by platform over time
Source: InsideMobileApps.com.
(Accessed July 19, 2013: http://www.insidemobileapps.com/2012/09/26/trackinggrowth-
the-itunes-app-store-vs-google-play/.)

an explosion of entry by app developers, a strong reflection of how low the incremental cost of app development has fallen.

The pace at which the industry creates social value depends not only on those low technical costs but also on the rate at which new, high value markets are founded and on the pace at which innovation and competition drive up product quality and drive down prices in new products enabled by the technical opportunity in the industry.

8.3 Matching Apps to Customers: App Store Rankings

The mobile apps marketplace serves primarily consumers, and thus needs institutions to match applications products (and their sellers) to mass-market buyers. The design of a market institution for matching appears to be very difficult. New apps are introduced at a very high rate, and the problem of categorizing apps is largely unsolved. Moreover, the extremely rapid growth of the user base means that the distribution for demand for apps can be volatile. Within the broad established games category, products have short lives. Thus the search problem for a particular consumer is difficult. The solutions offered by the online stores, which we discuss in detail in this section, are of a "top list" form and do not appear to be very effective at matching demanders to desired apps.

8.3.1 iTunes Store

Apple tightly bundles the services of the iTunes Store as a distribution and app-discovery market with the technical products that make up its platform. This is consistent with Apple's overall plan of providing a controlled "stack" of components and services and to vertically integrate into all of the general purpose components.

From one perspective, the iTunes Store is a roaring success. The applications store was added to iTunes with 500 distinct apps in July 2008; by May 2013 users had downloaded fifty billion apps.[7] Apple's tight control over distribution gives it control over developer revenues, and Apple charges distribution fees of 30 percent of revenues. Through May 2013, Apple had paid developers over $8 billion.

From another perspective, the iTunes Store is much more problematic for developers. The mechanisms by which users and apps are matched are limited. A user can search for apps by keyword. A user can look at "top lists," such as "top free apps" (or top paid or top grossing, i.e., highest revenue). A user can also look at "top lists" within broad applications categories, though it appears very difficult to categorize new apps. Finally, a user can arrive at the iTunes Store knowing the app they want, either by following a link from another website or by remembering the app's name. The "top lists" are, from a user perspective, a collaborative filter. As each user arrives at the site, the top lists show those apps earlier users have chosen to download. Our understanding is that Apple deals with these difficult trade-offs by displaying apps that have been most downloaded in the previous twenty-four hours on the "top" lists.[8] This makes the collaborative filter very responsive to current demand conditions. The iTunes Store also offers user-opinion-based collaborative filters, with comments and ratings (one to five stars) written by earlier users and read by later ones.

An important implication of the top lists structure of the iTunes Store is that there is a strong winners-take-all flavor.[9] An app can be popular because it is highly visible on the top lists, but it is on the top lists because it is very popular.

8.3.2 Google Play Store

Google, the supplier of the Android market, similarly runs an online app store called Google Play.[10] However, comparatively open-systems Google does not bundle services of Google Play with the Android platform, so developers can distribute apps themselves or use a different online store, such as Amazon, which has an Android apps store. Despite the possibility of developers going elsewhere, Google Play has caught up to iTunes' number of app offerings (see figure 8.1).

Google Play has a slightly more complex set of top lists than does the iTunes Store, mostly because there are more lists, including top paid, top free,

---

7. See http://www.apple.com/pr/library/2013/05/16Apples-App-Store-Marks-Historic-50 -Billionth-Download.html. Accessed October 15, 2013.

8. At this writing (summer 2013), Apple was considering changing away from this ranking system to one also based on user reviews.

9. See Sorensen (2007) for analysis of this form of distribution mechanism.

10. Before March 2012, Google maintained separate app, music, and book markets. Afterward, the former Android Market merged with Google Music and Google eBookstore and became Google Play.

new apps, and trending apps both overall and within categories. Another important difference is that the downloads-based collaborative filter is based on a longer window: our understanding is that "top" apps are the most downloaded over the preceding eight days (vs. one day for the iTunes Store). Otherwise, the structure is basically the same: a mixture of a user-action-based collaborative filter (recommending apps recently downloaded) and a user-opinion-based collaborative filter (comments and ratings).[11]

Like the iTunes Store, Google Play has been very successful on a volume basis. But it is also problematic as a matching mechanism, for much the same reason. At this early stage of the industry, it is very hard to run an effective collaborative filter.

8.3.3 Problems Facing Collaborative Filters

Developers, especially developers who do not have a preexisting connection to their customers, express considerable frustration with the effectiveness of the app store collaborative filters. Several features of the environment make effective implementation of a collaborative filter difficult. Users typically search from mobile devices, limiting their ability to browse through long lists. New apps are constantly being submitted, so early users would have to choose the best apps quickly if the collaborative filter can find them for later users. Most users do not have a large number of apps on their device, so there is limited information to fine-tune the recommendations tailored to a particular individual user. Apple is the only channel for distribution for its mobile devices, which exacerbates the problem of ranking being the dominant matching mechanism. Google's openness and the existence of alternative app stores does not seem to be the solution to the ranking lists provided by iTunes. The matching problem has proved an as yet unsolvable challenge despite Google's expertise in search. The same platforms that lowered the cost of entry are also the source of the largest costs to developers: the platforms have not been able to provide adequate institutions to help match consumers to the overwhelming product offerings.

The problem of matching buyers and sellers is an industry-wide problem, not just a marketing problem for developers. Instead of competing with other apps in their same market, apps are competing with *unrelated apps* for consumer attention. Our empirical analysis will show that the store rankings fail to reflect the value created by the app and match heterogeneous consumers to apps. Although the entrepreneur could benefit from rapid adoption by the huge mass of end consumers, the ability to capture the "right" consumers in that dynamic is unclear, leading to lots of expensive

---

11. Apart from the star ratings and reviews, Google Play also has apps recommended by members of a user's Google Plus "circles," that is, their Google Plus social network. The importance of this improved collaborative filter is constrained by the limited penetration of Google Plus into widespread use as a social network.

investment in capturing all consumers. As a result, there is a curious duality about the new mobile platforms: they have lowered the *technical* costs of entering and supplying application products dramatically but have left very high *marketing* costs for entrant app developers. This duality is familiar from earlier ICT development platforms.[12] Further, it is familiar that marketing costs can affect industry structure.[13] What is striking here is that the technical entry costs are so low because the platforms can offer so much accumulated and recombinable technology, while the marketing costs are so high, at least at this early stage, because the problem of matching app buyer and seller is particularly difficult when the seller is an end consumer.[14]

The incentive then arises for developers to game the rankings by purchasing downloads to rise in the rankings, which then makes interpretation of downloads as demand suspect. We will later discuss the emergence of a number of institutions as a result of the influence of collaborative filters. Alternative ranking and rating systems outside the app stores have emerged to better help match consumers to apps, based on niche interests or social networks: Facebook is now well placed to combine social networking with app advertising and app filtering. The platform providers are also tinkering with the structure of the online stores ranking systems to address the collaborative filter problems. For now, the rankings strongly influence the development of the app ecosystem.

8.4 Data

Our tables are built on three main data sources. The sampling frame for existing industry data sources are built around *products*, a peculiarity of our particular industry. There are no industry-wide data with the more usual economic sampling frames of firms or markets. We have thus worked to build firm-level and market-level data sets to pursue some parts of our analysis.

8.4.1 comScore

We utilize the "mobile metrix" data set from comScore. Like other comScore products, this is based on a panel of users, in this case approximately 10,000–12,000 US adult users of mobile devices. The bulk of the panel is

---

12. See, inter alia, Bresnahan and Greenstein (1999).

13. Fixed marketing costs have played a large role in the economic analysis of market structure and entry (see Sutton 1991). Sutton's concept of "endogenous fixed costs" is closely related to the use of mass-media advertising by firms in competition with one another in product markets. We are examining a very different institutional structure in which products not in competition with one another in markets compete for the limited attention of consumers in a ranking system. These costs, too, are endogenous, in that the efforts of one firm to get attention for its products raises the costs of other firms' getting attention for their products.

14. This is both like and unlike the problem of founding markets in earlier ICT technologies. There, the customer could typically be located, but the use of the technology involved invention of new organizations or work practices by the customer.

two subpanels, each approximately 5,000 users, one with Android phones and the other with iPhones.[15]

The underlying fundamental data are about each panelists' possession and usage of apps.[16] However, the data come to us aggregated to the product*platform*month level.[17]

The sample of products (apps) on which data are available meet a minimum usage test for each month. For each platform, iOS (iPhone) or Android (smartphone), comScore includes data on the app only if it is used on that platform by more than five (at least six) unique users.[18] As a result, apps enter and leave the reported sample month to month.

It is very difficult to identify the same app on the two different platforms in this industry. In this data set, comScore staff members (working with developer clients when available) manually identify the same app on iOS and Android. Three interesting issues arise with this. First, this is a continuing project, so not all apps have been processed yet. Second, comScore does not depart from its used-by-enough-users standard platform by platform. Thus an app may be *available to users on both platforms* but only included in the comScore data on one platform because it does not have many users on the second. Third, comScore assigns a common, sensible name to each app, which is common across platforms. This means that a multiplatform app can have a comScore name that is different from its name in the iTunes Store and also different from its name in Google Play. An app that is only listed on one platform on comScore can also have a comScore name that is different from its name in that platform's app store. We have linked the comScore names of the apps to product and firm information.

For each month*platform*(included) app, comScore compiles a number of metrics. These include estimates of the total number of unique users for each of those apps during that month, the total number of minutes for which the app was used by all users that month, the average minutes per visitor (which is the ratio of total minutes to unique users), and the average daily

15. There is also a smaller subpanel of iPad users, which we do not use in this chapter. The sample sizes change somewhat over the period in which the data exist.

16. There are also data on panelists' visits to mobile websites, but we do not use these data in this chapter.

17. An app is not exactly a product (observation in the data set), but it is close. For some apps, comScore has aggregated distinct versions into a single "property," typically because (a) they view the apps as different version of the same thing, (b) the app supplier views them as the same thing, or (c) the app supplier sells ad space in the different apps as a single-ad product. Often, the merged "property" includes both the free part and the paid part of a pair of "freemium" apps, though sometimes free and paid are two separate properties.

18. A second criterion could also lead to inclusion if the app developer has implemented a comScore provided software development kit (SDK) that includes a piece of software in every copy of the app. This piece of software then reports to comScore whether (inter alia) the user has used the app. If more than 11,000 unique US users have used the app during the month, it is included. In the months we examine, this second criterion does not appear to lead to the inclusion of any apps that would fail the first test.

Table 8.1   **Number of unique apps by month**

	Total	AOA	EOA	AOI	EOI	Multihomed
September 2012	1,301	715	439	862	586	276
October 2012	1,243	710	434	809	533	276
November 2012	1,202	691	408	794	511	283
December 2012	1,203	727	433	770	476	294
January 2013	1,231	793	507	724	438	286

Source: comScore.

Notes: AOA = all on Android, EOA = exclusive on Android, AOI = all on iTunes, and EOI = exclusive on iTunes. The table classifies apps in the comScore sample according to their availability on Android and iOS. In January 2013, users in the comScore panel have used 1,231 distinct apps in total, 793 and 724 of which were available on Android and iOS, respectively. Of the Android apps, 507 were exclusive to that platform, while 438 of the iOS apps were exclusive on that platform. Multihoming apps comprised the remaining 286 apps.

visitors. The average daily visitors is calculated by taking the average, over all days in the month in question, of the number of unique visitors that the app had in a single day. For example, if the universe of mobile app users consisted of two users, one of whom used a particular app every day in April, while the other used it every other day, then the "average daily visitors" measure for that app for April would be 1.5. All of these data are projected to the entire United States based on a set of comScore weights for their panel.

We have looked at the joint distribution of all of these different demand metrics and concluded that there is little more than two dimensional variation in the cross section of apps in a given month. The size metric, unique visitors per month, and the engagement metric, average minutes per visitor, together can explain almost all of the variation in the rest of the size measures. Both measures are correlated with the two remaining ones, average daily unique visitors and minutes per month, but are very close to uncorrelated with each other (0.06). Thus the raw data yield a simple two-factor model of product "size" that we use below.

Sometimes the total unique users measure is converted by comScore into the "reach" of the mobile app. An app's reach is defined as the percentage of all "potential users" of the app who actually used it during the month in question. This is given by the ratio of comScore's estimate of the total number of unique visitors to the app on a particular platform during that month to the total number of users on the platform during that month.

Table 8.1 shows monthly statistics on app developers' platform choices as reported by comScore between September 2012 and January 2013. Since the comScore panel, census projection methodology, and apps who use the comScore software development kit (SDK) are in flux over time in this very new industry, we are skeptical of reporting any changes in the data over time. That said, that five-month period was one of comparative stability in the definitions, and there does not appear to be much change over time in

Table 8.2   Summary statistics of key variables

	Mean	Sd.	Min.	Max.
Unique monthly visitors (000)	1,289.92	4,324.60	11.12	60,805.94
Avg. unique daily visitors (000)	300.51	1,739.01	0.46	38,657.62
Total monthly minutes (millions)	1,993.24	16,289.40	0.26	391,802.25

Source: comScore.

Notes: There are 1,517 app-platform observations, January 2013. Unique monthly visitors provides a count of the number of comScore panel members who visit an app in the sample at least once per month, weighted to be representative of the US population. Average unique daily visitors provides the average of the daily unique visitors per day of an app over all days in a month. Total monthly minutes records the total minutes an app was used in month.

the relative magnitudes across columns of these figures within a month, so hereafter we will focus on the data from January 2013 in our analysis.

The table reports six metrics for each month. "All" is the total number of distinct apps in the data, 1,301 in September 2012. "All on Android" (AOA) is the number that are available on the Android platform, and "exclusive on Android" (EOA) is the number of apps that are available only on the Android platform. "All on iTunes" (AOI) and "exclusive on iTunes" (EOI) are defined similarly. Finally, "multi" is the count of apps in the comScore sample found on both platforms in the month.

In table 8.2, we report simple descriptive statistics for the size measures. One can immediately see that all variables are heavily skewed. For instance, on average an app is used for 1,993.24 million minutes per month. However, the standard deviation for minutes used is more than eight times larger than the mean. The maximum unique monthly visitors for an app is three to four orders of magnitude larger than the minimum.

8.4.2 App Annie

App Annie is a market research business. They write, "App Annie tracks your apps' metrics and has the best app store data to help you make smart business decisions." The App Annie app metrics we use in this chapter are download metrics for the online stores (iTunes and Google Play). App Annie copies the rankings of top apps from each of the online stores each day. For each day we obtain the ranking of the top 500 free, top 500 paid, and top 500 grossing (revenue) apps for iPhone and Android phones. For Android, there is also a "top new free" and "top new paid" ranking each day.

The two online stores clearly have slightly different ranking algorithms. Neither online store publishes its ranking algorithm, but there is, of course, constant discussion of the algorithms among industry participants. Our understanding is that the top apps on iTunes is based on downloads in the past twenty-four hours, whereas the top apps on Google Play are based on downloads in the past eight days.

8.4.3 App Questionnaire

Finally, we have ourselves undertaken a survey of apps. We created a questionnaire, focused on apps' monetization strategies, categories, content and presence on multiple platforms, and asked research assistants to download apps onto their mobile devices and then fill out the questionnaire. The main point here is that there are a number of critical questions about app developers and apps—especially the monetization strategies, which are public because they are visible in the app to any users but are not collected anywhere else (they are not observable on any publicly available websites).

Our app questionnaire sample includes nearly 5,000 free apps (2,281 Android and 2,713 iOS). We have two sample inclusion criteria. First, we attempt to survey the most popular apps. This is simple on Android, since Google Play reports, for each app, what total cumulative downloads "bin" it falls into. We define popular on Android as an app that has had at least 500,000 downloads. It is more difficult on the iTunes Store, which has no similar data publically visible. For iOS, we define a popular app as one that has been on the App Annie rankings list for at least ten days at any time. We continue to survey popular apps today, but the samples used in the tables in this chapter focus on three periods: September 2012–January 2013, January 2013, and June 2013.

We also include apps in the questionnaire if the app appears in the comScore data. This adds a surprising number of apps (about 1,000). This could arise because comScore is a measure of current usage while our definitions of "popular" are based on downloads, or because of other gaps between comScore's sampling frame and the marketplace. However, we think it is more likely that it reflects the natural sampling uncertainty. To be popular in our definitions includes a large number of quite unpopular apps, as we have pushed our definition of popular far out into the long tail of apps. For these observations, comScore's definition of "widely used" as 1/1,000 in a sample of 5,000 has considerable sampling variability.

For each app, the questionnaire asks whether the app utilizes advertising and a number of questions about the products and services that are advertised, as well as a number of questions about the frequency, duration, and format of the ads. We also ask a number of other questions not used in this chapter.

We see an app on either iOS or Android. We ask the research assistant to look at the developer's website to see if the app is present on the other platform.[19] While on the developer's website, we ask whether the developer has a different app on the other platform. Similarly, the website tells us whether the app is part of a "freemium" pair. Our examination of the developer websites is the closest thing to a firm-level data set on app suppliers in this industry.

---

19. Since developer websites were not always available, this part of the questionnaire data set is smaller than the rest.

There are two problems with our questionnaire sampling frame; neither is solvable. First, we do not know exactly what function of downloads we are using as a cutoff for iOS apps. Second, if advertisements are targeted, then our data on advertising is conditional on our survey respondents' behavior (i.e., students who download many apps per week) and thus may not be representative of the market. At this stage of industry development, our conversations with advertising firms suggest that targeting at the individual level is not commonly used, so this second sampling problem is less important.

8.5 App Success is Highly Concentrated

Many theoretical models of platform success simply count apps.[20] Mobile industry sources do as well, noting that there are approximately three-quarters of a million apps available for each of the two largest platforms, with significantly fewer for Windows Mobile or Blackberry (and for the tablet format of both iOS and Android). Yet it is not obvious as an empirical proposition that simply counting apps is a good way to think about the contribution that aggregate app supply makes to either the competitive success of a particular platform or to the overall value of all platforms to the economy. Some apps may be much more important in delivering value to consumers and thus much more important in creating (competitive or growth) value for the platform. In this section, we examine the cross-section distribution of app demand; in the next section we look at some simple short-run dynamics.

Mobile apps compete in a wide variety of markets: games (including many subcategories), entertainment (music, television, books, or magazines in any of a wide variety of business models), content (news, weather, mapping services, financial information), Internet services (browsers or maps), travel services (airline and hotel online resellers like Expedia), online advice (like tripadvisor), communications services (text-message managers, Skype, social networks), and so on. This heterogeneity in purpose across the hundreds of thousands of apps has several implications for our analysis. From an industry life cycle perspective, the apps world does not call for one set of experiments to find the best app technically and in terms of business model, it calls for dozens of parallel sets of experiments. Second, most apps are not in *market* competition with a large number of other apps. In contrast, all apps are in competition with all other apps for users' *attention* in the app store.

We looked in the comScore data to examine the size distribution of app demand. For each of the two larger platforms, the comScore data contain a number of size measures.[21] As we noted in the data section, there is a very

---

20. Farrell and Klemperer (2006) give an overview of these models.

21. The data we are looking at are for smartphone apps. The comScore data also have, for a somewhat shorter time series, data on iPad apps. A number of other industry sources attempt to measure more size variables than we have here, notably app revenue.

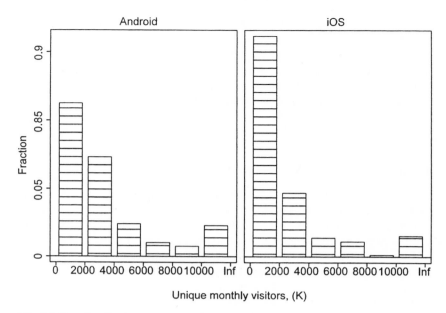

Fig. 8.2 Size distribution of apps by platform

Source: comScore, January 2013. This histogram of app usage has equally spaced bins, except for the rightmost one, which has no upper limit. The vertical axis is the log of the fraction in the bin.

substantial correlation among a number of these measures, and we are thus drawn to examine only two of them: (a) how many users have the app on their device and use it, and (b) how many minutes a typical user has the app open. We begin by looking at the first, measured by "total unique monthly visitors" for January 2013.

This investigation leads us to two facts, each about a different portion of the cross-section distribution of app demand. First, both platforms exhibit a high degree of skew in the size distribution of app demand. We noted above that apps compete in hundreds of markets. With only a couple of exceptions, the top twenty apps on each platform are not in competition with one another in those markets. We will return throughout this chapter to understanding the economic forces behind this high degree of concentration across all categories.

A similar phenomenon arises across the range of app demand high enough to be observed in comScore, that is, showing up on about 1/1,000 phones. To see this, we examine the histogram of the size distribution of apps (still measured by January 2013 "total unique monthly visitors," but in log scale) shown in figure 8.2. As you can see, the frequency of each fixed-width-bin of projected total US users drops rapidly as you make the number of users larger. What does the histogram *not* show? Stronger versions of the same

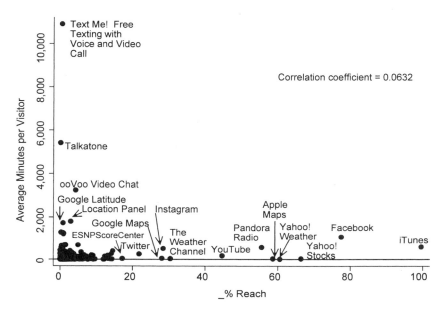

Fig. 8.3 Joint distribution of minutes and reach
Source: comScore, January 2013.

phenomenon. The rightmost cell contains the very few apps that have the most usage. To the left of the leftmost cell lie the overwhelming majority of apps (all but about 700 of the 750,000 or so on each platform) that have too little usage to appear in comScore.

We can learn a bit more about this considerable heterogeneity in app demand by looking at the joint distribution of the number of users and of the number of minutes of typical use. For this purpose, we look at iOS smartphone apps in the same month, January 2013. We use comScore's reach metric and average minutes per visitor as the two size measures in figure 8.3.

In this figure, every app ("property") is a dot. The two measures of size are not highly correlated ("reach" is very highly correlated with "unique visitors" and with "total monthly minutes," and none of these is very highly correlated with minutes per visitor in the cross section of apps).[22] Industry participants, not surprisingly, think of this as a distinction between apps, which deeply engage the user (y axis) and apps that have mass appeal to a very large number of users (x axis). It is clear that these are not the same object. Nonetheless,

---

22. This feature of the figure would stay the same if we switched to Android or to another month. One important detail would change: this figure is taken from Apple's ill-fated experiment with kicking Google Maps off of iOS. Earlier or later time periods and the Android platform would show that app much farther to the right.

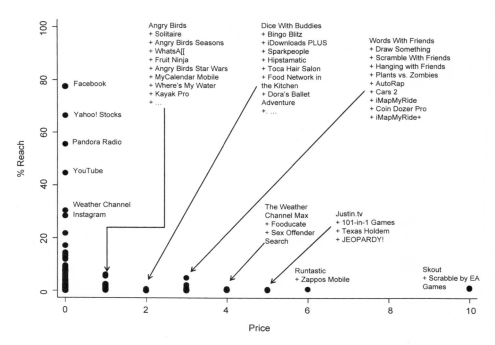

Fig. 8.4 Joint distribution of price and reach
Source: comScore, January 2013, and iTunes.

it is easy to see that there would be a very highly skewed distribution of the attractiveness of apps whether that was measured by reach, minutes, or both.

At least one source of variation in the size distribution, pricing, can be descriptively examined. In figure 8.4, we look at the joint distribution of reach and of price for the iOS platform in the same month. This is once again based on the comScore data, which we have linked to the iTunes Store in order to collect the prices.[23] As a threshold point, one can see from this figure that free apps make up a very large fraction of total app distribution. Paid apps, if they are high enough up in distribution to be labeled in the figure, are most frequently games.

Why the enormous skew in app demand? First, some apps are simply more useful, cheaper, or better programmed and marketed than other apps. Second, among equally attractive apps, some got to market earlier than others—and quite possibly, some very important apps are still growing in our snapshot figures. Finally, the most important distribution and app-discovery mechanisms

23. This matching process leads us to drop a small number of apps for which we simply cannot figure out which comScore app is which iTunes Store app. When there is both a free and a paid app in a "freemium" property in comScore, we use the price of the paid app in the figure. When comScore reports the free and paid halves of a "freemium" pair separately, they both show up in the figure.

are the online stores, and those add an element of positive feedback to app demand through the online ranking system. No reader of Sorensen (2007) will be surprised by this last point: the app stores are a "greatest hits" recommendation system, and the size distribution of apps shows a few greatest hits. We will return to this third, institutional, cause over and over throughout this chapter.

8.6 Short-Run Dynamics

If the reason for the highly skewed distribution of app demand were that there are only a few apps that have been creating most of the value for consumers, we would expect those apps to dominate over a period of time. In this section, we examine that proposition through analysis of the churn in app rankings at different time scales and for both Android and iOS apps.

To examine churn, we adopt the simple short-run dynamic measure of Waldfogel (2013), that is, "the number of apps that appear on the top fifty list seen at two dates." This is unlike the question of how many top fifty economics departments or business schools there are: at any given date there are only fifty apps on the top fifty list. However, if we look at two dates, there may have been some turnover. If, for example, (exactly) seven apps that were on the top fifty at time s are off it at time t, fifty-seven apps total will have been on the top list at the two times. The higher this number, the more churn there is.

We measure churn in the flow-of-apps-downloaded based on administrative records gathered from the iTunes Store and Google Play by the recommendation site App Annie.[24] App Annie has been collecting the daily download rankings data over time since the app stores opened. While the iTunes Store opened before Google Play, in the figures that follow we restrict attention to a more recent period in which both stores were open.

More precisely, using daily rankings from App Annie, we determine the volatility of a ranking from two dates s and t as follows: take one of the ranking methods from the online stores, "top paid," "top free," and so on. Locate the top fifty apps in both rankings, R_s and R_t. Next, we count the number of apps in the union of the two sets, D. Table 8.3 contains the average values of D for several interval lengths ($t–s$) over a long period of time and for a number of different "top" lists for each platform.[25]

To read the first number in the first row (Android top paid, $t–s = 1$) for Android paid apps, there are 51.35 apps on the Android "top fifty paid"

---

24. The distinction between stocks and flows may be smaller in this industry (at its present rate of development) than it might be in a more mature industry. The rate of growth of new users and of new replacement devices for existing users is very high, so the stock does not lag all that far behind the flow.

25. The "top paid" and "top free" lists rank downloads. The "top grossing" apps are ranked by revenue. Most top grossing apps have in-app payments (IAP), and the IAP revenues plus the original payment for the app (if any) are used to create revenues. The "top grossing" list is thus a bit closer to a stock measure. No analog of the two Android "new" lists exists for iOS. Since there are no ties in the data, the reported means all lie between fifty and one hundred.

Table 8.3	Average number of distinct apps in top fifty daily lists at different lags					
Days in between rankings	1	7	30	90	180	360
Android						
—Top paid	51.35	56.59	61.05	64.56	67.77	72.11
—Top free	51.26	56.00	63.11	68.99	74.60	78.15
—Top grossing	51.56	56.89	64.22	74.32	81.38	91.08
—Top new paid	53.59	69.28	99.69	100.00	100.00	100.00
—Top new free	53.98	72.25	99.91	100.00	100.00	100.00
iOS						
—Top paid	53.64	61.30	69.02	75.29	80.24	83.86
—Top free	55.65	68.18	79.58	85.31	88.24	91.24
—Top grossing	53.96	57.64	62.91	69.18	75.96	84.38

Source: App Annie rankings between January 1, 2012, and June 28, 2013.

Notes: This table shows, for a number of different daily top fifty rankings, the average number of apps that make up the rankings on two different days. The columns index the number of days between the rankings. The top-left figure shows that the Android top fifty paid ranking on two consecutive days contains on average 51.35 apps.

list over the average two-day period. That is, if we look at Monday and Tuesday, we would expect 1.35 apps, on average, to have fallen off the top fifty list.

The first fact visible in table 8.3 is that there is a good deal of churn in all the lists. This is perhaps what you would expect from a very new market. A second fact can be learned by looking at the churn that happens within a month (third column) versus the churn that happens within a year (last column) for all the lists except the Android "new" lists.[26] The churn is in general high frequency, with quite a bit of the turnover that happens in a year occurring within a month.

A third fact is that after either a day or a week the churn on each of the Android "new" lists is significantly higher than the overall top paid and top free app lists. This is not all that surprising: new apps are experiments, and they mostly fail. Many developers tell us that the Android top free lists are based on an eight-day history of downloads. Thus, after about a week even a developer who has been trying to "buy" downloads may well know the experiment is not working.

Fourth, if we look at either the top free or top paid list, there is more turnover of top apps on iOS than on Android. This difference is evident at t–s = 1, and continues to grow as we increase t–s. After a year, for example, just over 50 percent of the top free Android apps (78.15–50) will have departed, while over 80 percent of the top free iOS apps (91.24–50) will have departed the top lists. There is an obvious explanation for this. Developers tell us it is more important (and more effective) to "buy" downloads on the iTunes Store, because it uses a shorter history of app downloads to form the top list.

---

26. By construction, all the turnover on the Android "new" lists happens within a month.

Had we looked at a shorter or longer top list—say top ten apps or top 300—these findings would have been qualitatively, if not quantitatively, the same.

There are, of course, a number of possible explanations for these simple descriptive statistics about churn. New and better apps could be constantly introduced, displacing old (ninety days) outmoded competitors. Alternatively, new apps, not necessarily competitors of the apps that used to be on the top list, could be most of the flow of downloads while earlier hot apps sit on the stock of phones, having already completed their diffusion. These explanations are unpromising. Developers and other market participants have not proposed these explanations, and they would need *extremely* rapid dynamic competition or diffusion processes: on the iOS side, 40 to 60 percent of the top fifty at any given date are out of the top fifty a month later.

The churn figures are also consistent with another type of very powerful dynamic competition, competition across all apps, whether substitutes in functionality or not, for consumer attention on the app stores. Developers seeking to "buy" a position on the top lists for their apps by advertising or "incentivized" downloads are competing with one another for a limited resource: the top of the top lists and the attention it brings from potential users.

8.7 "Top List" Implications for Market Development

We have established that the collaborative filter of the online apps stores is an important barrier facing apps. Many entrepreneurs entered this business hoping to take advantage of the technical opportunity to attract consumers and then sell advertisements to firms who were hoping to acquire those consumers as customers. Difficult search and matching makes the customer acquisition part of commercialization difficult for an entrepreneur who has written an attractive app. If consumers who would value the app cannot find it, demand will be small. This gives app developers an incentive to build volume quickly. The high marketing costs have led to a market response, with a host of services that have arisen to try and solve commercialization and monetization problems for apps struggling to become visible in the clutter of apps.

This can lead to the irony, (now) frequently noted by industry observers, of an app that plans to be supported by advertising in the future but spends on mobile advertising today to get more downloads despite zero revenue. More generally, any app without an external body of customers has some incentive to keep up its rate of downloads at all times, even at those times when that might be difficult, such as right after introduction, just before a significant upgrade, and so on. App developers are demanders of advertising space in apps. One way for app developers who have a budget to advertise is to buy space through an ad broker. There are also ad-exchange clubs, in which app developers agree to show one another's ads for apps. Finally, multiproduct app developers can run an ad for their own products, either another app, an in-app purchase of a virtual good, or some other kind of product. Regardless of the channel, developers have allocated resources to purchase advertising to promote apps instead of attracting marketing dollars from products outside of the industry.

Above and beyond advertising to build volume, app developers can pay to have users download their app. These are called "incentivized" downloads in the industry, and a thriving and changing business has grown up to supply them. Firms such as Tapjoy have flourished in this space. Suppose app firm A has a successful app with a virtual currency used for in-app purchasing (IAP). App firm B wants downloads. Then B pays Tapjoy to pay A to offer virtual currency to A's customers in exchange for downloading B's app. Industry sources tell us that the strategy of buying users tends to be more prevalent in apps that monetize through the sale of virtual goods, subscriptions, or apps that are an arm of an existing consumer-oriented commerce firm. It appears that the strategy of buying users is closely related to those apps that have an anticipated high average revenue per user.

Like advertising focused only on mass downloads, "incentivized" downloads may not lead to users who are good customers (i.e., who use the app and monetize well). This is a recurring complaint among industry sources we interviewed. This conundrum has created further innovations, including firms that incentivize users to watch movie-style trailers of apps, letting the user decide if she should download the app, and firms that incentivize but give the user a choice of apps to download. We anticipate that ad-tech firms will continue to innovate new ways to "buy" users in the coming years, potentially discovering new solutions to incentivize users with a high likelihood of using apps and spending money inside apps.

The high marketing costs of using the app stores fall more severely on entrepreneurial app developers than on existing firms with consumer connections. The costs of inducing many consumers to download an app are much smaller for those app developers who already have a marketing connection to their customers, typically because those are already *online* or *product* or *service market* customers. If this is correct, it suggests a market equilibrium shift away from "disruptive" entrepreneurs to existing firms.

8.7.1 Evidence of Buying Users

We can examine some of this behavior of "buying users" by looking at the identity of advertisers. To do this, we return to our app questionnaire data. We asked students to download apps and answer questions about both the location and the content of the ads.[27] We count all the different ads that show up in one use of the app—defined as, at a minimum, passing all the

---

27. To the extent that there is targeting, the resulting sample of ads is composed of either mass-market ads trying to hit more or less all users, ads targeted to people who use their phones like students, or ads targeted to people who download many apps (since our students download a significant number of apps to fill in our survey).

Table 8.4   **Banner ad content**

	Android	iOS
Apps with banner ads	$N = 1,106$	$N = 1,583$
Any app ad	.802	.847
Any non-app ad	.370	.371

Source: App questionnaire as of January 2013.

Note: Column totals sum to more than one because an observation is an app and the measures are for *any* advertisement of a given type, so that multi-ad apps may show up in both rows.

Table 8.5   **Percent of apps with banner ads in each category**

	App ad	Non-app ad
Same firm	0.348	0.091
Different firm	0.623	0.295

Source: App questionnaire as of January 2013.

Note: Column totals sum to more than one because an observation is an app and the measures are for any advertisement of a given type, so that multi-ad apps may show up in both rows. Figures are weighted average over platforms.

places where ads might be shown. Thus the probabilities sum to (considerably) more than one. The dominant location is banner ads, so we report the results only for those. This exercise treats paid and unpaid (same firm, or bartered through an ad exchange) the same.

A surprising finding from table 8.4 is that if an app has a banner ad, there is an 80 percent chance or greater that it will serve an ad for an app! In table 8.5, we see that there is quite a bit of self-advertising in this industry. Over a third of the apps with banner ads advertise their own apps. The proportion of apps with banner ads that advertise other firms' products outside of the mobile app industry is quite low at under 30 percent. The demand for ad space in apps is demand for space to advertise other mobile apps.[28]

In table 8.6, we report four categories of ads, based on the kind of product advertised and the identity of the advertiser. For this purpose we have only two kinds of products, mobile apps and all others, and only two identities for the advertiser, the owner of the app in which the ad is running and all others. We recorded data for more categories than this, but once the table is viewed it will be obvious why this is what we report.

This is a complex enough table that it is useful to walk through the numbers. Table 8.6 reports the products and advertisers seen in Android and iOS

28. When we first talked with app-oriented venture capitalists about this fact in late summer 2012, many thought it was simply false. Today (summer 2013) savvy VC accept the necessity of buying downloads.

Table 8.6   **Advertised products and vertical integration in advertising**

	Android banner ad apps $n = 1,106$		iOS banner ad apps $n = 1,153$	
	App ad	Non-app ad	App ad	Non-app ad
Same firm	0.102	0.0171	0.486	0.133
Different firm	0.727	0.357	0.565	0.26

Source: App questionnaire as of January 2013.

platform apps that displayed at least one banner ad. Within the Android banner apps, about 10 percent (.102) display a banner ad for another app from the same firm, while just about 2 percent (.017) display a banner ad for another product from the same firm that is not an app. An example of the latter would be a media company advertising a television show in its app, a common ESPN behavior, for instance. Continuing with table 8.6, over 70 percent of the Android apps (.727) that have a banner ad have a banner ad for an app from a different firm. Finally, just a bit over a third of these apps displaying banner ads on Android (.357) display an ad for a non-app product or service from a different firm.

In contrast, we find that for iOS, about 50 percent (.486) display a banner ad for another app from the same firm, while over 10 percent (.133) display a banner ad for another product from the same firm that is not an app. Only 57 percent of the iOS apps (.565) that have a banner ad have a banner ad for an app from a different firm. Finally, just a bit over a quarter of these apps displaying banner ads on iOS (.26) display an ad for a non-app product or service from a different firm.

Self-advertising is much more important on iOS because (a) a developer can more easily influence an apps placement in the rankings on iTunes relative to Google Play since the rankings only use the last twenty-four hours rather than the last week of downloads, and (b) since iTunes is the only distribution channel for iOS apps, influencing iTunes rankings is relatively more important to an app's success than Google Play rankings (which marginally competes with alternative Android app markets). Apps with banner ads advertising a different firm's apps are less likely on iOS since those apps might represent direct competition in the rankings. Note that since the percentages do not have to add up to one, there is no mechanical reason why self-promoting app ads would be higher on iOS and competitor-promoting app ads would be lower on iOS.

These facts come from an early stage in the development of the industry, and they also come from our students' phones, that is, not necessarily from the most valuable advertising audience. Given that potential oddity of the sampling frame, we think that there is a strong conclusion and a weak conclusion. The strong conclusion is robust to our sampling frame: app developers today have a powerful motivation to buy downloads in order to become visible on the online app lists.

This conclusion is reinforced by the different behavior of corporate app developers (who have that motivation much less) and entrepreneurial app developers (who have it much more) since these two groups of developers have very different exposure to the costs associated with the collaborative filter of the online app store. See figure 8.14.

Our weaker conclusion is that exports from the entire sector of mobile app advertising to the rest of the economy are growing slowly. Here our sampling frame may matter, to some degree. We want to point out that robust revenue numbers for app advertising does not rebut this finding. Those numbers come from summing the revenues across app developers,[29] without netting out the within-sector sales.

The rate and direction of application innovation is being affected by the need for developers to devote resources to solving the matching problem and getting noticed out of clutter, rather than devoting resources to monetization efforts based on creating value for customers and rather than trying to gain money from marketing products outside of apps.

8.8 The Economic Return to the Development of New Apps

In this section, we take up the factual question of how developers seek to earn an economic return on their development effort. Our primary concern is understanding the formation of new markets, a key step in the creation of value out of a new platform industry. As an incidental payoff, we will be able to address some management-normative questions about the "monetization" of apps.

While mobile apps are a general purpose technology and thus might have a very wide range of uses, three main ideas about the way they might become valuable informed much early app development:

1. Many apps might be part of a cluster of entertainment services consumed on mobile devices. Games played on the mobile device are an obvious example.

2. Many apps might remove life's annoyances from users' lives. Maps are an obvious example.

3. Apps provide a dramatic new advertising medium, with the ability to condition advertising on a user's location as well as on many other targeting data.

Within all of these three broad categories, early app development was heavily influenced by the idea that app developers would be *entrepreneurs* and that

---

29. "Revenues" in the industry sources are themselves not the strongest numbers in the economy.

they would seek to *disrupt* existing industries.[30] The scope of the anticipated disruption was dramatic. Entertainment, advertising, and media markets would be disrupted. So, too, would industries that could take advantage of a very localized advertising and information service; in retail trade, for example, attracting consumers into the store might depend less on the brand capital of a chain store or on the locational attractiveness of a mall, and more on sending an ad to someone walking down the sidewalk. In parallel, industries characterized by transaction costs of aligning buyers and sellers—say taxicabs—might be disrupted by new services to match buyers and sellers over mobile devices.

There are at least two reasons to anticipate some disruption. First, earlier advertising and communications technologies have changed industry structure in advertising-intensive industries and in communications-intensive industries.[31] Second, the creation of the mobile app development platforms has certainly disrupted mobile telephony, as influence over customers and over the direction of industry standard setting has shifted from the mobile carriers to Google and Apple. One can only hope for more disruption as mobile devices compete with established forms of mass market computing such as PCs. It is worth noting, however, that essentially none of this earlier disruption came from entrepreneurs.

The broadly anticipated categories of app value creation also suggested some variety in how apps would earn their economic return. Early conjectures focused on two main ideas: consumers might pay to be entertained or to avoid the hassle of transaction costs, or advertisers might pay to gain consumers' attention.

Given the heterogeneity in apps, it is not surprising that there is a good deal of heterogeneity in the way the developer seeks to earn an economic return if the app is successful. It should be noted that although some monetization strategies like freemium have been featured prominently in the press, many entrepreneurs are uncertain about which methods are most effective for their apps. For example, one iPhone game developer echoed the sentiment of many entrepreneurs we spoke with:

> In all honesty, we still haven't figured out whether ads, freemium, discounts, or in-app purchases are the best way to make money off our loyal customer base, so we've committed to a period of experimentation with these different methods. It may come down to how these strategies work in combination with our different games. Maybe we can have a menu of payment options, but I suspect we'll have to commit to one or two because our customers may revolt if it gets too complicated.

As may be apparent, this is already significantly broader than that suggested by the "who pays" dichotomy between advertising-supported and

---

30. See, for example, Christensen, Johnson, and Horn (2010).
31. See, for example, Bresnahan (1992) and Sutton (1991) for the impact of mass communications on industry structure.

paid-by-the user apps. We now review a number of "monetization" strategies taken up by developers. Once we have discussed the main forms of app monetization, we will present some simple statistics on their prevalence.

8.8.1 Charging the User

Some apps, so-called "paid" apps, are bought by the user at the time of download. These apps pick up on the themes of providing valuable entertainment or removing an annoying hassle from users' lives. For a number of reasons, however, straight-up paid apps are not the most common revenue-gaining strategy, even when the user ultimately pays.

Two different strategies for charging the user have a try-before-you buy structure. "Freemium" apps have two versions. The "lite" version is free but limited in some way. A "pro" version is available at a positive price and is typically recommended with the "lite" version and sometimes even purchased within it.

Similarly, in-app purchasing (IAP) lets users buy complements to the app as they are using it; in a war game, IAP could be used to buy a bigger sword or more lives for the player's avatar, for example. The IAP and freemium models are not mutually exclusive. Many "pro" versions also entice in-app purchases. The widest use of IAP is in games—if the game is diverting, or even addicting, the user pays.[32]

The use of try-before-you-buy strategies in a new product area is not surprising. It is even less surprising when you consider the weak institutions of the mobile platforms for matching buyer and seller; buyers can be much more easily enticed if they do not have to pay up front. Furthermore, apps appear to be experience goods, as evidenced by the large number of apps in our comScore data that get downloaded but are not used.

Other apps use a subscription model, a recurring monthly or yearly IAP that could be managed by the developer's payment system or the app stores itself. It is common for media-related firms to offer apps such as Pandora, Hulu, and Netflix on a subscription model that include complete passes to unlimited media consumption. Matchmaking, personal finance, antivirus, and even navigation apps have also called for users to subscribe.[33]

---

32. It is not uncommon for apps that monetize through IAP to have dual-currency types in an effort to spur greater levels of monetization. For example, one form of currency may be easy to obtain through continued gameplay while another form of currency may be more difficult. Paying the developer through an IAP makes acquiring the scarce currency much easier. Further investigation could help us better understand the in-game currency dynamics, especially as they relate to dual-currency games.

33. App developers may be able to skirt the rules of transferring a portion of their app revenue to the app marketplace if they ask for the payment of the subscription dues outside the app and use a preestablished username/password for authentication. The carriers have even sponsored some subscription apps that allow users to pay through mobile phone bills. However, this practice is much less prevalent in the United States than in other countries, including Japan, that generally have higher rates of app monetization.

8.8.2 Advertising

There are also a number of ways that apps can be advertising supported. Both Google and Apple offer "ad brokerage" services to app developers. The developer creates a space in which an advertisement can run—before the app loads, across the top of the screen, between stages of use, and so forth—and the broker sells that space to potential advertisers. The app developer can be paid either like traditional media firms, that is, by the number of users who are shown the ad or like traditional online firms, by the number of users who immediately take action in response to seeing the ad. In addition to the in-house advertising brokerage services, there are a number of independent ad brokerage networks.

The wide variety of organizational structures for advertising is accompanied by a wide variety of ways to display advertising. A partial list would include text ads, banner ads, "click-to" ads, expanding banner ads, video ads, rotating banner ads, and interstitial ads (ads that are displayed full-screen during a pause in use of the app, such as between rounds of a game). Interstitial ads are like TV or radio ads, taking over the entire user interface for a brief time, while the other forms of ads are like newspaper or magazine ads, appearing over or under or beside or in the middle of the still-visible app.[34]

Additionally, as mobile phone screens are limited in size, erroneous clicks become more prevalent. This gives developers the opportunity to undertake less-than-consumer-friendly strategies; some have placed ads near the portion of the screen where the user might touch in the course of playing a game. This leads game players to erroneously click on ads, boosting the click-through rate and making the ad more lucrative. Other developers go down a very different path, allowing one full use of the app before showing ads—an indirect version of try-before-you-buy.

Finally, ads may be more or less targeted to the user. In principal, targeting could be based on a large number of different actions on attributes of the user including location, past purchases, the content of communications, and so forth. Some observers believe that targeting will lead to a dramatic reduction in trade frictions, as users are presented with valuable product offers at just the right time, place, and so forth to create gains from trade. Other observers believe that targeting is a large scale loss of privacy for consumers and represents an important loss of consumer sovereignty. Both sides are largely forecasting the future, not describing the present. The scale, scope, and form of targeted mobile advertising are changing; targeting today is

---

34. As innovation by ad-tech companies continues, regular interstitials evolved to offer brands an immersive near-virtual reality experience for consumers to explore new products. Additionally, simple banners evolved into pop-up push notifications appearing in the status bar of a phone, which some in the industry have termed "malware." Ad options vary by particular platform. Google allows push notification advertising, while Apple's tight control over the mobile phone software development process leads them to shun this advertising practice.

typically quite primitive. While some advertising is integrated with the users' past behavior, location, and the like, many other ads are much closer to broadcast than targeted. A number of complementary inventions are going to be needed before there is much effective targeting in mobile advertising. App developers are going to have to define and create audiences, the way traditional media do today.[35] Ad networks, including those that are part of the large platform firms, are going to need to work out what to measure about users, how to sell that knowledge, and how to divide the rents associated with targeted advertising with developers. Users will need to decide whether to respond to offers they get from mobile advertisers. The mobile advertising world is today not as sophisticated as, say, the online advertising and product-recommendation world. All of this inventive activity is today (summer 2013, the reader may be at a much later point in the development) closer to creating a new market than fine-tuning a targeting formula.

We should not be surprised at this slow development in light of the historical experience with inventing advertising-supported industries. Television broadcasting and receiving were tremendous technical inventions. The economic return to those inventions was dramatically increased by, for example, changes in the nature of certain existing content (e.g., the NFL), the creation of effective means of delivering that content to attract a specific audience (e.g., CBS sports for adult men), and the creation of advertisers with very large budgets to reach those specific audiences (e.g., Budweiser, a 50 percent share brand that emerged out of a very fragmented US beer industry). For mobile advertising, those coordinated commercial inventions (as opposed to technical inventions) still lie largely in the future.

8.8.3 No (Current) Revenue Stream—Corporate Apps

Finally, there is a very important class of app that is not paid, not subscription, has no IAP, no freemium, and shows no ads—in short, which has no current revenue stream. Many of these are "corporate" apps, which offer a product or service that is complementary to paid products from a consumer-oriented firm. A banking app, for example, lets a consumer check balances or take a picture of a check for deposit. An airline app similarly lets a traveler display an electronic boarding pass or check the seat map for his or her flight.

These corporate apps are large and growing as a portion of all app downloads. By the time this chapter is published, this class of app will likely be widely recognized. But after we first discovered it, leading industry figures told us for many months that we were simply mistaken (they have now stopped). These corporate apps represent a change in the mechanism by

---

35. A number of brokers and advertisers have told us that making the target audience part of app design (to the extent an advertiser cares about the message recipient) lies largely in the future.

which the mobile app economy delivers value to the broader economy. The original idea was that apps would offer advertising services to bring in new customers to consumer product and services companies. The corporate apps are typically given by the corporation to existing customers, and offer a wide variety of forms of improved customer service. That, of course, may raise demand, but not (directly) through bringing in new customers.

Monetization of the corporate app is most direct for mobile commerce retailers. Their apps make purchasing opportunities more accessible as impulsive purchasing messages surround consumption-driven mobile users. A push notification from a daily deal site could easily induce a consumer riding the subway to make an impulse purchase in a physical environment where monetary transfer may otherwise be limited to giving a beggar spare change. Even without push notifications, apps allow firms to sell tangible goods from locations where traditional e-commerce would never have been possible; eBay has been particularly effective in letting both buyers and sellers access from anywhere. Banks have picked up on this anytime, anywhere theme strongly in their messages to their customers. Without quite saying that banking is an annoying chore, they have suggested one can complete it in times or places where consumers' value of time might otherwise be low.

It is, of course, entirely possible that the strong turn to corporate apps is temporary; indeed, we argue in this chapter that it is driven, to a considerable extent, by the ineffectiveness of the online app stores (corporate apps go to existing customers and escape any problems with app discovery in the app stores). For now we note, however, that the race between *disruption of everything* and incremental improvements in quite a few things is not being won by disruption.

8.8.4 Other (Currently) Zero-Revenue Apps

Seller monetization strategies may be dynamic. An app developer may therefore choose to have no revenue in the present in anticipation of having revenue in the future. We have talked with many industry participants about this, and four main themes arise. Our interpretation of these themes is that the value of delaying revenue applies particularly to entrepreneurial developers, and within entrepreneurial developers with particular characteristics. Of course, the ability to finance a delay in revenue is also particularly challenging for entrepreneurs who have not been able to secure venture capital or angel financing.

As a threshold point, in some circumstances an app may be building up a large volume of users with a plan to somehow monetize in the future. Industry participants emphasized a short list of variants on this theme.

First, an app that is going to be monetized may first want to reach efficient scale. The structure of the online stores means that minimum efficient scale is measured, at least in part, against the rate of downloads. Thus a firm that seeks to have a paid version in the future may have only a free one in the present to avoid sliding up the demand curve. Similarly, the supplier of advertising-supported apps may seek to avoid annoying users in early stages, that is, show them few or no ads while the app is gaining volume.

Second, some app developer firms may be looking forward to a merger or an initial public offering (IPO) and thinking about their (equity) market valuation at that point. Many, many developers and their financial advisers believe that the equity markets have valuation models that depend on user headcount.

Third, at this very early stage of the industry, the ultimate point of writing an app may not be to have a successful app. For example, the point of the app may be to advertise the development capabilities of the team. Members may hope to be hired into established businesses. The entire app development team may seek to be hired as a group to convert its app into one that creates value post-takeover.

Fourth, some entrepreneurial developers simply have no commercial interest at all. We have met a substantial subset of developers with no interest in profit whatsoever. There are a number of tech savvy developers who build apps as "art" or tools to make their own lives easier. Their app solved a problem for them, and they are quite pleased that it solves a similar problem for others. These developers may be anticapitalist or simply too well paid by their day job developing software to care about the app profits. They are primarily motivated by the validation of their users using the app they built, the experience of building an app, or the simple utility (including oddity) of the app itself. Our industry sources tell us to expect such "hobbyists" more (today) on Android than iOS.

8.8.5 Some Statistics

In this section, we quickly examine some statistics about app developer monetization strategies. We first look at figure 8.5 and figure 8.6, which come from an industry source.

Note that the category of corporate apps is not present, as the Developer Economics data source follows current industry practice of focusing on "monetization" as a topic in management-normative analysis. The survey shows that there is considerable heterogeneity in the monetization strategies used by developers. While advertising is the most popular, there is also considerable use by developers of paid apps, in-app purchases, and/or freemium. Subscription payments used significantly less.

In contrast, when we look at the revenue-per-app figures, subscription apps are the highest, with in-app payments second. Advertising-supported apps have the lowest revenue. The management-normative conclusion that developers should shift out of advertising-supported apps and into subscription apps is likely not warranted. Subscription models, as we pointed out above, are only suitable for certain kinds of apps, most obviously those that distribute content such as music.

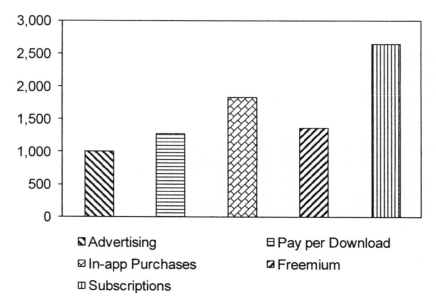

Fig. 8.5 Revenue per app by monetization strategy
Source: Developer Economics survey.

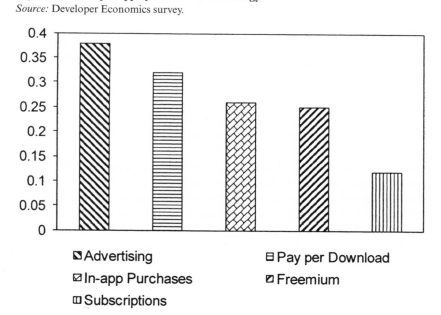

Fig. 8.6 Frequency of use of monetization methods
Source: Developer Economics survey.

Table 8.7	Frequency of use of IAP and advertising in free apps (wide sample)			
	(1)	(2)	(3)	(4)
Ads used?	No	No	Yes	Yes
IAP used?	No	Yes	No	Yes
	Neither	**IAP only**	**Ads only**	**Both**
Android	.394	.0772	.449	.0798
iOS	.221	.108	.394	.277

Source: App questionnaire (July 2013) sample is all apps. N: Android 2,281, iOS 2,713.

On the other hand, these figures do suggest that the question of who pays for mobile apps—users or advertisers—is not yet settled. Since there is some overlap between categories (e.g., an app can have both advertising and IAP), it is a slightly odd calculation to look at the share in total revenue represented by a category. Still, it appears that the total revenue across apps that have advertising is smaller than the total revenue of apps that use any other method. Advertising-supported apps are not taking over, at least not in a revenue sense. Given that a good portion of the advertising revenue of the typical app comes from ads for other apps, the overall net advertising revenue of the app industry must be even smaller.

We can get another estimate of the relative size of the different categories of revenue strategies, and include the important "corporate" form, by examining our app questionnaire data. In table 8.7, we show revenue strategies for the most popular free apps on the Android platform, and for a sample of iOS apps that are of comparable popularity. The table focuses on the joint distribution of any use of advertising and of any use of IAP in free apps. This yields four categories: (a) neither advertising nor in-app purchases; (b) no advertising, but in-app purchases; (c) some advertising, but no in-app purchases; and (d) some advertising and in-app purchases.

On both platforms, the most widely used monetization strategy for free apps is advertising without IAP, which was used in 2012 for 44.9 percent of free Android apps and 39.4 percent of free iOS apps. The most visible difference between the platforms in this table is that in-app purchases are considerably more widely used on iOS compared to Android, especially in conjunction with advertising (27.7 percent on iOS compared to 7.98 percent on Android). The higher use of IAP on iOS is consistent with industry surveys and likely stems from the tendency of iOS users to be richer than Android users.

This table also reveals that more than a third (39.4 percent) of free Android apps neither use advertising nor in-app purchases to users, and similarly for iOS, about a fifth (22.1 percent.) This highlights the significance of the class of apps that do not yet pursue a stand-alone commer-

Table 8.8   **Frequency of use of IAP and advertising in free apps (narrow sample)**

	(1)	(2)	(3)	(4)
Ads used?	No	No	Yes	Yes
IAP used?	No	Yes	No	Yes
	Neither	**IAP only**	**Ads only**	**Both**
Android and iOS	.258	.135	.357	.251

Source: App questionnaire (July 2013) sample is apps in comScore.

cialization strategy that is tied to the app economy.[36] These are mostly corporate apps.

As a robustness check, table 8.8 presents the same set of frequencies as table 8.7, but for a subsample of the app questionnaire data. Recall that the app questionnaire sampling frame is comprised of the union of two criteria: either an app is identified in the comScore data, or it satisfies the criteria for being a top app on iTunes or Google Play. The subset in table 8.8 are the apps that satisfy the intersection of the two criteria: both these apps are top apps and appear in the comScore data. Furthermore, the subset is limited to those apps in this intersection that actually had been processed by our research assistants as of January 2013. The result was 431 free apps over both platforms. Note that despite this more limited sampling frame, we still get the same results as we do in table 8.7.

We have also checked the "neither" column against a measure of firm type. Overwhelmingly, these apps are associated with a firm that, in addition to having the mobile app, is a nonmobile consumer products and services (preexisting, "corporate") firm. At this stage this is a limited subsample, since we are in process of defining firm types after linking each app to its sponsoring firm.

The emergence of a large category of "corporate" apps was a surprise. Most industry figures expected a primarily entrepreneurial form of app supplier. Further, the most common articulation of the broad value proposition of mobile was that an app developer's knowledge of where the consumer was located would permit selling location-specific advertisements. The mobile opportunity, in this advertising-centric view, was a more granular version of zip codes. However, mobile devices and services are a general purpose technology. Consumer product and services corporations have adopted this GPT and used it to provide customer service. This is an important, unanticipated growth category and represents a potentially large-scale increase in the economic value of the mobile opportunity.

36. It is possible that the lower incidence of these apps on the iOS side is an artifact of our sampling frame for apps, which, because of the iTunes Store's higher-frequency collaborative filter, may have picked up more apps that were "buying" distribution. These apps would tend, systematically, *not* to be corporate apps. We are investigating this using an alternative sample based on comScore.

Table 8.9   **Asymmetries between platforms**

iOS and iTunes Store	Android and Google
Early: more devices in use	Now: caught up in total devices in use (mostly phones)
More tablets	Tablets rapidly growing
More commercial infrastructure including payment processing tools	Absent, but catching up
Richer users	Less rich users: may not buy IAP, for example
More restrictions on developers	
Development environment on Macintosh	Develop anywhere
Apple dictates tools to be used (e.g., flash)	Use Java, popular with developers
Limited range of devices	Fragmentation
"Managed" change from year to year: Porting an app to the newest iPhone/ iPad devices from older ones usually simple	Changes in environment from year to year (e.g., substantial UI changes) Different hardware manufacturers use different OS versions
Distribution restricted to iTunes Store	Open; multiple distribution channels

Source: Discussion with industry participants.

8.9 Developer Behavior: Platform Choice and Multihoming

In this section, we examine the platform choices of smartphone app developers on iTunes and Google Play. We take up three topics. First, we examine the relative attractiveness of the two main platforms. While much industry discussion suggests "iOS-first" as a strategy, developer platform choice behavior is close to evenly balanced between iOS and Android at this time. Since the installed base of iOS and Android phones is approximately equal, this is consistent with the theory that the most important determinant of developer behavior is the installed base. Second, we examine developer decisions to multihome, that is, write for both platforms. Here the basic facts are more complex. If we look at the decision to *write* the app for both iOS and Android, multihoming is quite common. However, if we look at *marketing* the app to a mass market, multihoming is rarer, about one-third as common. Our interpretation of this dual result is consistent with the theme of this chapter: at this stage of industry development the technical entry barriers to either platform are low for developers, but marketing barriers remain high because of the problem of matching app demand and supply. Third, we consider the possibility of future changes in the platform market.

8.9.1 Relative Attractiveness of the Two Platforms

There is a lively debate in the industry about how the two main platforms differ and about ideal strategies for a developer in choosing among them. The debate emphasizes the asymmetries between the two platforms, summarized in table 8.9.

The industry discussion summarized in this table does reflect elements of the usual economic theories, that is, the size of the installed base of devices on each of the competing platforms. At present, an app developer can reach approximately half of all US smartphone users by writing only for iOS and distributing through the iTunes Store, and approximately half of all mobile users by writing only for Android and distributing through the Google Play store, Amazon, and so on. A much smaller installed base of users can be reached by writing for some other platforms, such as Windows mobile, Blackberry, and so on. Thus, one attractive theory is that developers ought to be approximately indifferent between the two platforms.

The dynamics of platform competition might lead to different expectations by developers of future installed base sizes. Apple's iOS started with a large lead, but Apple left an opening by not having iOS devices universally available. Android phones then grew more rapidly. Today, there is some discussion of possible future Android growth, but hardly a wide consensus on the part of the developers that either platform will be important in the future. On the other hand, developers do appear not to expect that a third platform will catch up to today's top two, though both Microsoft and Blackberry have proposed one.

There is also a great deal of discussion of the per-customer profitability for a developer on each platform. A broad consensus in the industry appears to favor iOS over Android here, primarily because iPhone users spend more through their mobile devices. It is thus less than obvious whether the initial lead for iOS plus the richer potential customers or the open systems approach of Android will have the advantage.

Examining developer platform preference by examining developer platform choices is, in general, a difficult problem.[37] We undertake a simple version based on the comScore sampling frame. In comScore, an app appears if 1/1,000 of panelists use it. The iPhone panel and Android phone panel are the same size. One simple approach to examining developer choice is to count the number of apps that appear in the iPhone and Android phone panels. This is shown in figure 8.7, which shows a very interesting trend. In September 2012, at the beginning of the time period shown, which is also likely the earliest date at which the comScore data can be used for this purpose, there appears to be more developer preference for iPhone apps than for Android apps. By January 2013, the most recent data that we have in cleaned form, this preference appears to have been reversed. This pattern is, of course, consistent with the idea that Android is catching up or even overtaking iOS in attractiveness with app developers, as it clearly is in attractiveness to users. The small differences between the two platforms are also consistent with the theoretical prediction that, with approximately equally

---

37. Rysman (2009) surveys the literature very well.

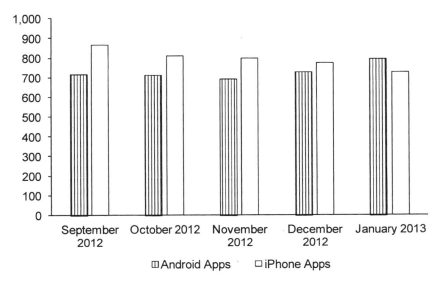

Fig. 8.7 Number of apps by platform by month
Source: comScore. Data are a subset of those in table 8.1.

many users, the two leading smartphone platforms should be approximately equally attractive to developers.

As of this writing, it appears that Apple's original lead has disappeared. Whether the mild trend to Google visible in the total number of apps (figure 8.1) or in the number of apps with reach over .001 (figure 8.7) is the start of a platform shift depends on the attractiveness of the trending apps, a topic for the future.

8.9.2 Multihoming

Multihoming—supplying the same app for more than one platform—is another aspect of developer supply behavior with important implications in the economics of platform markets.[38] With two platforms of approximately equal installed base, multihoming is also important for commercialization. Any app developer attempting to supply a mass market must multihome to both iOS and Android (Blackberry and Windows mobile do not contribute much installed base at this juncture). It is clear that whether, when, and how to multihome are critical decisions facing new ventures and established firms producing mobile applications. For example, one developer of productivity applications summarized the trade-offs facing his firm during multihoming:

38. There is a large literature about competition for developers and completion for users by platform firms. One behavior emphasized in this literature is multihoming. See, for example, Armstrong and Wright (2007).

Like many companies, we started on the iPhone, but with the explosion of the Android market, we are seriously considering a move. We don't have the infinite resources of a Zynga or Rovio, so the real decision is whether to make a new iPhone app for our existing customer base, or get in on the ground floor with Google Play by porting our existing app to Android. The problem is: there is a lot more uncertainty about whether Android will ever make money, but it's possible that iTunes is tapped out for us.

Analysis of multihoming also calls for matching apps across the two platforms and across data sources. Matching apps across the two platforms is significantly more difficult than one might at first expect. The same app can have different names in the iTunes Store and Google Play, and different apps can have the same name. The supplier has only one name on Google Play, and that can be the same as any or none of the three supplier names on the iTunes store.[39] For the limited set of apps it covers, comScore has linked across platforms, but unfortunately comScore does not use the same firm and product names as do developers. Our best results have come from looking at developer websites one by one and capturing all the apps—and their unique identifiers on each platform—listed by the developer.[40] Since this website is found through our app questionnaire, this hand-match lets us employ our app questionnaire and the comScore data to examine the current state of multihoming.

Finally, we define two different definitions of multihoming and use two different data sets to examine them. One definition of multihoming is "written for both platforms." This technical definition of multihoming is observed in our developer website survey. Another definition of multihoming is "marketed successfully on both platforms." We introduce this second definition because we suspect that the fixed costs of marketing a mass-market app may, along with the fixed costs of writing it, limit entry. This technical + marketing definition of multihoming is observed in the comScore data. Finally, we should point out that all our data sets condition on app success; in what follows, we will be looking at multihoming among apps that have had considerable success on at least one platform.

We begin with the January 2013 comScore sample.[41] In that month, there were 1,231 apps in the comScore data set. In figure 8.8, we show the distri-

39. Computer scientists have built sophisticated name-matching software for this and other purposes (and enthusiastically recommend it to us regularly). The best software solutions for matching in this context correctly identify just under half of true matches—that is, have more false negatives than true positives—and also identify a significant number of false positives, about one-tenth of the number of true positives.
40. When we are working with only the comScore data, we can use comScore's matching. But comScore's names for apps differ from both the iTunes Store and Google Play. Accordingly, we have also handmatched comScore "apps" to unique identifier app ids. We very much appreciate the help comScore has given us in resolving the last few difficult cases.
41. The comScore mobile product is new and had an initial period of rapid improvements in coverage; January 2013 is after the comScore mobile sample and definitions settled down.

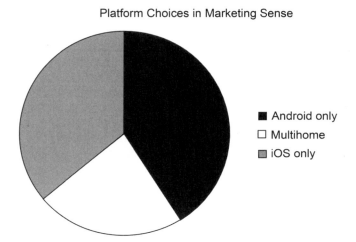

Fig. 8.8 Multihoming, iOS exclusive, and Android exclusive apps in comScore
Source: comScore, January 2013. "Multihoming" here means appearing in both comScore panels.

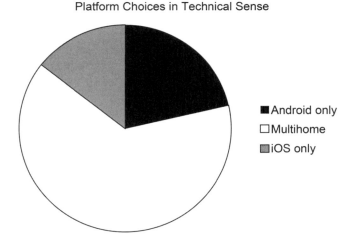

Fig. 8.9 Multihoming, iOS exclusive, and Android exclusive apps in comScore
Source: App questionnaires (platform choice based on listing of app on developer website).

bution of apps among iOS only, Android only, and both categories. As you can see, there is a low rate of multihoming.

We can take a broader view by drawing on our questionnaires. Figure 8.9 is based on the same sample of 1,231 apps as figure 8.8. Here, however, we define multihoming based on the developer-website information. An app is multihomed if the developer has links to download it both from the iTunes Store and Google Play. A single-homed app is one that appears on comScore on one platform, but the developer does not list a link to the other platform on its website. As you can see, the rate of multihoming on this "technical" definition is dramatically higher.

The big difference here is between the tendency to *supply* an app for both platforms, which yields a rate of 64 percent multihoming (figure 8.9) in our survey of developer websites, and the tendency to *widely market* the app on both platforms, which yields a 23 percent multihoming rate in the comScore data (figure 8.8). We interpret this large difference as a clue that the barriers to porting to a second platform are not merely the technical costs of rewriting the app, but also the marketing costs of finding an audience.[42]

8.9.3 Weighted Multihoming Rates

We noted previously that the size distribution of app demand was highly skewed. The larger the demand for an app, the more likely it is to be profitable for the developer to bear the incremental fixed costs of multihoming, that is, the technical costs of porting to the second platform and the marketing costs of establishing a connection to consumers on the second platform.[43] That would imply a higher rate of multihoming for more popular apps, a hypothesis that can be examined by looking at different *weighted* propensities to multihome. We examine this hypotheses once again in the January 2013 comScore data.

To associate weights with each app, we now turn to a frame in which each app has a "base" platform and the definition of multihoming is that the app is also observed on the other platform. In figure 8.10 the base platform is iOS, so "multihoming" means availability for Android. Figure 8.11 is symmetric; the base platform is Android and multihoming means we observe the app on iOS.

In both figure 8.10 and figure 8.11 we display the rate of multihoming under five different weighting schemes.[44] We first ask which weighting schemes make a difference, that is, which weights are correlated with the firm's decision to multihome? For this purpose, the fifth column, "unweighted," is the base case. On both base platforms, we see the same pattern. Column (2), weighting by "minutes per visitor," predicts about the same rate of multi-

---

42. Of course, a more careful analysis of the exact nature of the marketing costs of finding an audience on the second platform and of the impact of those costs on supply would be necessary to make this inference complete. An important part of such an analysis would be to distinguish between app quality and marketing costs as alternative theories. Sorensen (2007) has an interesting analysis of precisely such a problem in a related context, "best-seller" lists. The problem of finding an audience for an app on a platform is, to a considerable degree, one of breaking into a "best-seller" list.

43. Bresnahan and Reiss (1990).

44. The schemes are daily "visitors" (i.e., users), minutes per "visitor," that is, the average amount of time the app is open when it is being used, total minutes (over all users), unique "visitors" over the month, and unweighted. Definitions of all these different weights are found in the footnotes to Error! Reference source not found..

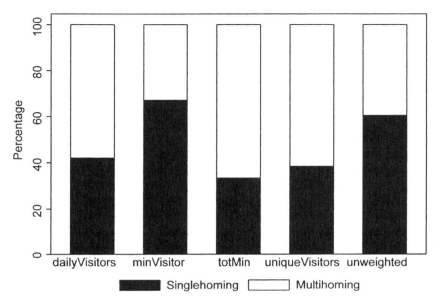

Fig. 8.10 Weighted multihoming rate on iOS
Source: comScore, January 2013.

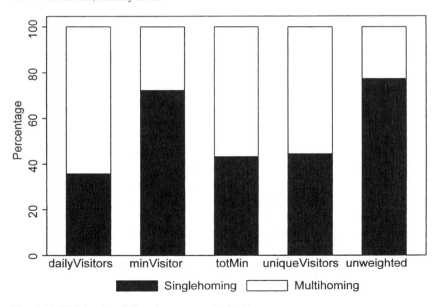

Fig. 8.11 Weighted multihoming rate on Android
Source: comScore, January 2013.

homing as the base case. The other three columns, unique monthly visitors, daily visitors, and total visitor minutes are all similar to one another and all, on both base platforms, predict a higher rate of multihoming than the base case.

The second fact is easy to interpret. When we weight by any of the three "size of the market" variables, we see more multihoming. That is, those developers who have a larger market size on the base platform are also likely to be observed on the other platform. This sounds very much like a familiar result from the entry literature. If (plausibly) market size on one platform is correlated with market size on the other platform, our weighting fact is no more than this: firms that expect a large market bear the technical and marketing costs of entering. If this is the right interpretation, even the weighted multihoming rates in columns (1), (3), and (4) are surprisingly low, another clue that the total fixed costs (technical plus marketing) of entering to serve consumers on a platform are substantial.

The other fact, based on the second bar (minutes per visitor), shows results much closer to the unweighted case. Now, industry figures are likely right that this measure is related to the intensity of preference at the individual user level—typically called "user engagement" in the industry. One might have thought that higher user engagement would mean more per-customer value for the app developer, and thus that there would be a significantly higher rate of multihoming than the base case. However, this does not appear to be the case. There are two obvious explanations for this, and we are not yet able to distinguish them. First, high-engagement apps may be more difficult to port from one platform to another, either because it is harder technically to write them for the other platform (e.g., they may be games with much use of the user interface) or because it is more expensive to find an audience for them. Second, having the same user for twice as many minutes may not be as close to getting twice as much profit as is having twice as many users. This would certainly be the case for paid apps or subscription apps, or if high-engagement apps cannot show advertisements (profitably) twice as often if the app is open twice as long.

These weighting results establish clearly that more popular apps (larger market size) are more likely to multihome. The obvious interpretation is that a firm with a more popular product can pay the incremental fixed costs of supplying a second platform. Since the observable definition of multihoming in these results is appearance in both platforms on comScore, we cannot, by these results alone, distinguish marketing fixed costs from technical fixed costs. Other indicia, however, suggest that the marketing costs are high.

8.9.4 Multihoming at the Firm Level

One theory of platform choice frequently articulated in the early stages of the industry is the importance of firms' technical capabilities. Some firms might have engineers trained in iOS or in the programming languages

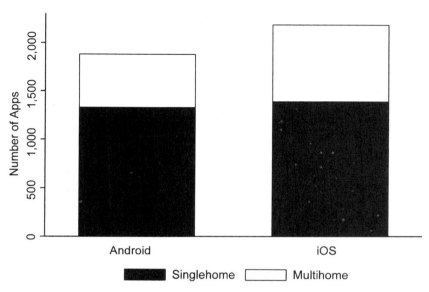

Fig. 8.12 Developer websites: This app available on the other platform
Source: App questionnaire, n: Android = 1,844, iOS = 2,117.

approved by Apple, while other firms might have the corresponding Android technical capabilities. We examine this hypothesis by looking at multihoming at the firm level.

We examine this hypothesis in figure 8.12 and figure 8.13. These figures are based on new samples, that is, the largest groups of apps for which we have observed developer websites. Once again, we use the "base" platform concept, that is, we examine the behavior of writing for the "other platform for all the apps we survey in the 'base' platform." This sample goes considerably deeper into the long tail of apps on both platforms than does the sample of apps we were just examining. Also, the definition of multihoming here is writing an app for a second platform, not writing and marketing it to achieve success.

Figure 8.12 looks at multihoming at the app (product) level. The sample for the first column is the most popular apps on base platform Android, so "multihoming" means writing that same app for iOS. The second column is symmetric, an observation is a popular app on base platform iOS, and "multihoming" is writing that same app for Android. Figure 8.13 has the same structure and sample, but here "multihoming" refers to the same firm supplying any app to the other platform.

There are three interesting facts in these figures. First, the rate of multihoming at the app level (figure 8.12) has fallen in these broader samples compared to the data we saw in figure 8.9. Above, we had the most popular 1,231 apps and about a 64 percent rate of multihoming. Here, we are looking

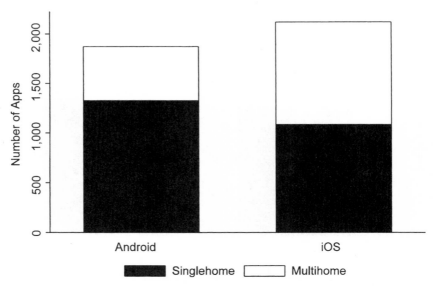

Fig. 8.13 Developer websites: Some apps available on the other platform
Source: App questionnaire, n: Android = 1,874, iOS = 2,117.

at the most popular 4,000 apps or so, and the rate of multihoming has fallen: 23 percent of Android apps and 29 percent of iOS apps are multihomed. Second, while the rate of multihoming has fallen, it has not fallen to zero. This encourages us, in future work, to examine the relationship between market size and multihoming more quantitatively to learn more about the incremental fixed costs of writing for a second platform.

The third fact comes from the contrast between figure 8.12 and figure 8.13. Simply put, there is much more multihoming at the firm level than there is at the app level. On both platforms, there is a considerable probability that a firm that does not write the same app for both platforms nonetheless has apps on both platforms. The difference is not small. Also, it is not symmetric. When the base platform is iOS, meaning that the firm has at least one popular app on iOS, there is a very large, 49 percent, probability that the firm will offer some app on Android.

This is strong evidence against a technical capabilities theory of platform choice. We see a large number of firms that already multihome today; they know how to develop for the other platform. They choose not to port all apps to the other platform; they are not constrained by lack of capabilities. This is interesting in itself about mobile firms; more importantly, it suggests market forces will be able to influence future developer platform choices should platform market structure move away from the current top-two-tiered structure.

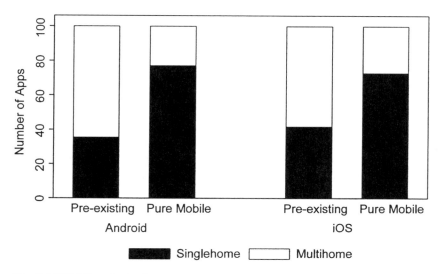

Fig. 8.14 Multihoming by firm scope

Source: App questionnaires. Based on the question: "Is the app produced by (or in partnership with) a larger company (one with a presence outside the mobile economy)?"

8.9.5 Multihoming by Firm Type

In the app questionnaire data set, we have also constructed variables for the type of firm associated with the app developer (purely mobile app firm or a preexisting, "corporate" firm offering nonmobile products). We report the rate of single homing and multihoming for each observable type of firm in figure 8.14. This question is based on our linking the developer websites to standard firm-information sources; linkage involves more hand work, as not all developers are all that specific about their parent firm.

The value of this linkage lies in investigating a hypothesis about the incidence of the marketing costs associated with the mobile industry's market institutions for matching users to apps. A firm that was founded in the mobile industry, an entrepreneurial mobile app firm, must acquire customers on each platform using the mobile industry's market institutions. A preexisting firm may already have customers, and can market its mobile app to them. Thus the marketing costs of porting to a second platform may be significantly lower for a corporate app than for an entrepreneurial app.

Figure 8.14 provides considerable evidence for the hypothesis that the costs of porting and multihoming *for entrepreneurial firms* are comprised, in no small part, of the marketing costs of finding an audience on a second platform. Top Android apps whose developers have a connection to their customers from outside the mobile world, such as existing consumer product and services companies, companies with an online presence, and so forth are, at 65 percent

(0.645), quite likely to multihome. For iOS, the probability of a top app multihoming if the developer is preexisting is 58 percent (0.58). These probabilities are two to three times higher than the probabilities for apps from purely mobile developers. Such firms have no need to find a new audience on the new platform, and thus have far lower incremental marketing costs of multihoming.

There are a number of limitations to this discussion. The difficulties in our matching apps across data sets means that we do not have a fully integrated data set yet. Furthermore, we have not filtered out apps for whom the porting decision may not be interesting or may already be predetermined (wireless carriers, platform providers, etc.). A great deal of interesting analysis may arise from augmenting this data with information on categories (Davis, Muzyrya, and Yin 2013). Finally, we as yet do not have a serious model for understanding first and second platform choice by developers.

Still, after looking at a number of different indicia, we have reached what we think is an important interim conclusion. There appear to be substantial marketing costs of reaching both halves of the mass-market audience in the mobile industry, and these marketing costs fall far more heavily on entrepreneurial firms than on established consumer product and services firms. That suggests once again that market institutions may also be limiting the range of economic experiments about value creation. Another implication of our interim conclusion is that the market institutions of the young mobile industry are slowing the rate of innovation effectively delivered to consumers and biasing its direction toward corporate apps.

8.10 Alternative Equilibrium Scenarios

8.10.1 Competition among Platforms

The implications of all this developer behavior, including single homing, multihoming, and platform choice for the single homers for the competition between platforms is difficult to discern at this juncture. No tip to a single platform seems imminent, even though full multihoming is rare. The Android/Google Play and iOS/iTunes Store platforms are approximately equally attractive to developers today, which suggests temporary stasis near the unstable platform market equilibrium of approximately even market shares for the competing platforms.

Modern platform competition theory emphasizes the critical role played by expectations in resolving which of the many platform market equilibria could arise.[45] There does not appear to be a strong expectation—in any direction—among consumers.

---

45. See, for example, Farrell and Klemperer (2006). This handbook chapter particularly focuses on the distinction between expectations that track efficiency (developers and users expect the better platform to be adopted) and inertial expectations.

To the extent that developers have expectations, those would appear to favor the original incumbent iOS/iTunes Store platform. (See table 8.1.) Money-oriented developers are directed toward the richer customers of the iOS/iTunes Store platform, even today, and of course for a while that platform had a strong lead in installed base in smartphones (and then, for another while, a strong lead in installed base in tablets).

However, many developers appear to be treating the two platforms as broadly equal and largely skipping the Blackberry or Windows Mobile efforts. As we saw in table 8.1, the count of larger developers in the com-Score data with Android-only exceeds the count with iOS-only, with a bit of a trend toward Android. For now, our interpretation is that the market is *not* tipping to early leader iOS. Whether a more careful investigation of developer behavior would reveal a tilt of those apps more influential on users toward either platform remains to be seen.

Other entertainment uses of smartphones and tablets have a more multihomed structure today than do apps. Media (books, newspapers, music, etc.) are significantly more likely to multihome than are app developers. One reason may be that media have buyer-seller matching market institutions that are outside the mobile economy and media have low technical costs of porting to a new platform. Media are natural multihomers.

Within media, however, new issues arise. The Amazon Kindle platform and the related Amazon Kindle online store represent an important media-distribution and media-reading system, certainly a third alternative to iPads and Android tablets. Indeed, some media multihoming is partial, as when iPad users can run an Amazon Kindle app in order to read books they have obtained from the Kindle store. Still, there are powerful forces toward concentration in the media-reader function, just as there are in the app-platform function. These forces were demonstrated by the recent exit of the Barnes and Noble Nook.

Whether there will continue to be significant differentiation between special-purpose media reader devices (Kindle Fire) and general purpose mobile devices that also support media reading (iPad, Galaxy) remains to be seen. This is a second question parallel to, and analytically very similar to, the question of a platform "tip" within the applications platform function.

In short, we are in the "before" period in which a large number of alternative solutions to the same problem are racing in the marketplace. Whether the ultimate important platform(s) will be open (Android/Play store) or proprietary (iOS/iTunes Store), whether they will be general purpose devices like those or (also?) special-purpose media readers, all remain to be seen. So, too, does the boundary between all these mobile devices—today primarily supporting consumption—and PCs—today primarily supporting work—in markets served. Perhaps the most important development so far in the mobile devices running iOS or Android is that they have drawn enough users and app developers to create significant momentum, so that there is a real question not only about the competition for ultimate leadership in what is now the "mobile space," but also about competition between these platforms and existing e-reader and PC platforms.

8.10.2 Divided Technical Leadership (DTL)

It is a fact of information technology industry life (if not a feature of "two-sided market" models) that many application platforms partially or sometimes fully overlap with one another. Today, for example, there is a positive feedback loop around not only the iOS/iTunes Store platform, but also around the Facebook applications development platform and around Google Maps—while both a Facebook app and a Google Maps app run on iOS devices. This structure, which causes divided technical and market leadership of the positive feedback loop in any one of the partially overlapping development platforms, has implications for the analysis of platform competition.

Bresnahan and Greenstein (1999) argue that this structure has been historically important in platform competition in the computer industry; DTL was important in the early PC industry and in the fall of IBM's hegemony in corporate computing. The structure has been important both in enabling new platforms to compete against an established platform and in enabling competition within a platform, even with positive feedback around a particular standard.[46] We have recently seen another very important example in the mobile space itself. Competition between mobile carriers such as AT&T and Verizon has been changed forever by the partially overlapping positive feedback loops around mobile devices and mobile platforms we are studying in this chapter. In light of its past importance, it is reasonable to examine DTL as one force that will affect future events in mobile platform competition.

The Android/Google Play mobile platform was set up by sponsor Google as if to enable divided technical and market leadership. Android can be altered by other firms, and has been, for example, to create e-readers that are now outside Google's control, to create special versions of Android for particular manufacturers, for particular screen sizes, and so on. Similarly, app distribution is permitted to flow through alternatives to Google Play, such as Amazon. Accordingly, Google cannot prevent the widespread distribution of apps, and thus permits app developers wide leeway. There are, as a result, a number of Android apps offering "infrastructural" functionality, functionality that many

---

46. Entry and competition against the established platform of IBM mainframes was, for example, greatly facilitated by the existence of the partially overlapping platform of database management systems, and firms such as Oracle were important complementors of entrants competing against the established IBM platform. A second implication of divided technical and market leadership is that it weakens, sometimes dramatically, the control of a platform sponsor over technical progress within a platform, permitting innovation competition among the different firms sponsoring the different layers. The influence of firms other than IBM on innovation in the "IBM PC" is a famous example.

operating system providers would seek to offer only in the OS itself. Some hardware manufacturers, such as Samsung, have made impressive efforts to turn Android for mobile devices into their own platform, not Google's.

Apple set up the iOS/iTunes Store platform in a much more centralized and controlled way, and thus has the capability to block the emergence of apps it feels are bad for consumers or not in its own interest. As a result, divided technical and market leadership on the iOS/iTunes Store platform cannot emerge as easily. (Apple's enthusiasm for enforcing restrictions on developers may well have declined over time after the success of Android.) However, a number of partially overlapping platforms, notably from Facebook and Google Maps, have created some divided technical and market leadership for Apple. Apple's unimpressive effort to replace Google Maps reflects the value to a proprietary platform sponsor of preventing divided technical and market leadership from emerging.

One significant influence on the direction of the mobile platform competition among Google, Apple, and potentially Microsoft, Blackberry, or others arises from highly influential suppliers of partially overlapping platforms. Should, for example, Facebook's influence with mobile users grow rapidly, that firm could have considerable influence on the market equilibrium for mobile operating systems.

While the current platform industry structure seems unlikely to persist for a long period of time, it is less than obvious what ultimate market structure will arise. Obviously a tip to either iOS/iTunes Store or to Android/Google Play could occur. Less likely, one of the newer platforms (Windows mobile, Blackberry) could get over the network effects hump and the market could tip to it. Given Microsoft's position in PCs and its extremely impressive capabilities as a "strong second" imitator, it seems certain that this firm will continue to invest in Windows Mobile for a long time (à la Bing) even if it cannot move beyond a second place platform. Another potentially important long-run equilibrium scenario is platform product differentiation. Many industry participants imagine a future in which Android is stronger in selling to poorer customers than iOS, possibly poorer customers in the rich countries, or possibly becoming the dominant platform in poorer countries. Like a platform tip, a platform product differentiation equilibrium involves a substantial change in app developer behavior. At this stage, we do not see any strong precursors of any of these long-run scenarios. The average market participant who comments on this issue suggests a tip to Apple, a scenario that seemed to us more likely a few years ago than it does today.

Finally, the OS and online store platforms we see today could be commodified by the movement of control of standards to a different layer in the value-creation "stack." Control of mobile development standards by Facebook is one possible scenario here, with the authors of the mobile device OS becoming less influential on developers (à la the shift from "the IBM PC and clones" to "the Windows PC").

8.11 Conclusion

We have been writing, for much of the last few sections, about the institutional and conceptual bottlenecks to successful exploitation of a tremendous technical and market opportunity. This is interesting for two reasons. First, it illuminates an important general point about the creation of economic value out of technical progress: the "last step" in the chain of invention that leads from new basic science to new economic growth can be quite difficult, and can, even though it is "merely" the discovery of the most valuable uses for a new technology and the creation of markets to serve those uses, call for profound innovation itself. In this regard, with the need for "mere marketing" to be successful for the industry to expand, mobile is like earlier rounds of ICT invention such as the mainframe computer, the superminicomputer, the PC, or the widely used Internet.

Second, the sheer size of the mobile opportunity illustrates an important change of twenty-first century innovation from earlier rounds of innovation, at least in ICT. There are three-quarters of a million apps on each of the major mobile platforms: a great deal of experimentation can fit into three-quarters of a million new products! There has been much success at the firm level, but there is still an opportunity for some of those experiments to create new rounds of innovation by finding important general sources of value for mobile applications. This is an important difference from earlier ICT industries. At the stage when the PC industry found its first big hits—white-collar work apps like the spreadsheet and the word processor—there were far fewer app innovations (in the hundreds) and a vastly smaller (<1 million) device market. We are seeing the benefits of cumulative technical progress joined to the benefits of successful exploitation of social scale economies through the platform model of industrial organization.

The large potential markets created by the rapid growth of mobile device usage and the low costs of entry created by the platform technologies are an invitation to potential app inventors with a wide variety of knowledge bases and a wide variety of incentives. The resulting entry, even at this early stage, has been heterogeneous as well as numerous. The creation of a global communication infrastructure has meant that the sources of supply of apps have also been global. We can once again draw the analogy to the early PC industry, in which international app supply meant primarily the "localization" of applications first written in English; this time it means drawing on a global talent pool to create firms such as Rovio and Distimo.[47]

---

47. There were, of course, exceptions to the marked US-ness of the early PC industry, but they had no important influence on the industry's technical or (other than enabling global sales through "localization") market development. See Bresnahan and Malerba (1999) for a discussion of national versus international supply in the creation of a number of earlier computer markets.

Similarly, high device penetration has made it economically feasible in the mobile era to have app experiments that are very heterogeneous with regard to the developer's business model. What Hayek (1945) called the economic function of the entrepreneur, finding overlaps between supply opportunity and demand need, can be taken up by any organizational form—and it has been. Right now, the online app-discovery mechanisms tend to favor existing businesses that give their existing customers an app. That is, the entrepreneurial function is being taken up by firms that do not take the entrepreneurial form. But there is no strong reason to believe that this is a permanent situation; the economic returns to the creation of new app-discovery mechanisms are simply too large to believe there will not be new invention in that area.

While many industry participants talk about universal "disruption" of existing economic institutions and markets resulting from mobile systems development, that remains largely in the future. A number of very interesting experiments are being tried, and a number of potentially important applications areas are entering a very early stage of a long diffusion process. Industry participants routinely speak as if they already know the results of the experiments and as if everyone is already doing what only the earliest adopters are trying out. This is a familiar situation in the applications of information technology, which economists labeled "the problem of the tenses" three decades ago.[48]

Why has the app exploration taken so long? This question falls into "the problem of the tenses"! A better question is: Why will app exploration take a long time to create all the markets and other institutions this industry will need for long-term value creation? This question leads us to consider a very simple consumer problem, one where mobile devices and apps have already created a lot of value, of finding a coffee shop in an unfamiliar neighborhood. "Maps" is a good application for this, and it is possible that there will continue to be a very large market in which maps are free to the consumer and (some) retailers pay for ads to be displayed in maps or some other product supplied jointly with maps. Or, since coffee shops are consumed socially, either a general purpose social network app or a special purpose meet-and-greet social app, such as those used frequently by young people to find a bar, could guide consumers to a coffee shop their friends like, one their friends are in, one where potential new friends are sitting, and so forth. Those different solutions are all technically feasible, and they also all entail different visions of consumer behavior and retailer behavior as well as of app development. Without experimentation, do we really know what, other than caffeine, consumers will want from the system that helps them find a coffee

---

48. As we write, Google faces the difficult problem of continuing to talk about what everyone "is" doing with their Android phones and tablets while moving on to talking about what everyone "is" doing with Google Glass. This is "the problem of the tenses" in the present.

shop?[49] Even more difficult to foresee without experimentation is the market equilibrium balance between consumer interests (Do I find the coffee shop I like? Do shops compete for my business?) and retailer interests (Do the ads bring me customers I would not otherwise have had?). The (nonmobile) online world continues to explore these equilibrium questions almost two decades on; that pace is determined by the pace of market exploration, not by the pace of technical change.

To continue to use the very simple coffee shop example just for a moment, there is another possibility, which is that rather than a general app like maps or Facebook or a special third-party app like Coffee Shop Finder that helps the consumer actively search, a retailer app comes to play a very important role in this area, tying the consumer more tightly to a particular chain of coffee shops. Starbucks, for example, implements a significantly more successful volume discount (loyalty program) through its mobile app than it ever could through prepaid discount cards, and reports that its customers are much more "sticky" when they use the mobile app to pay. They also report that over 10 percent of their (US) sales are now (early summer 2013) paid by mobile app, which might lead you to think that (some) retailers are not entirely powerless in the struggle with Google, Apple, Facebook, and new "disruptive" entrepreneurs over who will get the rents from the mobile opportunity.

There is no particular reason to think that this struggle over rents will play out the same way in all markets; an enormous market-creation cluster of parallel experiments in a large number of consumer markets awaits. There is no particular reason to believe that the momentary advantage given to large, preexisting firms (like Starbucks) by the app discovery process today will persist. That too, as we have said, could be changed by new innovation. What there is every reason to believe is that the incentives for new commercial innovation over the next decade created by the opportunity—incentives for a wide variety of new and existing firms—are enormous.

We are not arguing that there is no widespread prospect for disruption, rather the reverse. The mobile developments have already made a considerable impact on a few areas. Music and other media, already going through a dramatic change because of the Internet, see that accelerated by mobile. (There is likely more to come, as firms like Spotify, Last.fm, and Pandora are in competition with the music portion of the online stores.) Mobile tele-

---

49. Some grumpy observers have already said that the high weight that app development so far has put on social features (the coffee shop where your friends are or where new friends might be found) arises from twenty-something app developers thinking about the concerns of twenty-something customers, in this case the incomprehensible tribal and mating behaviors of twenty-somethings. This is uncharitable; market experiments are heterogeneous because different experimenters have different knowledge, goals, and powers of conjecture. Heterogeneity is very good in markets as large as the mobile industry; ultimate economic value creation does not turn on how many experiments are wrong, but on whether any are right.

phone carriers have found their business radically changed and are (mirabile dictu) embracing open systems at long last. But many of the other obvious loci for disruption from mobile are still in the future. For example, radical change in advertising markets is still to come.

We have written much about the app discovery bottlenecks holding back this progress. But it is clear that the market process is working rapidly to resolve these problems. Already, we see a tremendous market response to the needs of app firms. The broader problems of value creation will be solved. Right now, the bottlenecks in the system favor established firms over entrepreneurs, so there is an immediate advantage to value creation from established firms. That, too, could easily change through new innovation. Study of that awaits occurrence.

This gap between technical opportunity and market value creation is characteristic of information technology innovation over the last sixty years. Another element of continuity from the past is that the early uses of an important platform initiative need not be the ultimately most valuable ones, and that important interim innovation, even after early success, is important. The twenty-first century has brought a series of important changes to this. Some arise from scale, and the sheer size of the opportunity has drawn remarkable resources to the mobile area. Others arise from the quick entry of a second platform in competition with iOS/iTunes Store, so that there is technical and market heterogeneity in even the general purpose components at a fairly late stage.

References

Armstrong, M., and J. Wright. 2007. "Two-Sided Markets, Competitive Bottlenecks and Exclusive Contracts." *Economic Theory* 32 (2): 353–80.

Bar, F. 2001. "The Construction of Marketplace Architecture." In *Tracking a Transformation: E-Commerce and the Terms of Competition in Industries*, 27–49. Washington, DC: Brookings Institution Press.

Bresnahan, T. F. 1992. "Sutton's 'Sunk Costs and Market Structure: Price Competition, Advertising, and the Evolution of Concentration.'" *RAND Journal of Economics* 23 (1): 137_52.

Bresnahan, T. F., and S. Greenstein. 1999. "Technological Competition and the Structure of the Computer Industry." *Journal of Industrial Economics* 47 (1): 1–40.

Bresnahan, T. F., and F. Malerba. 1999. "Industrial Dynamics and the Evolution of Firms' and Nations' Competitive Capabilities in the World Computer Industry." In *Sources of Industrial Leadership*, edited by D. Mowrey and R. Nelson. Cambridge: Cambridge University Press.

Bresnahan, T. F., and P. C. Reiss. 1990. "Entry in Monopoly Markets." *Review of Economic Studies* 57 (4): 531–53.

Bresnahan, T. F., and M. Trajtenberg. 1995. "General Purpose Technologies: 'Engines of Growth'?" *Journal of Econometrics* 65 (1): 83.

Christensen, Clayton, Curtis W. Johnson, and Michael B. Horn. 2010. *Disrupting Class, Expanded Edition: How Disruptive Innovation Will Change the Way the World Learns*. New York: McGraw-Hill.

Davis, J., Yulia Muzyrya, and Pai-Ling Yin. 2013. "Experimentation Strategies and Entrepreneurial Innovation: Inherited Market Differences from the iPhone Ecosystem." Working Paper, INSEAD.

Farrell, J., and P. Klemperer. 2006. "Coordination and Lock-In: Competition with Switching Costs and Network Effects." Working Paper, Competition Policy Center, University of California, Berkeley.

Greenstein, S. 2001. "Commercialization of the Internet: The Interaction of Public Policy and Private Choices or Why Introducing the Market Worked So Well." In *Innovation Policy and the Economy*, vol. 1, edited by Adam Jaffe, Josh Lerner, and Scott Stern, 151–86. Cambridge, MA: MIT Press.

Hayek, F. A. 1945. "The Use of Knowledge in Society." *American Economic Review* 35 (4): 519–30.

Helpman, E., and M. Trajtenberg. 1998. *Diffusion of General Purpose Technologies*. Cambridge, MA: MIT Press.

Klepper, S. 2002. "Firm Survival and the Evolution of Oligopoly." *RAND Journal of Economics* 33 (1): 37–61.

Rysman, M. 2009. "The Economics of Two-Sided Markets." *Journal of Economic Perspectives* 23 (3): 125–43.

Sichel, D. E. 1997. *The Computer Revolution: An Economic Perspective*. Washington, DC: Brookings Institution Press.

Sorenson, A. T. 2007. "Bestseller Lists and Product Variety." *Journal of Industrial Economics* 55 (4): 715–38.

Sutton, J. 1991. *Sunk Costs and Market Structure: Price Competition, Advertising, and the Evolution of Concentration*. Cambridge, MA: MIT Press.

Waldfogel, J. 2013. "And the Bands Played On: New Music in the Decade after Napster." Working Paper, Carlson School of Management and Department of Economics, University of Minnesota.

9

State Science Policy Experiments

Maryann Feldman and Lauren Lanahan

State governments in the United States have experimented with programs that fund science. Many of these programs mimic or complement federal programs, while others attempt to increase industry investment within their borders. Forty-four states in the United States have adopted policy instruments that leverage university resources. Fiscal federalism dictates that different levels of government have specific obligations, with each state responsible for funding its public universities while also influencing private institutions within their borders. The demonstrated and growing interest among state governments in the basic research enterprise suggests that public support for R&D no longer rests solely at the federal level. State science policy actions over the past thirty years illustrate an evolution toward multilevel funding of US science, with states often motivated to facilitate the commercialization of academic science to capture the returns within their borders. The federal government's role as the source of public support—as outlined by Vannevar Bush's 1945 report *Science: The Endless Frontier* and discussed by Stephan (chapter 10, this volume), has been adopted by US states, which have great latitude in adopting new initiatives that may be particularly suited to their local circumstances or responsive to specific conditions. In contrast to the sizable literature that examines federal investment in research and development (R&D) (David, Hall, and Toole 2000; Feller 2007; Payne 2001; Ruegg and Feller 2003; Santoro and Gopalakrishnan 2001), there are few studies that consider *state* R&D investments. The magnitude of state

Maryann Feldman is the Heninger Distinguished Professor in the Department of Public Policy at the University of North Carolina at Chapel Hill. Lauren Lanahan is a PhD candidate in the Department of Public Policy at the University of North Carolina at Chapel Hill.

For acknowledgments, sources of research support, and disclosure of the authors' material financial relationships, if any, please see http://www.nber.org/chapters/c13046.ack.

efforts to leverage science and the reasons behind the adoption of specific state programs are an underappreciated aspect of science policy explored in this chapter.

Individual states within the United States have flexibility to build capacity to either influence firms' R&D location decisions or leverage federal programs. Since 1980, state government expenditures for university R&D programs have increased threefold to $3.13 billion, and now account for 5.8 percent of all university research in the United States in 2011.[1] Placing this level of support in context, state and local government investment in academic R&D is larger than support provided by industry (National Science Board 2012, chapter 8). Moreover, these amounts do not consider state initiatives that fund capital investments in science or programs that support the recruitment of faculty and the promotion of research. To attract firms, states have simultaneously offered R&D tax credits and attempted to create good business climates (Wilson 2009; Hearn, Lacy, and Warshaw 2014). The adoption of science policy is another means to influence industry location decisions. Moreover, federal funding still accounts for the majority of academic R&D expenditures. Federal awards are competitive and subject to peer review. States invest in science as a means to increase their share of federal R&D expenditures (Sapolsky 1971). In addition, with federal funding agencies focused on developing regional centers of innovation often focused on technology-intensive sectors, state science policy is a way to build capacity to participate in cluster initiatives.

Viewing state science policies as experiments may help guide policymakers at the US federal level as well as from other states and countries. The US's federalist multilevel structure was intentionally put in place to create checks and balances on the national government. This structure places state governments in a position to experiment and vet the efficacy of varying programs as they seek to maximize their intended goals (Karch 2007). Scholars and policymakers have an opportunity to evaluate the successes and failures of these state experiments, consider the competitive nature of state actions, and arrive at more enlightened policy recommendations. While federal policy actions are accountable—as mandated by the Government Performance and Results Act of 1993—the same standard does not hold for state governments. There is less accountability and data are more difficult to access in a uniform format—this particularly pertains to fragmented programs. State-level policy analysis and accountability, nonetheless, is not only critical to constituents, but analysis has the potential to improve policy.

1. Data retrieved from NSF WebCASPAR; NSF Survey of Research and Development Expenditures at Universities and Colleges/Higher Education Research and Development Survey. University R&D estimates are adjusted for inflation using the Fiscal yr GDP Implicit Price Deflators—base year 2005. State activity is derived from the State/Local Govt Financed Higher Education R&D Expenditures for S&E metric.

Before we can consider the implications of these state science policy experiments, however, it is critical to consider which policies are most appropriate in different circumstances or to even understand what motivates the state adoption of certain policies. To this point there is little guidance except for broad discussions of the economic renewal of states (Fosler 1988; Eisinger 1988; Feller 1997) and some early descriptive studies (Combes and Todd 1996). The creation of a recent typology (Feldman, Lanahan, and Lendel 2014) allows us to examine the factors associated with the adoption of state science programs. Moreover, it allows us to test if states are attempting to promote an enterprise that complements federal efforts and to assess if states promote these programs to catch up or to lead in terms of R&D activity.

This research is part of a larger research portfolio that first set out to classify state efforts (Feldman, Lanahan, and Lendel 2014); second, to understand motivations relating to state government involvement (the focus of this current chapter); and third, to consider their efficacy and contributions within the national context. This chapter is not an evaluation of these state science programs, but rather considers the circumstances by which states adopt these policies. This is a critical step to consider before examining their efficacy.

The next section provides background on state science policy, with emphasis on state university R&D programs. This section highlights the trends in the progression of adoption for each of these programs, which includes the Eminent Scholars, University Research Grants, and Centers of Excellence programs. The following sections present the methods and empirical analyses assessing a series of factors associated with the adoption of each of these three programs. The final sections discuss the results for each program, consider the broader state policy portfolio, and conclude with considerations for further research.

9.1 Background on State Science Policy

Sapolsky (1971) argues that governors' attention to science and technology resulted from the tripling of federal appropriations in response to Sputnik from 1957 to 1963. The local economic effects of federal expenditures along Route 128 and what was to be later named Silicon Valley were already notable. Many governors sought to replicate that success, with an initial objective of increasing their share of federal science funding. In 1963, New York and North Carolina established entities to parallel the president's science advisor and created state science and engineering foundations modeled after the National Science Foundation (NSF). The US Department of Commerce's State Technical Service Program (STS) and the NSF state science advisor's initiative encouraged active engagement with science policy (Berglund and Coburn 1995). By 1967, twelve governors had science policy advisors (Sapolsky 1968).

In the 1970s, revenue sharing between the federal and state governments and the devolution of authority from the federal government provided states with the resources and political freedom to experiment with R&D programs (Vogel and Trost 1979). From 1977 to 1979, forty-nine out of fifty states participated in the NSF State Science, Engineering and Technology (SSET) program, which encouraged states to develop and implement science and technology (S&T) related strategic plans (Berglund and Coburn 1995). The funding that had been promised for implementation was not subsequently provided. However, the idea that states could strategically leverage science was established (Feller 1990).

During the 1980s academic research was increasingly seen as instrumental in economic growth. The 1980 passage of the Bayh-Dole Act, which granted universities the rights to commercialize results from publicly funded research, coupled with the monetary success of the Cohen-Boyer patents encouraged state legislatures to view universities as engines of economic development (Cozzens and Melkers 1997). Concurrently, the decline of federal and industry support for university R&D created uncertainty and resulted in a search for alternative sources of revenue (Teich 2009). In response, states began to actively experiment with new programs that involved university science.

State science programs are typically announced with great fanfare and given colorful names. There is a tendency to describe each program as unique and innovative. In reality, however, there are only a few policy levers available to state policymakers. In an effort to build a typology of similar programs, Feldman, Lanahan, and Lendel (2014) identify commonalities across state science initiatives.[2] Through this effort, three consistent state initiatives aimed to promote innovation capacity through university research institutions were identified: Eminent Scholars, University Research Grants, and Centers of Excellence programs. Table 9.1 provides the year that each state initially adopted each of the three programs, illustrating the variation in the order of adoption and in the combination of programs adopted.

The Eminent Scholars program provides funding for a chaired position to attract world-class senior researchers to public and private universities located within the state boundaries. This program can be conceptualized as an investment in human capital through the attraction of what Zucker and Darby (1996) term "star scientists." This program demands substantial upfront costs, often ranging between $3–6 million per scholar to support the scholar's salary, lab materials, graduate students, administrative support, and overhead. Despite these notable costs, this program is centrally premised on the idea that these scholars will recover the state's investment by the following: (a) building research capacity within the university, (b) leveraging addi-

---

2. Data collection efforts to identify the portfolio of state R&D university programs includes the following: (a) state-funded, (b) codified in a policy document, (c) focus on university R&D, and (d) administered by a state agency (Feldman, Lanahan, and Lendel 2014).

Table 9.1   **Year of state science policy adoption**

	Eminent Scholars	University Research Grants	Centers of Excellence
Alabama		1983	1975
Alaska			
Arizona	1991	2006	
Arkansas	2002	1983	1990
California		2005	
Colorado			1983
Connecticut	2006	1993	1965
Delaware		1984	1994
Florida	2006		1982
Georgia	1990	1990	1990
Hawaii			
Idaho			2003
Illinois			2003
Indiana		1999	1983
Iowa			
Kansas	2004	2000	1983
Kentucky	1997	1997	2003
Louisiana	1987	1987	
Maine		1990	1988
Maryland			1985
Massachusetts		2004	2009
Michigan		1999	1981
Minnesota			2005
Mississippi			1999
Missouri	1995		1986
Montana		1999	1988
Nebraska		1988	1987
Nevada			
New Hampshire		1991	1991
New Jersey		2007	1984
New Mexico			1983
New York	1999	2000	1983
North Carolina	1986	1984	1980
North Dakota			2006
Ohio	1983	1998	1984
Oklahoma	2006	1985	1989
Oregon			
Pennsylvania	2006		1988
Rhode Island			1996
South Carolina	1997	1983	1983
South Dakota		1987	2004
Tennessee	1984		1984
Texas	2005	1987	
Utah		2006	1986
Vermont			
Virginia	1964		1986
Washington	2007	2005	
West Virginia		2004	
Wisconsin	1998	2007	
Wyoming	2005		2008

Source: Feldman, Lanahan, and Lendel (2014).

tional federal and private funds, (c) serving as research magnets for industrial recruitment, and (d) ultimately generating revenue from commercialized research (Bozeman 2000; Feller 1997). By providing funds for endowed chairs at research-university campuses, states seek to increase innovative activity by cultivating a rich knowledge economy rooted by these individuals.

Virginia was the first to adopt this program in the 1960s; however, other states did not begin to introduce the program until the 1980s. With Ohio serving as the second adopter in 1983, only five additional states implemented the program within the following decade—these include Tennessee, North Carolina, Louisiana, Georgia, and Arizona. During the latter part of the 1990s, only a handful of states selected to adopt the program. However, this program gained the greatest traction after 2001 with nine states introducing it within a six-year period between 2002 and 2007. Arguably, this recent surge may have resulted from state reports published in the late 1990s highlighting the notable benefits of the state programs. As of 2009, twenty-one states were identified as having an Eminent Scholars program. State and local officials interviewed were very enthusiastic about the potential of the program to build academic resources (Feldman, Lanahan, and Lendel 2014).

The Georgia Research Alliance (GRA) and Kentucky's "Bucks for Brains" stand out as exemplary Eminent Scholars programs (Bozeman 2000; Youtie, Bozeman, and Shapiro 1999). One illustrative example of the program's benefits lies with a distinguished IBM researcher who was recruited to the GRA program for $1.055 million and in return secured an NSF grant to establish an Engineering Research Center in Electronic Packaging worth a total value of $40 million over a three-year period (Combes and Todd 1996). Kentucky's "Bucks for Brains" initiative increased the number of endowed chairs and professorships in the state by over fivefold from 1997 to 2010, while extramural research expenditures from two of Kentucky's research universities—the University of Kentucky and the University of Louisville—increased by roughly 250 percent over the same time period.[3]

The second state university-based program, the University Research Grants, provides state grants to support university science and engineering (S&E) research. Feldman, Lanahan, and Lendel's (2014) defining criteria for the University Research Grants programs are the following: (a) grants oriented toward basic scientific research, (b) grants available to all researchers at universities or research institutions within the state, (c) grants that do not fund physical infrastructure, and (d) grants that do not require supplemental funding by an industrial partner.[4] As of 2009, twenty-nine states were identified as having a University Research Grants program.

---

3. Source: http://cpe.ky.gov/news/mediaroom/releases/nr_110811.htm.
4. We consider research grants that require matching funds from firms as a separate category that creates collaboration and leverages university resources. See the later discussion of Centers of Excellence.

The first state to adopt a University Research Grants program was Arkansas in 1983. Named the Basic Research Grant Program, the primary aim of the program was to build "the state's scientific infrastructure and improve the ability of Arkansas research scientists to compete for awards at the national level by awarding grants to researchers at the state's colleges and universities."[5] This program targeted individual researchers who had not previously received federal funding and required a 40 percent cash or in-kind contribution match by the individual's home institution. The primary intention of this program, as stated in the research objectives, was "to use state funds as an incentive to get scientists interested in new areas of research and to provide them with a track record that will help them to compete for federal monies, thereby bringing more research funds to the state" (Berglund and Coburn 1995, 84). The idea of improving the ability of scientists to compete for federal funds is consistent for these programs, suggesting that states perceive themselves to be lagging in federal R&D funding.

The Center of Excellence—the third state university-based program—is geared for later-stage university research activity by focusing on university and industry collaboration. This program aims to build capacity by investing in physical infrastructure and strengthening research partnerships with industry. Connecticut adopted this program in 1965, followed by Alabama in 1975. As of 2009, thirty-seven states were identified as having a Centers of Excellence program. These programs include state initiatives alternatively called University Research Centers, Advanced Technology Centers, and Centers of Advanced Technology. The important differentiating criterion of this program lies with the more central, active role of the university's industrial partners. Given the breadth of organizational forms and research foci across Centers of Excellence programs, both in terms of research scale and scope, scholars have struggled to reach a consensus on the definitive features that characterize these unique research organizations (Aboelela et al. 2007; Mallon and Bunton 2005; Youtie, Libaers, and Bozeman 2006; Friedman and Friedman 1982).

Feldman, Lanahan, and Lendel's (2014) review identified four common features of Centers of Excellence programs. These include: (a) a directed research mission focused on basic and applied research, (b) emphasis on graduate training, (c) collaboration between universities and industry, and (d) a strong research orientation directed toward a specific industry sector or technology. Despite these common features, some states place greater emphasis on the partnership with industry, while others are more concerned with the research program. The Massachusetts' Centers of Excellence (2004) serves as an exemplar of the latter, placing a concerted aim on improving emerging technologies such as biotech and nanotech. The Florida Technology Development Initiative, however, exemplifies the former. This Cen-

---

5. Source: ASTA's website, http://asta.ar.gov/.

Table 9.2   Trends of initial policy adoption of Centers of Excellence, University
                   Research Grants, and Eminent Scholars programs

Policy	Number of states	Year adopt first program (state)		
		Mean	First to adopt	Most recent to adopt
Centers of Excellence	28	1987	1965 (CT)	2006 (ND)
University Research Grants	14	1992	1983 (AR, SC)	2005 (CA, WA)
Eminent Scholars	9	1988	1964 (VA)	2005 (WY)

ters of Excellence program promotes both functions of promoting research excellence and facilitating collaboration with industry for conduit building.

Among the portfolio of the three university programs the Centers of Excellence is not only the most widely diffuse; states tend to adopt it first. This suggests a prioritization of making investments in academic research directly linked to industrial activity over supporting more upstream efforts that are characteristic of the Eminent Scholars and University Research Grants programs. The descriptive statistics presented in table 9.2 show that twenty-eight states adopt the Centers of Excellence first, with Connecticut adopting first in 1965. Fourteen states initially adopted the University Research Grants, and nine initially adopted the Eminent Scholars program. These trends of adoption demonstrate a slightly different progression of state policy actions than presented by Plosila (2004). He groups the evolution of state S&E policy activity linked to economic development programs and practices into three stages—1960s to 1970s, 1980s, and 1990s—with the first focused on bolstering S&T programs, the second marking a shift toward university-based economic development initiatives, and the third directed to technology alliances and trade associations linking S&T to economic growth. Feldman, Lanahan, and Lendel's (2014) review of the portfolio of state university-based programs, however, finds little state university-based activity prior to the 1980s, with the pace of adoption remaining strong in the most recent decade after the turn of the century. Moreover, over the past thirty years states have adopted a range of programs from more upstream programs aimed to bolster the basic research enterprise within the university (Eminent Scholars and University Research Grants) to more downstream initiatives that link university research with industry (Centers of Excellence).

This descriptive analysis suggests that state science policy adoption is not random, but rather maps out in a systematic manner. Currently, our understanding of state science policy tends to rely on case studies that examine single programs and tends to provide more operational details rather than considering the motivation to adopt programs. While there is little theory to directly guide choices for state science programs, there are two broad literatures that we can draw from, specifically the state policy diffusion literature and literature on science policy. This analysis draws from these two distinct, yet complementary literatures to identify a series of factors that likely motivate state university R&D policy adoption.

9.1.1 What Motivates State Science Policy?

Since 1985, state governors have been convening annually at the National Governors Association (NGA). Each year, the chair presents a policy-based initiative, directing state attention to a range of issues from education, to healthcare, to economic development and R&D. Four initiatives have focused on the latter set—1988–89 NGA Chair Gov. Gerald Baliles's Initiative, *America in Transition: The International Frontier*; 1999–2000 NGA Chair Gov. Michael Leavitt's Initiative, *Strengthening the American States in a New Global Economy*; 2006–07 NGA Chair Gov. Janet Napolitano's Initiative, *Innovation America*; and 2011–12 NGA Chair Gov. Dave Heineman's Initiative, *Growing State Economies*. These national initiatives highlight that states need to invest in science for future economic growth. State policymakers operate with limited resources subject to bounded rationality in their policy decisions (Simon 1978). Given these limitations, they tend to rely on cues from other sources in their decision making. Directed attention on these issues at the annual governors meeting not only raises awareness to the initiative, but also may prompt governors to act within their own jurisdictions.

Moreover, these NGA initiatives point toward a broader economic restructuring that began in the 1980s, motivating many states to adopt science policy programs. Democratic governors pursued technology-based economic development as part of a new strategy. Widely known as "Atari Democrats" these democratic governors sought a contemporary equivalent of the New Deal that would revitalize the economy (Wayne 1982). Named after a then-popular consumer electronic game, the ideology favored R&D investment in growing industries and academic research figured prominently in their plans. Therefore, we expect that states are more likely to adopt one of the state science university policies when a democratic governor is in office. Additionally, there is considerable research to show that elected officials hold greater clout at the beginning of their term; thus we expect the state policy activity to take place during the first two years the governor is elected into office (Berry and Berry 1990).

The idea that states benchmark against one another is well established. Many times states that are lagging in terms of R&D expenditures or high tech capacity will be motivated to adopt science policy initiatives in order to catch up with their peers. Taylor's (2012) recent paper on the role of governors as economic problem solvers argues that a lagging economy or a low level of R&D may provide an incentive to implement S&T initiatives. While the precise referent group may be difficult to define, the literature has considered diffusion among contiguous states. These data do not support that pattern, but the prominence of the NGA suggests that benchmarking may be national. States that are behind the national average may be more likely to adopt science policies.

The ability to make these investments, however, will likely be related to the state's fiscal condition. In their influential study on state policy diffusion, Berry and Berry (1990) found that the fiscal health of the state budget influenced state lottery adoptions. While lotteries augment state budgets, S&T programs require slack resources and ability to fund programs that may be considered longer-term investments and discretionary. As such, we anticipate that states would be more likely to have science policy programs in years when they have fiscal growth.

In addition, certain states have demonstrated a commitment to science through previous efforts, suggesting that some states may have a proclivity toward supporting these types of science-based initiatives. In Sapolsky's (1968) review of science policy for state and local governments, he identified twelve states that had taken early action to establish science advisory units. These positions were created shortly after the federal science advisory position was established and reflect an early commitment by the states for science policy that has likely carried through our time frame of interest.

National trends of federal and industry R&D activity likely drives state actions. Historically, federal and industry R&D investments have been primary sources of support for S&T activity, overshadowing investment from state governments and other sources of funding. The federal government tends to lead in supporting more upstream activity, while industry is more prominent in supporting more downstream efforts. Moreover, research within the policy diffusion literature finds states rely on the federal government when making policy decisions. As an illustrative example, Baumgartner, Gray, and Lowery (2009) consider the nature of vertical policy diffusion between congressional activity and state lobbying actions and found the top-down influence to be considerable. The results suggest that federal R&D actions guide subsequent state policy activity. In this case, we expect that increased federal R&D spending will prompt greater state attention to science policy initiatives. This expectation is reinforced by a series of studies (Blume-Kohout, Kumar, and Sood 2009; David, Hall, and Toole 2000; Diamond 1999; Payne 2001) providing evidence that additional private support results from federal investment in R&D results. This literature finds evidence of a complementary or crowding-in effect between these two sources. Although these studies focus on the relationship between federal funding and private R&D, a complementary relationship likely holds for state governments as well as for the adoption of state policies designed to contribute to the R&D enterprise. Increased federal investment in science is likely to motivate state attention to science policy. We anticipate that state policy actions will complement federal and industry science investments.

Our understanding of state R&D activity is relatively nascent compared to federal R&D policy actions, thus the quantitative analyses in the next section serves as an exploratory effort toward understanding whether and how economic, political, and R&D-related factors influence state science policy-making decisions. We estimate the impact of the economic, political, and R&D-related factors associated with states adopting one of the three university state science programs, respectively.

9.2 Methods

We employ a Cox proportional hazard model (equation [1]), based on semiparametric assumptions about the distribution of adoptions.[6] In contrast to parametric models, this approach leaves the transition rate as unspecified as possible, relying on the proportionality assumption and appropriate specification of the functional form for the influence of covariates. As Blossfeld, Golsch, and Rohwer (2007) highlight, theory in the social sciences for selecting the appropriate parametric model is underdeveloped; thus, semiparametric models offer a useful alternative, particularly when primary interest is on the magnitude and direction of the observed covariates.

$$(1) \qquad \text{ADOPT}_{it} = h(t) * \exp(\beta_k(x_{it})\alpha).$$

Our empirical model is specific to state i, and year t. ADOPT is our primary outcome variable of interest, the transition rate of adoption for each policy respectively—Eminent Scholars, University Research Grants, and Centers of Excellence. This dichotomous variable is coded 1 in the year a state adopts one of the three respective science policies and 0 in the years leading up to the adoption.[7] While data on the dollar amount of expenditures for these programs would provide a more ideal dependent variable, the most information that we have is the year of adoption for each of the three programs. The unspecified baseline rate is $h(t)$, and $(\beta_k(x_{it})$ is a vector of covariates. Table 9.3 lists the variables, functional forms, and sources of the covariates considered in this analysis; each is considered in turn.

Building off Berry and Berry's (1990) study on state lottery adoptions, we include a series of economic and political variables in the analysis. First, we include Fiscal, which estimates the rate of growth in the state's revenue. This measure estimates the state's slack resources and ability to afford science

6. While hazard models are prominently used to estimate transition rates, scholars from a variety of fields have been using these methods to estimate other types of transitions prominent in labor market studies, social inequality studies, demographic analyses, sociological mobility studies, and state policy diffusion (Blossfeld, Golsch, and Rohwer 2007). We follow the literature on the latter and employ an event history model (Berry and Berry 1990; Mintrom and Vergari 1996; True and Mintrom 2001; Volden 2006; Karch 2007).

7. As is characteristic of hazard models, observations beyond the initial year of adoption for states with the policy of interest are dropped.

Table 9.3   Indicators used to measure independent variables

Variable	Metric	Source
Fiscal	$\dfrac{\text{total revenue}_{it} - \text{total revenue}_{it-2}}{\text{total revenue}_{it-2}}$	US Census State Government Finances
Dem Gov	Democratic Governor$_{it}$	Book of State: Council of State Governments; state government websites
Dem_Early	Dem Gov * Early Term[a]	Derived
Early science advisor	Early Science Advisor$_i$	Sapolsky (1968)
NGA initiative	NGA Economic Development Initiative$_{it,\,t+1}$	National Governors Association
EPSCoR	EPSCoR status$_{it}$	National Science Foundation
S&E degrees[b]	S&E Degrees Location Quotient[c] Quartile Rankings$_{it}$	National Science Board S&E State Indicators on Higher Education
High-tech industry[d]	High-Tech Industry Location Quotient[e] Quartile rankings$_{it}$	Bureau of Economic Analysis
University R&D[f]	University R&D Expenditures$_{it}$ (logged)	NSF Survey of R&D Expenditures

Note: i denotes state and *t* denotes year.

[a] Early Term is defined as the first two years in a term or consecutive term as Governor$_{it}$. Information was derived from state government websites.

[b] S&E degrees are defined by the National Science Board and include physical, life, earth, ocean, atmospheric, computer, and social sciences; mathematics; engineering; and psychology. Higher education degrees include bachelor's, master's, and doctoral degrees but exclude associate degrees. The quartile of states each year with the highest location quotient, Q4, serves as the referent category.

[c] S&E Degrees LQ $= \dfrac{(\text{higher ed S\&E degrees}_{it} \,/\, \text{higher ed degrees}_{it})}{(\text{higher ed S\&E degrees}_t \,/\, \text{higher ed degrees}_t)}$.

[d] High-tech industries were based on the BLS definition (Hecker 2005). The quartile of states each year with the highest location quotient, Q4, serves as the referent category.

[e] High Tech Industry LQ $= \dfrac{(\text{high tech employment}_{it} \,/\, \text{total employment}_{it})}{(\text{high tech employment}_t \,/\, \text{total employment}_t)}$.

[f] This includes federal and industry R&D expenditures in S&E fields, including direct and recovered indirect costs, respectively. The data are logged expenditures of university R&D (in real dollars).

programs. Dem Gov is a binary variable coded 1 in the years the state government inaugurates or has a democratic governor and 0 otherwise. In addition, politicians often have greater political influence when they are initially elected to office. We include a binary variable indicating whether the democratic governor is in his/her first or second year in office—Dem_Early. In the event the governor is reelected, we only code the first two years in nonconsecutive terms in office as 1.

The time frame for this analysis begins in 1982 and continues through 2009 for states that have not adopted the respective policy. We begin in the early 1980s, as this marks a time when increased federal and state attention was directed to university-based R&D activity. This coincided with the passage of the Bayh-Dole Act of 1980. With the exception of three state policy adoptions, it is after 1980 that we witness the diffusion of university-based R&D policy activity. State attention to science issues, however, did not begin at this time. In earlier reviews of state and local science policy actions, Sapolsky (1968, 1971) identified twelve states that established general science advisory units.[8] While other states were considering similar positions in the late 1960s, these states demonstrated an active interest in state science policy by implementing a science advisory position that mirrored the position in the federal government. We have created a binary, time-invariant, variable for this subset of states—Early Science Advisor—and anticipate that states with this demonstrated record supporting science issues are more likely to adopt these university-based R&D programs.

We have created a binary variable—NGA Initiative—coded 1 in the year the first three initiatives focused on economic development and R&D (mentioned above) were presented and the following year, as we expect there may be some lag in policy implementation. The most recent initiative, presented by Governor Heineman in 2011, falls outside the time frame of this analysis.

In addition, we include three state-level benchmarking measures that capture the S&T capacity of the state—EPSCoR, High Tech Industry, and S&E Degrees. EPSCoR is a dichotomous variable that denotes the status of the state in the Experimental Program to Stimulate Competitive Research (EPSCoR) program. Administered by NSF, EPSCoR is a federal program that began in 1980 to support and encourage disadvantaged states to improve their research and development activity (Hauger 2004). As of 2009, twenty-five states have received EPSCoR status. The first cohort of EPSCoR states in 1980 included Arkansas, Maine, Montana, South Carolina, and West Virginia. A second cohort was added in 1985: Alabama, Kentucky, Nevada, North Dakota, Oklahoma, Vermont, and Wyoming. The third cohort of states was added in 1987 and included Idaho, Louisiana, Mississippi, and South Dakota. In 1992, Kansas and Nebraska joined. Between 2001 and 2009, seven additional states have been added: Alaska,

8. These states include: CT, GA, HI, KS, KY, LA, MD, MA, NY, NC, OK, and PA.

Hawaii, and New Mexico in 2001; Delaware and Tennessee in 2003; New Hampshire in 2005; and Utah in 2009.[9]

High Tech Industry measures the annual high-tech employment for a state. To compute this indicator, we use the Bureau of Labor Statistics definition of high-tech industries (Hecker 2005),[10] and compute the ratio of high-tech employment to total employment. S&E Degrees measures the extent to which the state's higher education graduates are concentrated in the fields of science and engineering (S&E). This measure is drawn from the National Science Board's S&E State Indicators on Higher Education activity and estimates the ratio of S&E graduates to total graduates. To estimate benchmarking activity, we compute quartile rankings of the state location quotients for these two state R&D-related measures (High Tech Industry and S&E Degrees). For computation of the location quotient, the national ratio serves as the reference base in the denominator. The fourth quartile of the location quotients—the cohort of states with the highest rankings— serves as the referent category for both sets of variables. These values vary by states and year.

We also include a set of federal and industry metrics to account for national R&D-related trends. We expect that the larger, external spending environment will influence state science policy activity. University R&D denotes the sum of federal and industry investment in each specific state's university research activity. Given that the three university programs have different aims, we adjust the source of R&D for this measure. Eminent Scholars and University Research Grants programs are designed to support earlier stage, more upstream R&D activity; therefore, for the variable University R&D we include the federal investment in university R&D for these two sets of models. The Centers of Excellence program aims to support later-stage university R&D activity that should be more responsive to industry R&D investment. Thus, we include industry investments in university R&D for this model. Although we are unable to discern the precise direction of causality in this analysis, these measures approximate whether federal or industry university R&D investment in a given state complements or substitutes that state's university science policy adoption activity.

Table 9.4 provides descriptive statistics for the covariates for the years leading up to and including the initial year of adoption. While the same set of covariates is used for each of the three models, we present three sets of descriptive statistics given the variation in the cohort of states that have

---

9. Three additional states have received EPSCoR status after 2009: Rhode Island in 2010, Iowa in 2011, and Missouri in 2012.

10. Hecker's classification of high-tech industries is used for the National Science Board's definition of high-tech sectors, and therefore serves as a valid source for defining the list of NAICS and SIC codes that constitute high-technology industries. Using employment data from the Bureau of Economic Analysis (BEA), we matched Hecker's list by industry title to the BEA's LineCode classification scheme at the three-digit industry.

Table 9.4 Descriptive statistics

Variable	Eminent Scholars				University Research Grants				Centers of Excellence			
	Mean	S. D.	Min.	Max.	Mean	S. D.	Min.	Max.	Mean	S. D.	Min.	Max.
Fiscal	0.140	0.155	−0.754	1.164	0.143	0.162	−0.788	1.164	0.144	0.159	−0.788	1.164
Dem Gov	0.490	0.500	0	1	0.524	0.500	0	1	0.521	0.500	0	1
Dem_Early	0.152	0.359	0	1	0.159	0.369	0	1	0.160	0.367	0	1
NGA initiative	0.202	0.401	0	1	0.196	0.397	0	1	0.188	0.391	0	1
Early science advisor	0.206	0.405	0	1	0.197	0.398	0	1	0.195	0.397	0	1
EPSCoR	0.366	0.482	0	1	0.261	0.439	0	1	0.371	0.483	0	1
University R&D (federal)	18.524	1.389	14.660	22.165	18.658	1.358	14.660	22.102	16.045	1.467	11.925	20.043
University R&D (industry)	0.998	0.145	0.687	1.469	1.003	0.146	0.591	1.469	0.992	0.153	0.578	1.469
S&E degrees (LQ)	0.882	0.264	0.250	1.882	0.884	0.267	0.250	1.882	0.839	0.277	0.250	1.474
High-tech industry (LQ)	0.366	0.482	0	1	0.261	0.439	0	1	0.371	0.483	0	1

Notes: The time frame of interest spans 1982–2009. The number of observations varies between policies given that the number of states and adoption years for each of the policies varies. Twenty Eminent Scholars state adoptions are considered in this time frame—1982 to 2009; the total number of observations is 1,146. All adoptions are considered for the University Research Grants analysis given that the first adoption was in 1983; the total number of observations is 992. Thirty-four adoptions are considered in this time frame of interest for the Centers of Excellence; the number of observations is 707.

adopted each of these three programs over the time frame of interest. As of 2009, twenty-one states have adopted the Eminent Scholars program; however, twenty adoptions are considered in this analysis. Virginia is left-censored with initial adoption of the program in 1964. The total number of observations for Eminent Scholars is 1,146. As of 2009, twenty-nine states have adopted the University Research Grants program. All adoptions are considered in this analysis given that the first adoption was in 1983; the total number of observations is 992. As of 2009, thirty-seven states have adopted a Centers of Excellence program; however, four adoptions are left-censored—AL (1965), CT (1975), NC (1980), and MI (1981). Thirty-four adoptions are considered in this time frame of interest; the number of observations is 707.

Table 9.5 provides the correlation coefficients for the covariates. University R&D (Industry) and University R&D (Federal) have a high correlation, 0.8959, however, only one of these measures is used in each of the models to control for external R&D—University R&D (Industry) for the Centers of Excellence and University R&D (Federal) for the Eminent Scholars and University Research Grants programs.

We estimate three sets of Cox proportional hazard models for the policies—one for the Eminent Scholars, University Research Grants, and Centers of Excellence programs, respectively. Additional specification tests were run to ensure validity of the proportionality assumption and goodness of fit. Regarding the former, the results hold for the covariates in the three sets of models with the exception of the first quartile ranking of S&E Degrees (Q1) and the third quartile ranking of High Tech Industry (Q3) for the Centers of Excellence program. Both of these variables were significant in the specification tests with p-values of 0.048 and 0.037, respectively. Given this limitation, we exert caution in our interpretation of these coefficients in the Center of Excellence model as the values for these two are likely problematic.

9.3 Empirical Results

Tables 9.6, 9.7, and 9.8 present the results of the Cox proportional hazard models for the three state science policy programs. For ease of interpretation and discussion of results across the three policies, the coefficients are reported indicating the direction of the hazard rate to adopt.[11] Model 1 provides a baseline with economic and political covariates (Fiscal, Dem Gov, and Dem_Early). Model 2 includes state science policy variables (NGA Initiative and Early Science Advisor) and Models 3, 4, and 5 add in the benchmarking and R&D-related covariates (EPSCoR, University R&D, S&E Degrees, and High Tech Industry). The empirical results for each policy are discussed in turn.

---

11. Conversion of the coefficient, β_k, to the hazard rate: Hazard Rate = $(\exp(\beta_k)-1) * 100\%$.

Table 9.5   Correlation coefficients for covariates

	Fiscal	Dem Gov	Dem_ Early	NGA initiative	Early science advisor	EPSCoR	University R&D (industry)	University R&D (federal)	S&E degrees (LQ)	High-tech industry (LQ)
Fiscal	1									
Dem Gov	0.0325	1								
Dem_Early	0.0336	0.4253	1							
NGA initiative	0.0172	-0.0356	-0.0101	1						
Early science advisor	-0.0154	0.1335	0.0395	-0.0000	1					
EPSCoR	-0.0463	-0.0144	-0.0275	0.0321	-0.0643	1				
University R&D (industry)	-0.0876	-0.0281	0.0247	0.0597	0.2531	-0.42	1			
University R&D (federal)	-0.089	-0.0404	0.0144	0.0618	0.2566	-0.4688	0.8959	1		
S&E degrees (LQ)	0.0021	-0.002	-0.0366	-0.0029	0.0099	-0.2098	0.0142	0.0855	1	
High-tech industry (LQ)	0.02	-0.102	-0.0323	0.0053	0.0748	-0.5017	0.5044	0.5597	0.1563	1

Note: The time frame of interest spans 1982–2009. The number of observations is 1,450 (29 years * 50 states).

Table 9.6   **Empirical results of Eminent Scholars policy adoption**

Variables	ES (1)	ES (2)	ES (3)	ES (4)	ES (5)
Fiscal	2.376**	2.799**	2.860**	3.727***	3.717***
	(1.033)	(1.095)	(1.114)	(1.359)	(1.386)
Dem Gov	0.515	0.533	0.450	0.366	0.250
	(0.482)	(0.485)	(0.489)	(0.499)	(0.518)
Dem_Early	−0.825	−1.048	−1.002	−1.063	−1.051
	(0.795)	(0.802)	(0.806)	(0.804)	(0.806)
NGA initiative		−0.476	−0.515	−0.595	−0.596
		(0.541)	(0.547)	(0.573)	(0.579)
Early science advisor		1.556***	1.831***	2.243***	2.162***
		(0.459)	(0.508)	(0.536)	(0.554)
EPSCoR			−1.137*	−1.994***	−2.017***
			(0.625)	(0.735)	(0.762)
University R&D			−0.444*	−0.429	−0.445
(federal)			(0.251)	(0.275)	(0.294)
SE degrees Q1				2.666***	2.427***
				(0.922)	(0.934)
SE degrees Q2				1.913**	1.599*
				(0.869)	(0.897)
SE degrees Q3				0.619	0.541
				(0.865)	(0.871)
High-tech industry Q1					−0.049
					(0.957)
High-tech industry Q2					0.689
					(0.804)
High-tech industry Q3					0.554
					(0.686)

Notes: Cox proportional hazard model was run with adoption of the Eminent Scholars program serving as the transition indicator. The coefficients, β_k are reported indicating the direction of the hazard ratio. The number of observations is 1,146. Twenty state adoptions are considered in these models; Virginia was left-censored due to early adoption in 1964. Standard errors are in parentheses.

***Significant at the 1 percent level.
**Significant at the 5 percent level.
*Significant at the 10 percent level.

9.3.1 Eminent Scholar Results

Table 9.6 reports the coefficients from equation (1) for the Eminent Scholars program. The effect of the state fiscal growth (Fiscal) is positive and statistically significant. States exhibiting a growth in revenue increases adoption activity. The political covariates (Dem Gov and Dem_Early) are not statistically significant. Among the two state science policy covariates, the coefficients for Early Science Advisor are positive and significant. The cohort of twelve states that initially adopted state science policy units in the 1960s is more likely to adopt the Eminent Scholars program than those that did not have this position.

Turning to the set of benchmarking and R&D-related measures, the EPSCoR coefficient is statistically significant and negative demonstrating that lagging states in terms of R&D performance are less likely to try to attract eminent scholars. To reiterate, states receive EPSCoR status if their R&D performance falls below a minimum threshold. In other words, states without an EPSCoR status are more likely to adopt the Eminent Scholars program than those with the status. The negative coefficient for University R&D is weakly significant in Model 3 (p-value < 0.1) and not robust across Models 4 and 5.

The coefficients for S&E Degrees are positive and statistically significant for Quartile 1 and Quartile 2. In contrast to the referent category—Quartile 4, the cohort of states with the largest location quotients—states with lagging concentrations of S&E graduates are more likely to have this program. The size of the coefficients from the quartile dummies, notably Q1 and Q2, indicate that the likelihood to adopt the Eminent Scholars program increases as states fall in rank. Taken at first glance, this stands in contrast to the implications from the EPSCoR coefficients; this is discussed in greater detail following the discussion section. Lastly, the coefficients for the quartile rankings of High Tech Industry are not statistically significant.

9.3.2 University Research Grant Results

Table 9.7 reports the coefficients for the Cox proportional hazard model for the University Research Grants program. A number of the state R&D coefficients mirror the results from the Eminent Scholars model. Most notably, the coefficients for Early Science Advisor are positive and statistically significant across Models 2–5. Moreover, states with an EPSCoR status are less likely to adopt the Eminent Scholars program—as indicated by the negative and statistically significant coefficients. As for the benchmarking measure S&E Degrees, Quartile 1 is positive and statistically significant, though the effect is statistically insignificant for the two other quartiles. Regarding similarities in terms of insignificant results, the political variables (Dem Gov and Dem_Early) and NGA Initiative are statistically insignificant for this set of models.

In contrast to the Eminent Scholars results, however, the coefficients for Fiscal are not statistically significant. Results for University R&D are robust—negative and statistically significant. As federal investment in university R&D within states increases, the likelihood of states adopting the University Research Grants decreases. While we anticipated state policy decisions to complement external—notably federal—investment in university R&D, these results suggest a substitutive relationship, or crowding-out effect; this is discussed at greater length in the discussion section. Turning to the last set of benchmarking measures, the coefficient for Q1 for High Tech Industry is negative and statistically significant. Again, the fourth quartile—those states with the highest location quotients—serves as the referent cate-

Table 9.7   Empirical results of University Research Grants policy adoption

Variables	URG (1)	URG (2)	URG (3)	URG (4)	URG (5)
Fiscal	1.267	1.317	1.269	1.358	1.412
	(0.804)	(0.839)	(0.860)	(0.919)	(0.973)
Dem Gov	0.270	0.214	0.109	0.121	0.316
	(0.389)	(0.392)	(0.400)	(0.408)	(0.419)
Dem_Early	−0.350	−0.311	−0.244	−0.346	−0.466
	(0.576)	(0.577)	(0.581)	(0.584)	(0.588)
NGA initiative		0.117	0.091	0.091	0.108
		(0.408)	(0.416)	(0.420)	(0.424)
Early science advisor		0.794**	1.322***	1.467***	1.571***
		(0.400)	(0.430)	(0.454)	(0.455)
EPSCoR			−1.292***	−1.631***	−1.226**
			(0.498)	(0.532)	(0.545)
University R&D (federal)			−0.710***	−0.779***	−1.113***
			(0.187)	(0.196)	(0.237)
SE degrees Q1				1.063**	1.229**
				(0.510)	(0.577)
SE degrees Q2				−0.027	−0.075
				(0.609)	(0.633)
SE degrees Q3				0.103	0.452
				(0.545)	(0.573)
High-tech industry Q1					−2.097***
					(0.723)
High-tech industry Q2					−0.754
					(0.644)
High-tech industry Q3					−0.727
					(0.533)

Notes: Cox proportional hazard model was run with adoption of the University Research Grants program serving as the transition indicator. The coefficients, β_k are reported indicating the direction of the hazard ratio. The number of observations is 992. Twenty-nine state adoptions are considered in these models; every state is initially considered in this model. Standard errors are in parentheses.

***Significant at the 1 percent level.
**Significant at the 5 percent level.
*Significant at the 10 percent level.

gory. In contrast to states that lead in terms of high-tech industrial capacity, states ranked in the lowest quartile are less likely to adopt the University Research Grants program. In other words, this suggests that states leading along this measure are more likely to adopt the program than those that lag.

9.3.3 Centers of Excellence Results

Table 9.8 presents the results of equation (1) for the Centers of Excellence program. There is notable overlap in the results with the University Research Grants program, though the results are not as robust with this program. The effect for the Early Science Advisor is positive and statisti-

Table 9.8   **Empirical results of Centers of Excellence policy adoption**

Variables	CE (1)	CE (2)	CE (3)	CE (4)	CE (5)
Fiscal	1.011	1.061	1.087	1.196	1.408
	(0.877)	(0.879)	(0.920)	(0.975)	(1.107)
Dem Gov	−0.048	−0.040	−0.273	−0.308	−0.051
	(0.404)	(0.405)	(0.419)	(0.425)	(0.464)
Dem_Early	0.489	0.449	0.405	0.344	0.251
	(0.499)	(0.501)	(0.502)	(0.505)	(0.505)
NGA initiative		−0.375	−0.258	−0.236	−0.249
		(0.464)	(0.471)	(0.472)	(0.474)
Early science advisor		0.294	0.504	0.593	0.984**
		(0.409)	(0.416)	(0.435)	(0.473)
EPSCoR			−1.074**	−1.290***	−0.322
			(0.421)	(0.454)	(0.513)
University R&D (industry)			−0.370***	−0.416***	−0.738***
			(0.118)	(0.126)	(0.160)
SE degrees Q1				0.665	0.910
				(0.498)	(0.567)
SE degrees Q2				0.303	0.755
				(0.503)	(0.550)
SE degrees Q3				−0.136	0.319
				(0.560)	(0.575)
High-tech industry Q1					−2.646***
					(0.758)
High-tech industry Q2					−0.661
					(0.655)
High-tech industry Q3					−0.607
					(0.543)

Notes: Cox proportional hazard model was run with adoption of the Centers of Excellence program serving as the transition indicator. The coefficients, β_k are reported indicating the direction of the hazard ratio. The number of observations is 707. Thirty-four state adoptions are considered in these models; Alabama, Connecticut, North Carolina, and Michigan are left-censored due to early adoption prior to 1982. Standard errors are in parentheses.
***Significant at the 1 percent level.
**Significant at the 5 percent level.
*Significant at the 10 percent level.

cally significant, though the effect only holds for the full model (Model 5). As for EPSCoR, the coefficients are negative and significant for Models 3 and 4; however, the results are not robust in the full model. Regarding external R&D—notably industry investment in university R&D—the coefficients for University R&D are negative and statistically significant across all models. Again, we had expected a positive effect; the implications of this are discussed in the next section. Lastly, the effect on High Tech Industry is negative and statistically significant for Quartile 1. As with the University Research Grants results, the results are not significant for the other rankings.

The remaining covariates are statistically insignificant. Notably, as with both the results for the Eminent Scholars and University Research Grants, the political variables and the NGA Initiative do not show signs of statistical significance.

9.4 Discussion

This analysis presents the results from three empirical models to identify factors associated with the adoption of the Eminent Scholars, University Research Grants, and Centers of Excellence programs. As mentioned above, this analysis serves as an exploratory exercise to identify trends with state policy-making activity rather than claiming causality. The results for each policy offers insight on the factors associated with the state adoptions.

For the Eminent Scholars program, the results provide evidence to suggest that states leading in terms of R&D interest and activity are more likely to adopt the program. This is made evident from the results on the Early Science Advisor and EPSCoR variables. The former offers evidence to suggest that states that have demonstrated an early interest in state science policy are more likely to adopt the program. The latter highlights an association between state R&D capacity and the program. To reiterate, the Eminent Scholars program is a more upstream policy—in contrast to the Centers of Excellence—requiring substantial up-front costs to invest in distinguished research scientists and their labs. These earlier-stage investments are made with the expectation that the researchers will recover the costs as they develop their research and commercialize. The results for Fiscal, therefore, are not surprising, as states with fiscal growth are more likely to adopt this program. States exhibiting an increase in revenue have greater slack resources and are able to afford the program with longer-term returns on investments. Taken together, these results suggest that states rely on their R&D capacity as they invest in this program. This policy adoption can be viewed as states building upon their demonstrated strengths when considering this more upstream policy.

The results for the benchmarking measure, S&E Degrees, at first glance appear to stand in contrast to these conclusions. The positive coefficient for Quartile 1 indicates that states with a lagging ratio of S&E higher-education graduates are more likely to have the program. We might expect the direction of the coefficient to have the opposite sign, given the implications from the other variables that suggest that R&D capacity is associated with this policy adoption. In looking at the distribution of the S&E Degrees measure more closely, these results affirm what scholars already know about the complexity of the R&D process: strength along one dimension of R&D does not necessarily ensure strength along another. To better understand the distribution of the S&E Degrees quartile rankings, we consider it in contrast to the distribution of High Tech Industry quartile rankings; the former measure offers

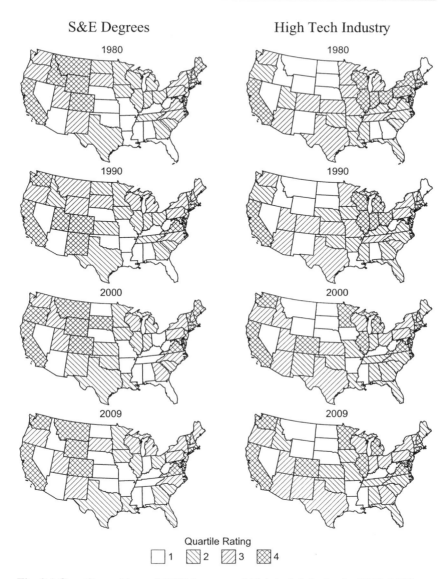

Fig. 9.1 Quartile rankings of S&E degrees and high-tech industry for 1980, 1990, 2000, and 2009

a more upstream measure of R&D capacity, while the high-tech measure is more downstream. Figure 9.1 presents a series of maps of these two measures showing how these measures have changed over the past forty years.

Most notably the leading cohorts of states for both measures vary. The left column in figure 9.1 illustrates a general trend that the states with leading ratios of S&E Degrees are concentrated along the West Coast and Rocky Mountain region, with a few located in New England and in the mid-Atlantic region. The lagging states demonstrate a greater concentration in the plains and Southeast. California, New York, Virginia, Maryland, and Washington are among the leaders along this measure, which is not surprising given their history of demonstrated R&D and economic performance. What is more unexpected, however, is the group of EPSCoR states—including Montana, Wyoming, Vermont, and Maine—that lead along this dimension as well. While the latter cohort lags in terms of its relative share of R&D activity— which qualifies them for EPSCoR status—these states produce a greater ratio of S&E degrees compared to the US average. This figure illustrates that the concentration of S&E Degrees is more varied and does not directly align with more traditional, downstream measures of R&D.

The more traditional, downstream activity is illustrated in the right column in figure 9.1. These maps of High Tech Industry more closely mirror the overall economic health of the state. This is illustrated by the notable concentration in the Rust Belt region in the 1980s and 1990s, followed by a shift to mid-Atlantic states and Washington, Colorado, Illinois, and Minnesota in the more recent decades.

Turning to the University Research Grants program, there is notable overlap for some of the coefficients with the results from the Eminent Scholars program. In particular, results for the Early Science Advisor and EPSCoR covariates similarly suggest that states with a demonstrated R&D capacity are more likely to adopt the University Research Grants program. Like the Eminent Scholars program, this is a more upstream program focused on university research with the expectation that the public investment will bolster capacity and yield returns in the longer term. This conclusion is additionally supported by the negative coefficient for Quartile 1 of the High Tech Industry covariate. State interest and investment in this program can be viewed as an effort to build upon their demonstrated strengths of more downstream R&D measures. As for the positive coefficient for Quartile 1 of the S&E Degrees covariate, it is important to take into consideration the distribution of the variable—as is highlighted above. It is important to recognize that the rankings along this more upstream measure do not directly correlate with common, more downstream perceptions of state R&D capacity.

While the negative coefficient for University R&D was only significant in Model 3 for the Eminent Scholars, these coefficients were robust for the University Research Grants models. Originally, we anticipated that the relationship between federal investment in university R&D would be positively associated with the adoption of the University Research Grants program. This relationship, in fact, suggests the opposite effect. This suggests that states are less likely to adopt the University Research Grants program when federal investment in university R&D increases. Another way to look at this is that states tend to adopt when federal investment in university R&D

decreases, pointing toward a crowding-out or substitutive effect. A substantial portion of the federal budget for university R&D is discretionary and historical trends of spending have been tenuous (Teich 2009). With annual threats to cut the federal R&D budget, this state science policy activity can be viewed as a proactive effort to provide public investment for university R&D. This suggests that states are not necessarily mirroring federal R&D actions, but rather are responding to decreases in federal spending by taking an active stance to support this program.

Results for the Centers of Excellence program are not as robust as the other two models. The coefficient for the Early Science Advisor is only positive and significant in the full model, though the coefficient for EPSCoR loses significance in the full model. While the effect is stronger for the other two policies, we interpret these weaker results to suggest that this program appeals to a broader cohort of states, not necessarily those with a demonstrated science capacity. States with an early, demonstrated interest in science and non-EPSCoR states are likely to have the program; however, the weak results indicate that even more states tend to adopt this program. The Centers of Excellence program stands out in contrast to the other two programs given that it is a later-stage, more downstream program. In other words, this program appears to attract broader appeal given that the investment is closer to more immediate, tangible, and economic outcomes. The negative coefficient on the High Tech Industry quartile (Q1) lends credence to this conclusion as well. States are cognizant of the strength of their R&D capacity—in terms of more downstream measures—when they consider adopting the Centers of Excellence program, which is centered on university and industry collaborations. Our interpretations of the negative coefficients on University R&D variables mirror the discussion with the University Research Grants program. While we anticipated a complementary association, the results indicate the opposite. We interpret this to suggest that states adopt this university-based R&D program when external support is lacking, demonstrating a bottom-up commitment from states.

Looking across the three programs, we see some overlap between the Eminent Scholars and University Research Grants programs and between the University Research Grants and Centers of Excellence programs. The former pair of policies is a set of more upstream investments in university R&D. Results suggest that states seek to build upon their strengths when making these investments. State commitment requires a stronger R&D capacity to offset the more upstream investment. Although there is overlap with the University Research Grants and Centers of Excellence programs, the results are not as robust for the Centers of Excellence. Both policies, in contrast to the Eminent Scholars, have a central research component— with the former focused on earlier-stage activity and the latter concerned with collaboration between the university and industry. The implications of the weaker results for the Centers of Excellence program suggest that the program attracts broader appeal to a larger cohort of states given that it is a program designed to produce more immediate tangible outcomes. In looking at the rate of adoption alone, this is not surprising, as 74 percent of states have adopted this program.

While there are commonalities across the portfolio of programs, the results highlight that state governments rely on a different set of incentives when adopting and maintaining these programs. We attribute these differences to the structure of the state programs with the Eminent Scholars and University Research Grants programs aimed at supporting earlier-stage, more basic university R&D activity, and the Centers of Excellence program designed to support later-stage R&D activity. Arguably the more downstream policy has greater appeal since this investment is closer to more tangible, economic outcomes.

9.4.1 Consideration of the Eminent Scholars, University Research Grants, and Centers of Excellence as a Portfolio

Thus far, we have focused on each program separately, taking into consideration some of their similarities and differences. Now we briefly consider these programs as part of a larger state university-based science portfolio. In looking at table 9.1, most states have adopted more than one policy. Figure 9.2 presents a series of snapshots of the continental US illustrating the path of diffusion of this portfolio of programs over the past three decades.[12] By 1990, marking one decade of state policy adoption activity, both North Carolina and Georgia established all three programs; by 2000, New York, Ohio, and South Carolina joined this cohort; and by 2009, five additional states adopted the entire portfolio.[13] These maps highlight a concentration of state-funded university R&D programs along the East Coast and Midwest with states in the Southeast, Rust Belt region, and lower Midwest demonstrating greater state policy efforts by adopting more programs. As of 2009, forty-four states had at least one university R&D policy. Of those, thirty-three had two policies and ten had adopted the entire portfolio.

Table 9.9 presents descriptive statistics for the years of adoption for having a first, second, and third policy, respectively. On average, states adopted one of these policies by 1989—with Virginia leading as the first adopter in 1964 and North Dakota serving as the most recent state to adopt their first state university-based policy in 2006. Of those states with more than one policy, on average, they adopted a second policy by 1996 and a third by 1999, respectively. For those states with more than one policy, table 9.10 provides information on the time lag between adopting a second and third policy.

---

12. A map for the baseline year was not included given the dearth of state university R&D policy activity at this time. In 1980, only four states—AL, CT, NC, and VA—had one of the three programs. Figure 9.1 is intended to reflect the diffusion of adoption, thus the first image of state policy activity is 1990.

13. These states include: AR, CT, KS, KY, and OK.

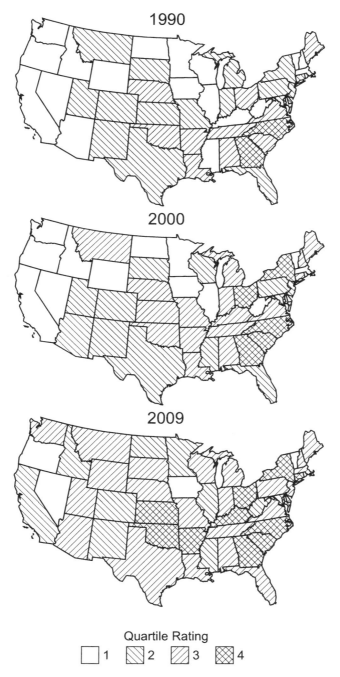

1990

2000

2009

Quartile Rating
☐ 1 ▨ 2 ▨ 3 ▨ 4

Fig. 9.2 Distribution of state university R&D portfolio—1990, 2000, and 2009

Table 9.9   Adoption years for portfolio of state university R&D programs

		Year adopt (state)		
Number of policies	Number of states	Mean	First adopter	Most recent adopter
1	44	1989	1964 (VA)	2006 (ND)
2	33	1996	1983 (SC)	2009 (MA)
3	10	1999	1986 (NC)	2006 (CT)

Table 9.10   Duration between policy adoptions within states

Policy time lag	Number of states	Mean (years)	Standard deviation	Min.	Max.
First to second	33	9.94	8.61	0	28
Second to third	9	8.56	6.67	0	17
First to third	9	17.11	11.37	0	41

On average, the time lag between adopting the first and second policy was roughly ten years; the average lag decreased slightly to 8.5 years for states between adopting the second and third policy.

9.5 Reflective Conclusions

Over the past thirty years, the fifty state governments have experimented with different programs that attempt to leverage academic science to create economic growth. Given the current economic climate and the federal government's tenuous commitment to R&D, we are likely to see more state efforts that attempt to leverage science resources for economic development. Most notably, for the more upstream Eminent Scholars and University Research Grants programs, states seek to build upon the strengths of their universities when making this investment. The more downstream Centers of Excellence programs have broader appeal, which is likely due to the fact that the program is designed to produce more immediate tangible outcomes through involvement with industry. It seems also that states that have a historical commitment to science policy have stayed true to this tradition and became early adopters while other states are motivated to adopt science policy as a means to enhance their ranking and leverage their strengths.

This analysis is part of a larger research agenda focused on understanding state science policy. This chapter not only offers greater insight on the typology of state science programs, it serves as an essential first step toward policy evaluation. It is a critical assumption that policy evaluations reliant on natural experiments depend on exogenous policy switches (Shadish, Campbell, and Cook 2002). This chapter takes this assumption seriously.

Only with this information, will researchers be equipped to begin to assess the efficacy of these programs.

The increased prevalence of these state-based science policies suggests that state policymakers have come to justify support of university R&D programs under the premise that R&D will stimulate innovation and thereby foster local entrepreneurship and economic growth. The adoption of these policies reflects the prevalence of new expectations for research universities. To the extent that states are investing in university research it is likely that policymakers will expect a return on their investments. Additional analysis and better data is required to see if these policies have had the desired impact. These quantitative analyses serve as an exploratory effort toward understanding whether and how state economic, political, and R&D-related factors influence the state science policy-making process. The logical next step is to examine their impact.

Public commitment to science research is evident by federal funding; what this research highlights, however, is that this public commitment stems from multiple levels extending to state governments as well. The large-scale adoption of these programs suggests that state experiments with science policy might fruitfully give way to national policies that would level the playing field between states. However, even when the majority of states have adopted the program there are some states that elect not to participate. We hope that with these preliminary efforts that we encourage scholars and policymakers to further investigate state science policies.

References

Aboelela, S. W., E. Larson, S. Bakken, O. Carrasquillo, A. Formicola, S. A. Glied, J. Hass, and K. M. Gebbie. 2007. "Defining Interdisciplinary Research: Conclusions from a Critical Review of the Literature." *Health Services Research* 42 (1, pt. 1): 329–46.

Baumgartner, F. R., V. Gray, and D. Lowery. 2009. "Federal Policy Activity and the Mobilization of State Lobbying Organizations." *Political Research Quarterly* 62 (3): 552–67. doi: 10.1177/1065912908322407.

Berglund, D., and C. Coburn. 1995. *Partnerships: A Compenduim of State and Federal Cooperative Technology Programs*. Columbus, OH: Battelle Press.

Berry, F. S., and W. D. Berry. 1990. "State Lottery Adoptions as Policy Innovations: An Event History Analysis." *American Political Science Review* 84 (2): 395–415.

Blossfeld, H.-P., K. Golsch, and G. Rohwer. 2007. *Event History Analysis with Stata*. Oxford: Taylor & Francis Group, Psychology Press.

Blume-Kohout, M. E., K. B. Kumar, and N. Sood. 2009. "Federal Life Sciences Funding and University R&D." NBER Working Paper no. 15146, Cambridge, MA.

Bozeman, B. 2000. "Technology Transfer and Public Policy: A Review of Research and Theory." *Research Policy* 29:627–55.

Combes, R. S., and W. J. Todd. 1996. "From Henry Grady to the Georgia Research Alliance: A Case Study of Science-Based Development in Georgia." *Annals of the New York Academy of Sciences* 798 (1): 59–77. doi: 10.1111/j.1749–6632.1996.tb24856.x.

Cozzens, S. E., and J. E. Melkers. 1997. "Use and Usefulness of Performance Measurement in State Science and Technology Programs." *Policy Studies Journal* 25 (3): 425–35. doi: 10.1111/j.1541–0072.1997.tb00032.x.

David, P. A., B. H. Hall, and A. A. Toole. 2000. "Is Public R&D a Complement or Substitute for Private R&D? A Review of the Econometric Evidence." *Research Policy* 29 (4–5): 497–529. doi: 10.1016/s0048–7333(99)00087–6.

Diamond, A. M. 1999. "Does Federal Funding 'Crowd In' Private Funding of Science?" *Contemporary Economic Policy* 17 (4): 423–31. doi: 10.1111/j.1465–7287.1999.tb00694.x.

Eisinger, P. K. 1988. *The Rise of the Entrepreneurial State: State and Local Economic Development Policy in the United States*. Madison: University of Wisconsin Press.

Feldman, M., L. Lanahan, and I. Lendel. 2014. "Experiments in the Laboratories of Democracy: State Scientific Capacity Building." *Economic Development Quarterly* 28 (2): 107–31.

Feller, I. 1990. "Universities as Engines of R&D-Based Economic Growth: They Think They Can." *Research Policy* 19 (4): 335–48.

———. 1997. "Federal and State Government Roles in Science and Technology." *Economic Development Quarterly* 11 (4): 283–95.

———. 2007. "Mapping the Frontiers of Evaluation of Public-Sector RD Programs." *Science and Public Policy* 34 (10): 681–90. doi: 10.3152/030234207x258996.

Fosler, R. S. 1988. *The New Economic Role of American States: Strategies in a Competitive World Economy*. New York: Oxford University Press.

Friedman, R. S., and R. C. Friedman. 1982. *The Role of University Organized Research Units in Academic Science*. State College: The Pennsylvania State University, Center for the Study of Higher Education, Center for the Study of Science Policy, Institute for Policy Research and Evaluation.

Hauger, J. S. 2004. "From Best Science Toward Economic Development: The Evolution of NSF's Experimental Program to Stimulate Competitive Research (EPSCoR). *Economic Development Quarterly* 18 (2): 97–112.

Hearn, J., A. Lacy, and J. Warshaw. 2014. "State Research and Development Tax Credits: The Historical Emergence of a Distinctive Economic Policy Instrument." *Economic Development Quarterly* 28 (2): 166–81.

Hecker, D. E. 2005. "Occupational Employment Projections to 2014." *Monthly Labor Review* (November):70–101.

Karch, A. 2007. "Emerging Issues and Future Directions in State Policy Diffusion Research." *State Politics & Policy Quarterly* 7 (1): 54–80. doi: 10.1177/153244000700700104.

Mallon, W. T., and S. A. Bunton. 2005. "Research Centers and Institutes in US Medical Schools: A Descriptive Analysis." *Academic Medicine* 80 (11): 1005–11.

Mintrom, M., and S. Vergari. 1998. "Policy Networks and Innovation Diffusion: The Case of State Education Reforms." *Journal of Politics* 60:126–48.

National Science Board. 2012. Science and Engineering Indicators 2010. Arlington, VA: National Science Foundation.

Payne, A. A. 2001. "Measuring the Effect of Federal Research Funding on Private Donations at Research Universities: Is Federal Research Funding More than a Substitute for Private Donations?" *International Tax and Public Finance* 8 (5): 731–51. doi: 10.1023/a:1012843227003.

Plosila, W. H. 2004. "State Science- and Technology-Based Economic Development Policy: History, Trends and Developments, and Future Directions." *Economic Development Quarterly* 18 (2): 113–26. doi: 10.1177/0891242404263621.

Ruegg, R. T., and I. Feller. 2003. *A Toolkit for Evaluating Public R & D Investment: Models, Methods, and Findings from ATP's First Decade.* Gaithersburg, MD: US Dept. of Commerce, Technology Administration, National Institute of Standards and Technology.

Santoro, M. D., and S. Gopalakrishnan. 2001. "Relationship Dynamics between University Research Centers and Industrial Firms: Their Impact on Technology Transfer Activities." *Journal of Technology Transfer* 26 (1): 163–71. doi: 10.1023/a:1007804816426.

Sapolsky, H. M. 1968. "Science, Voters, and the Fluoridation Controversy." *Science* 162 (3852): 427–33.

———. 1971. "Science Policy in American State Government." *Minerva* 9 (3): 322–48.

Shadish, W. R., D. T. Campbell, and T. D. Cook. 2002. *Experimental and Quasi-Experimental Designs for Generalized Causal Inference.* Boston: Houghton Mifflin.

Simon, H. A. 1978. "Rationality as Process and as Product of Thought." *American Economic Review* 68 (2): 1–16.

Taylor, C. D. 2012. "Governors as Economic Problem Solvers: A Research Commentary." *Economic Development Quarterly* 26 (3): 267–76.

Teich, A. 2009. *AAAS Analysis of Federal Budget Proposals for R&D in FY 2010.* Washington, DC: American Association for the Advancement of Science: Forum on Science and Technology Policy.

True, J., and M. Mintrom. 2001. "Transnational Networks and Policy Diffusion: The Case of Gender Mainstreaming." *International Studies Quarterly* 45 (1): 27–57.

Vogel, R. C., and R. P. Trost. 1979. "The Response of State Government Receipts to Economic Fluctuations and the Allocation of Counter-Cyclical Revenue Sharing Grants." *Review of Economics and Statistics* 61 (3): 389–400.

Volden, C. 2006. "States as Policy Laboratories: Emulating Success in the Children's Health Insurance Program." *American Journal of Political Science* 50 (2): 294–312.

Wayne, L. 1982. "Designing a New Economics for the 'Atari Democrats'." *New York Times,* September 26.

Wilson, D. J. 2009. "Beggar Thy Neighbor? The In-State, Out-of-State, and Aggregate Effects of R&D Tax Credits." *Review of Economics and Statistics* 91 (2): 431–36. doi: 10.1162/rest.91.2.431.

Youtie, J., B. Bozeman, and P. Shapiro. 1999. "Using an Evaluability Assessment to Select Methods for Evaluating State Technology Development Programs: The Case of the Georgia Research Alliance." *Evaluation and Program Planning* 22(1): 55–64.

Youtie, J., D. Libaers, and B. Bozeman. 2006. "Institutionalization of University Research Centers: The Case of the National Cooperative Program in Infertility Research." *Technovation* 26 (9): 1055–63.

Zucker, L. G., and M. R. Darby. 1996. "Star Scientists and Institutional Transformation: Patterns of Invention and Innovation in the Formation of the Biotechnology Industry." *Proceedings of the National Academy of Sciences* 93 (23): 12709–16.

IV

Historical Perspectives on Science Institutions and Paradigms

The Endless Frontier
Reaping What Bush Sowed?

Paula Stephan

10.1 Introduction

Science emerged from World War II triumphant. Its contributions to the war effort included the Manhattan Project, radar, DDT, and penicillin. Its triumphs were sufficient to cause one National Institutes of Health scientist to remark that from the end of the war on, "science was spelled with a capital 'S' and research with a capital 'R'" (Strickland 1989, 17).

The time was ripe for funding for scientific research to gain a firm national footing. No one understood this better, or was better positioned to promote it, than Vannevar Bush, President Roosevelt's science advisor and Director of the Office of Scientific Research and Development. Sensing that the moment was propitious for a public initiative, Bush maneuvered for Roosevelt to request a report laying out a federal course of action. The request was duly dispatched and in the late fall of 1944 Bush set about writing what was to bear the name *Science: The Endless Frontier*.

Paula Stephan is professor of economics at Georgia State University and a research associate of the National Bureau of Economic Research.

I have benefited from the comments of participants at the "Changing Frontier: Rethinking Science and Innovation Policy" preconference and conference workshops, especially those of the organizers Adam Jaffe, Benjamin Jones, and Bruce Weinberg, the discussant. I have also benefited from discussions with Gregory Petsko and comments by Bill Amis. Various individuals helped in supplying unpublished data and answering inquiries. They include Robert Buhrman, Ronald Ehrenberg, John Griswold, Vicky Harden, Barbara Harkins, Peter Henderson, Richard Mandel, Joseph November, James Schuttinga, Buhm Soon, Lori Thurgood, and Hui Wang. Research assistance was provided by Nicholas Heaghney and Rhita Simorangkir. For acknowledgments, sources of research support, and disclosure of the author's material financial relationships, if any, please see http://www.nber.org/chapters/c13034.ack.

The report, issued in July of 1945, recommended a three-pronged course of action for the federal government.[1] First, the government should fund basic research at universities and medical schools because these "institutions provide the environment which is most conducive to the creation of new scientific knowledge and least under pressure for immediate, tangible results" (Bush 1945, 7). Second, the government should provide scholarships and fellowships to promote training. Both the research and training initiatives, it argued, were essential for economic growth; both addressed the concern that, due in part to the war, the United States faced a scientific deficit in terms of basic research and the highly trained individuals required to conduct the research. Third, the report recommended that the government continue to conduct research of a military nature during peacetime.

Science: The Endless Frontier "established an intellectual architecture that helped define a set of public science institutions that were dramatically different from what came before yet largely remain in place today."[2] It also gave birth to and nurtured a university culture that, although initially a bit skeptical of federal support, quickly began to ask for more—not only from the federal government, but also from faculty and staff. In the process, the research environment at universities underwent substantial change.

This chapter sets out to examine how *The Endless Frontier* changed the research landscape at universities, the response of universities to the initiative, stresses that have emerged in the system, and the implications they have for discovery and innovation. To cut to the chase: *The Endless Frontier* set about to grow research capacity at universities and increase the supply of individuals qualified to do research. Initially the agencies it established were in missionary mode, recruiting research proposals from faculty and applications from students for fellowships and scholarships. By the 1960s, however, the tables had begun to turn and universities, having tasted federal fruit, aggressively began to push for more resources from the federal government in terms of funds for research, support for faculty salary, and indirect costs. Universities also began to demand more from faculty in terms of external support for their research and support for graduate students. The process transformed the relationship between universities and federal funders; it also transformed the relationship between universities and faculty.

The plan of the chapter is as follows: section 10.2 describes the university research enterprise at the end of the war. Section 10.3 focuses on the early days at the National Institutes of Health (NIH) and the National Science Foundation (NSF). Section 10.4 examines the universities' response to federal funding during the 1960s. Section 10.5 focuses on the years 1970 until 2012.

---

1. Bush assembled a staff to assist in drafting the report. One of its members, Paul A. Samuelson, wrote an account of his role in the report in 2009 (Samuelson 2009).

2. Adam Jaffe and Benjamin Jones, e-mail to possible participants of the NBER conference, "The Changing Frontier: Rethinking Science and Innovation Policy," April 5, 2012.

Section 10.6 takes stock of how the university research enterprise has evolved and changed since *The Endless Frontier*. Section 10.7 examines stresses to the system and the chapter ends with concluding thoughts (section 10.8).

10.2 The Scientific Landscape Circa the 1940s and *The Endless Frontier*

Despite the large number of universities and colleges in the United States at the time Bush authored *Science: The Endless Frontier*, only ten to fifteen could be considered top research universities.[3] The number of medical colleges doing research was even smaller. The typical medical school's faculty was largely composed of part-time clinicians with minimal interest in research.

Bush estimated that $31 million was spent on research at universities and medical schools in 1940 ($513 million in 2013 dollars—or less than 1 percent, in real terms, of what was spent on university R&D in 2012); almost all the funds came from endowments, private foundations, and donations (Bush 1945). The small amount of university research supported by the federal government came by way of contracts. Grants as a mechanism for supporting research were rare.

Expenditures for research equipment and materials were modest by today's standards. The 200-inch reflecting telescope that Caltech was building at the time—later named the Hale—cost approximately $6 million dollars, or $79 million in today's dollars. By comparison, the Thirty Meter Telescope (TMT) that is currently on the drawing boards, a joint project of Caltech and the University of California, has an estimated price tag of $1 billion. The first model for Lawrence's cyclotron, built with wire and sealing wax, cost approximately $25, not enough in today's dollars to pay for a minute of the electricity required to run the Large Hadron Collider built by the European Organization for Nuclear Research (CERN), estimated to have cost about $8 billion at the time it first came online in 2008. Labs in chemistry and the biomedical sciences were reliant on tabletop equipment. Organisms used in research were often of the garden variety—worms, fruit flies, and mice.

At the time of World War II, forty-seven institutions awarded the PhD degree in mathematics, fifty-five in physics, seventy-four in chemistry, thirty-nine in earth sciences, thirty-seven in engineering, and seventy-four in the life sciences (table 10.1). Production of PhDs in science and engineering (S&E) had grown steadily during the 1930s, going from 895 in 1930 to 1,379 in 1939 (figure 10.1).[4] By 1940, the number of degrees awarded in S&E

---

3. Based on the number of doctoral degrees conferred in science and engineering, the ten-to-fifteen included the University of Chicago, Columbia, Cornell, the University of Wisconsin, Harvard University, Johns Hopkins University, the University of Illinois, University of California, Berkeley, and Yale. Data provided by Lori Thurgood, unpublished.

4. Throughout this chapter S&E is defined to include engineering, geosciences, life sciences, math and computer sciences, and the physical sciences.

Table 10.1   Number of doctorate-granting institutions in the United States by five-year period, 1929–1974

Field	1920–1924	1925–1929	1930–1934	1935–1939	1940–1944	1945–1949	1950–1954	1955–1959	1960–1964	1965–1969	1970–1974
Mathematics	22	33	43	45	47	49	71	74	91	127	159
Physics	28	37	46	55	55	54	74	84	114	150	167
Chemistry	43	47	66	76	74	84	100	112	143	171	194
Earth sciences	24	24	37	39	39	38	50	59	74	96	121
Engineering	19	24	32	37	37	49	63	75	97	127	151
Life sciences	42	57	65	70	74	81	99	122	144	178	224

Source: National Academy of Sciences (1978, 95).

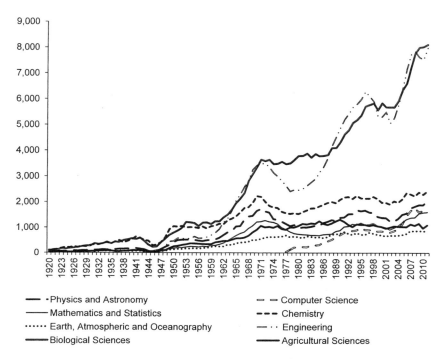

Fig. 10.1 PhD production in science and engineering, 1920–2010
Source: Unpublished NSF records and NSF Webcaspar.

was 1,618. As the war accelerated, however, the number of students enrolled in graduate school declined and PhD production in science and engineering fell to 1,030 in 1944 and 743 in 1945 (figure 10.1). Clearly, a deficit was in the making.

Time spent in doctoral training was considerably shorter than time spent in training today. Although data are sparse, Bush estimated that it took about six years from high school to get a doctorate. Bush completed his own doctoral training in electrical engineering in two years.

The principal objectives of *Science: The Endless Frontier* with regard to universities were to promote basic research through the provision of federal funds for research and to promote training of a future work-force by providing fellowships for doctoral and postdoctoral training and scholarships for undergraduate students. When it came to research, Bush not only wanted to support research at established universities and medical schools, but also wanted to build up less strong departments, especially at medical schools, which he saw as particularly lacking in terms of research capacity. With regard to training, while Bush advocated that training should occur in a research environment, he never suggested that the two should be jointly funded. Rather, he saw the two as separate activities.[5]

Bold for its time, the price tag was modest by today's standards. Bush envisioned that support for medical research would go from $5 million a year to $20 million a year in the fifth year "where it is expected that the operations have reached a fairly stable level" ($65 million to $260 million in 2013 dollars). With regard to the natural sciences, Bush saw funding going from $10 million to $50 million ($130 million to $650 million in 2013 dollars). Bush also saw stability of funding as key: "Whatever the extent of support may be, there must be stability of funds over a period of years so that long-range programs may be undertaken" (Bush 1945, 29).

The implementation of *Science: The Endless Frontier* was largely the responsibility of two federal agencies: the National Institutes of Health (NIH), which had been formally established in 1930, and a new federal organization for research, referred to in the report as the National Research Foundation.[6] Providing funds to the firmly established NIH proved much easier than establishing the new agency that Bush envisioned and the NIH clearly benefited from the political hurdles faced in creating the latter.[7] It was not until 1952 that the National Science Foundation opened for business.

10.3 Early Years of the NIH and the NSF

10.3.1 NIH

The NIH's budget in 1948 of $25 million was reasonably consistent with what Bush had envisioned for health-related research. However, by 1950, in nominal terms, the budget had almost doubled to $48 million. It doubled again by 1956, and again by 1958, and still again between 1958 and 1960, where it stood at approximate $400 million ($3.1 billion in 2013 dollars). Clearly, Bush had underestimated the amount of funds that would be directed to health research (National Institutes of Health).

In its early years, NIH was in missionary mode, encouraging institutions and individuals to submit proposals. To quote Fred Stone, circa 1950, an NIH official who later became the director of the National Institute of General Medical Science (NIGMS), "It wasn't anything to travel 200,000 miles a year" (Strickland 1989, 38). This was consistent with the NIH's view

5. See discussion in Teitelbaum (2014).

6. Health research became more consolidated in 1944 when the National Cancer Institute, established in 1937, was incorporated into the NIH.

7. A primary opponent of Bush's plan was Senator Kilgore of West Virginia, whose proposal to create a national science foundation, first introduced in 1942, had, as one of its objectives, the "geographic" distribution of the funds. It took five years to work out a compromise, which included among other things the provision that the new agency was to avoid an "undue concentration" of its funds. Finally, in 1952, the National Science Foundation became operative. See http://www.nsf.gov/pubs/stis1994/nsf8816/nsf8816.txt.

of its mission, which was not only to support top research but to build programs. The NIH also built capacity by supporting the construction of facilities at universities.

Grants were initially reviewed by sending them out to eminent scientists (National Institutes of Health 1959). But by 1946 the concept of study sections had evolved, and henceforth, peer review was to be organized around these. Success rates were high, by all accounts 65 percent or more (Division of Research Grants 1996). Requests were reasonably modest. The average grant, which was approximately $9,000 ($87,000 in 2013 dollars), lasted approximately a year (Munger 1960). This quickly changed. By 1951 the average duration of a grant was 1.8 years, by 1955 it was 2.5 years, and by 1957 it was 3.2 years (Munger 1960).

In its early years the NIH adopted the policy that the renewal award documents show the number of years of previous support for a particular project, a "high number portending a long-term commitment" (Appel 2000, 211). Not surprisingly, success rates for renewals were even higher and investigators became reluctant to change research focus.

Indirect rates were low: 8 percent. As early as 1951 various university and medical associations asked that it be raised to 15 percent. The request was refused (Division of Research Grants 1996). In 1956, however, the rate was raised to 15 percent; it was raised again to 25 percent in 1958 (Munger 1960). Although the goal was for grants "to add rather than replace support from the parent institution" (National Institutes of Health 1959), at some point in its early years, if requested, the NIH began to pay for a portion of faculty salary on the grants. Indeed, one reason that individuals reportedly preferred NIH grants over NSF grants in the early years was precisely for the ability to write off salary at the NIH (Appel 2000). While the NIH's extramural grants program focused on individual research projects, it also included funding for facilities and for equipment (Strickland 1989).[8]

Grants were heavily concentrated in the early years at a handful of institutions. Columbia University headed the list, receiving more than 5 percent of the funds, followed by Johns Hopkins, New York University, Harvard University, and the University of Minnesota (appendix table 10A.1). Taken together, the top ten institutions in 1948 received slightly more than one-third of all the NIH award funds; the top fifty received approximately 75 percent. Despite the heavy concentration, approximately 120 universities, medical schools, and colleges received one or more of the 795 research grants that institutions and hospitals were awarded that year.[9]

8. See November (2012) for a discussion of the conscious and directed effort on the part of the NIH in the 1950s and early 1960s to computerize the fields of biology and medicine.
9. A document dated 1948 lists names and amounts for the 198 institutions that received Public Health Service Grants in Aid in 1948. At least seventy-nine of these were independent research organizations, hospitals or, in a few cases, foreign institutions. See http://history.nih .gov/research/downloads/PHSResearchGrantsinAID-June30th1948.pdf.

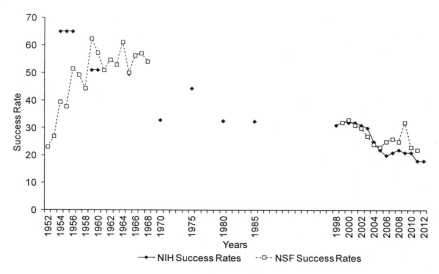

Fig. 10.2 The NIH and NSF success rates, available years

Source: Appel (2000) and various reports to the National Science Board on the NSF's Merit Review Process, various fiscal years. Data for NIH rates are from Chubin and Hackett (1990) and the NIH data book. See http://report.nih.gov/NIHDatabook/Charts/Default.aspx?showm =Y&chartId=124&catId=13.

Note: The NSF rates for 1952–1968 are for the Division of Biological and Medical Sciences; those for 1999 and thereafter are for all of the NSF.

Outreach was met with increased demand. The number of research projects reviewed by study sections almost tripled in the late 1950s, going from 2,750 to 7,975 (Division of Research Grants 1996). The average request also increased, going from $12,500 to $19,500 ($101,500 to $153,500 in 2013 dollars) (Division of Research Grants 1996). Approval rates fell in the 1950s from 65 percent to the low 50s (figure 10.2). It was not solely a question of the availability of funds. It was also a strategic decision to signal to Congress and the president that the NIH only funded quality research (Division of Research Grants 1996).

The NIH saw the shortage of talent to be a major bottleneck in getting the research done. According to Mary G. Munger, writing in 1960 on the history of the first twelve years of the NIH, "from the beginning of the extramural research grants programs, the lack of a sufficient number of qualified research investigators was a continuing bottleneck" (Munger 1960). To promote training, the NIH awarded predoctoral and postdoctoral fellowships, selecting applicants in house. However, it rapidly shifted some of the responsibility for selection to institutions, with the creation of training grants awarded to institutions to train individuals that they selected. Stipends started at $1,800 ($19,250 in 2013 dollars) for a first-year predoctoral fellowship and $4,500 ($48,000 in 2013 dollars) for first-year postdoctoral fellowships; most recipients of the latter in the early years were medical doctors. Allowances were also provided for dependents, travel, and tuition (National Institutes of Health 1959). When concern was raised in 1948 that "NIH fellows were being used simply as research assistants, as extra pairs of hands, as cheap labor" the NIH changed and strengthened the criteria for fellowships, trying to ensure that the fellow not "remain a sidekick to a senior scientist for an indefinite length of time"(Strickland 1989, 45).

10.3.2 NSF

The NSF's initial budget for 1952 was meager compared to that of the NIH's, starting at $3.5 million ($30.5 million in 2013 dollars). It grew rapidly, however, during the 1950s and by 1960 total obligations for the NSF were $158.6 million ($1.2 billion in 2013 dollars), or approximately 40 percent the size of the NIH's budget at the time (Appel 2000).[10] Although committed to quality, the NSF, like the NIH, made an effort to identify "atypically good researchers in underdeveloped institutions" (Appel 2000, 59).

Like the NIH, the NSF also awarded funds in the form of grants to assist faculty in doing research rather than award contracts for the purchase of research. Grants were reviewed and scored on a five-point scale by panels, populated through the "old boys network" (Appel 2000). In the early days, it was even possible to be a member of a review panel and have one's own research proposal reviewed and funded. Although success rates were initially below 30 percent, reflecting pent-up demand, by the mid-1950s success rates had grown, with but one exception, to over 50 percent (figure 10.2). In 1959 the success rate was 62 percent (Appel 2000).[11] Renewals (although the NSF, unlike the NIH, did not formally refer to them as such) had significantly higher success rates, always over 80 percent. Requests were generally for modest amounts. The median award in 1952 was $9,000 ($78,000 in 2013 dollars—identical to that at the NIH in the late 1940s); however, the average grant lasted for two years instead of one. By the late 1950s the duration of grants had lengthened, especially for strong investigators, who often received funding for three to five years. The size of the grant also increased. Leading researchers could count on $20,000 a year, and in some instances as much as $30,000 a year ($159,000 to $237,000 in 2013 dollars) (Appel 2000).

Indirect rates were initially set at 15 percent, but were raised to 20–25 percent by the mid-1950s (Appel 2000). From its beginnings, the NSF willingly supported two months of summer salary but resisted supporting academic-year salaries; the NSF leadership saw this as the responsibility of the university. Despite the opposition, in some instances support for academic-year salary was provided. Moreover, facile administrators and

10. This figure overstates the disparity between the two for support of university research because a goodly portion of NIH funds supported intramural research programs while NSF did not have an intramural research program.

11. Success rates are for the division of Biological and Medical Sciences (Appel 2000).

scientists could move money from one budget category to another after the award had been made (Appel 2000). The NSF also provided funds for the purchase of large instruments, supplies, travel, publication, educational projects, technicians, and facilities.

In its first year of operation, the NSF awarded ninety-eight grants totaling $1.1 million ($9.5 million 2013 dollars); sixty colleges and universities were recipients. The largest amount of funding was awarded to The California Institute of Technology (6.9 percent), followed by Indiana University, Bloomington (5.2 percent). The number of academic institutions receiving grants grew by 25 percent the next year; the number of awards increased to 172, and funding increased to $1.7 million ($14.5 million in 2013 dollars). The largest amount of funding went to Harvard University (6.5 percent), followed by Yale (6.3 percent). Taken together, the top ten institutions received 42 percent of the award funds (table 10A.2).

Consistent with Bush's vision and mission to build capacity, the Division of Scientific Personnel and Education was established within the NSF as part of the initial NSF act to award fellowships to students for graduate training. The selection process was overseen by the National Research Council. In the early years, the division awarded between 500 and 600 fellowships a year. The original stipend was for $1,600 ($13,900 in 2013 dollars), plus tuition and fees. The fellowship was usually awarded for three years (Freeman, Chang, and Chiang 2005). The division also awarded fellowships for postdoctoral training. From the beginning, graduate students and postdoctoral students were also supported on faculty grants. An audit of grants awarded by the division of Biological and Medical Sciences (BMS) in 1956 showed that 75 percent of one unit's awards supported predoctoral students; 20 percent of the units' awards included salaries for postdoctoral fellows (Appel 2000).

10.3.3 Other Federal Sources, Sputnik and the NDEA

Data are sparse to document in any detail the amount of research funds that came to universities from other federal agencies during the late 1940s and 1950s. Clearly, however, agencies other than NIH and NSF supported university research. Key among these was the Department of Defense (DOD), whose budget for research grew dramatically during the Cold War. The DOD funding, unlike that of the NSF and NIH, was highly concentrated at a handful of institutions. At the top was the Massachusetts Institute of Technology (MIT), which in the late 1940s had seventy-five separate contracts for defense-related work, totaling $117 million (Leslie 1993). Caltech was next with $83 million in contracts, and Harvard a far third with $31 million. (Assuming that these figures are for 1947, this represents, respectively, $1.25 billion, $888 million, and $331 million in 2013 dollars.) Throughout the Cold War, MIT maintained its dominant position, receiving more in contracts than many large industrial defense contractors. Unlike the NIH

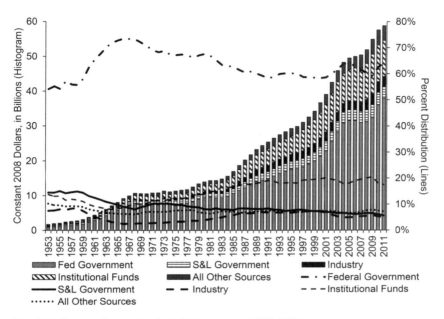

Fig. 10.3 Support for academic R&D by sector, 1953–2011
Source: NSF Webcaspar.

and NSF model, however, whose funds went primarily to individual investigators, DOD funds were directed to interdisciplinary research labs at universities. It is also notable that funds came in the form of contracts, not grants. Other universities learned from MIT's experience and used postwar defense contracts to propel themselves into research- university status. Stanford was an early example; more recently the Georgia Institute of Technology and Carnegie Mellon have benefited from defense-related research (Leslie 1993; Stephan and Ehrenberg 2007).

In 1957 the Soviet Union launched Sputnik. The United States responded in part by dramatically increasing federal support for university research, which nearly quadrupled during the period 1958–1968, going from $2,720 million in constant 2008 dollars to $10,685 million (figure 10.3). Universities also benefited from the scholarships and fellowships for students that the federal government provided post-Sputnik through the National Defense Education Act (NDEA). Retrospectively, the 1960s would be seen as the "golden age" of university research.

10.4 The University Response to Capacity Building: The 1960s

Universities were extremely responsive to the capacity-building initiatives of the NSF and NIH, increasing the number of PhDs they trained and the number of grants they submitted. But while the 1950s can be seen as a period where the federal government took the initiative in building the capacity of universities to do research, the 1960s can be seen as a transition period in which the tables began to turn. Universities not only responded to the government's capacity-building initiative, they began to aggressively push the government to cover salaries on grants and raise the allowable indirect rate. In short, before the 1960s, the federal government was pushing universities to develop research and training capacity and to perform research. After that, the roles were reversed and universities began to push the federal government for funds. Positive feedbacks of the system had begun to emerge, feedbacks that Vannevar Bush had not foreseen.

A number of metrics show the success with which capacity was built during the 1950s and 1960s. For example, the number of PhD recipients awarded in 1959 was 250 percent higher than its prewar high (figure 10.1). It was not just that traditional prewar programs were educating more PhDs, but that new programs were being created. Between the early 1950s and the early 1960s, the number of PhD programs increased by over 40 percent in all fields save math (table 10.1). Strong federal funding for students and research provided incentives for PhD production to continue to grow in the 1960s, tripling during the decade. Once again, it was not only that there were more PhDs. There were more programs. By the end of the decade, 27 percent of PhDs were being awarded by the top ten PhD-granting institutions in science and engineering, compared to 68 percent four decades earlier. American higher education was becoming more democratized (table 10.2).

The growth in PhD production was due in large part to the dramatic increase in federal support for PhD study after the war. The expansion, particularly in the late 1960s and early 1970s, was also encouraged by the availability of draft deferments for graduate study until 1968. In the short period between 1966 and 1970, the number of science and engineering doctoral degrees awarded per thousand thirty-year-olds in the US population increased by almost 50 percent, going from nine to thirteen (National Science Foundation 1994).

The federal government played virtually no role in the support of PhD students prior to World War II. During the 1950s, however, the federal government began to play a major role through the provision of fellowships by the NSF and NIH and also through the support of training programs. Moreover, a new use of federal research funds began to emerge in the 1950s— support of a graduate research assistant—on a faculty member's grant. By 1961, for example, research grants in BMS at the NSF supported 985 predoctoral students, 27 percent of all PhD degrees awarded in the bio sciences in the years 1959, 1960, and 1961 (Appel 2000). Across all NSF directorates, in 1966, a year for which data are readily available, the NSF supported almost 11,000 graduate students: 23.4 percent on fellowships, 35.9 percent on traineeships, and 34.6 percent as research assistants on faculty grants

(National Science Board 1969). The same year the NIH supported almost 10,000 graduate students—25.7 percent on fellowships, 47 percent on training grants, and 25.6 percent as research assistants.[12]

Other federal agencies also supported graduate students. The Atomic Energy Commission supported a substantial number of research assistants and NASA had a large training grant program. In addition, "other" federal agencies supported approximately 10,600 graduate students in 1966, the majority of whom (63 percent) were as research assistants (National Science Board 1969).

The federal government also built capacity by supporting postdoctoral fellows. Although the concept of postdoctoral study dates to 1919 (Assmus 1993), support for postdoctoral study before the war was minimal and the support that did exist was largely provided by private foundations such as the Rockefeller Foundation. From the very beginning, however, the NIH saw postdoctoral study as a major way to build research capacity. Throughout the 1950s and 1960s the number of postdocs supported on training grants grew, as did the number supported on fellowships. While some of these postdoctoral positions were for study at the NIH, many were for postdoctoral study at a university or medical school. Although data are sparse, the inference can be made that in 1969 the NIH was supporting about 6,050 individuals on postdoctoral fellowships and training grants.[13] The NIH supported additional postdocs on faculty research grants, although the number cannot readily be determined. The NSF also allowed faculty to pay the salaries of postdocs from research grants. The BMS in 1961, for example, supported 213 postdocs on research grants, approximately 9 percent of the PhDs awarded in biology during the two preceding years (Appel 2000).

The support of research assistants and postdocs on federal research grants meant that the government was now supporting graduate students and postdocs in order to get the research done now and not only supporting graduate students and postdocs on fellowships and training grants to build future research capacity. Perhaps because of this new role, the median time individuals spent in a PhD program (measured as "registered time") grew slightly, going from 4.9 in the physical sciences and 5.0 in the life sciences to 5.1 and 5.3, respectively, between 1958 and 1963 (table 10.3).[14] If Vannevar Bush's two-year degree is even remotely representative, time to degree had grown considerably since he received his degree in 1917. The observation is consistent with the finding that individuals supported on training grants

12. The importance of training grants and fellowships increased during the decade. By 1969, the NIH reported supporting 9,500 students in such positions. The 1969 number represents the peak of NIH support for students in the form of fellowships and training grants. By the end of the 1970s, the NIH was supporting fewer than 5,000 a year on training grants and fellowships.

13. Estimate based on the assumption that 37.9 percent of the individuals supported were postdocs, basing this proportion on data for 1992 (National Research Council 1994).

14. Table 1–3 (National Science Board 1969).

Table 10.2   Top twenty-five universities awarding PhDs in science and engineering (1920–1924, 1968, 2011)

Name of institution 1920–1924	Number of PhDs 1920–1924	Name of institution 1968	Number of PhDs 1968	Name of institution 2011	Number of PhDs 2011
University of Chicago	347	University of Illinois at Urbana-Champaign	409	Stanford University	512
Columbia University in City of NY	264	University of California, Berkeley	391	University of California, Berkeley	510
University of Wisconsin-Madison	215	University of Wisconsin-Madison	382	Massachusetts Institute of Technology	504
Cornell University	207	Purdue University	300	University of Florida	503
Johns Hopkins University	186	Massachusetts Institute of Technology	292	University of Illinois at Urbana-Champaign	494
Harvard University	152	University of Michigan	282	University of Michigan	487
University of Illinois at Urbana-Champaign	144	Stanford University	274	Purdue University	472
University of California, Berkeley	132	Cornell University	270	University of Wisconsin-Madison	470
Yale University	125	University of Minnesota, Twin Cities	241	Pennsylvania State University, main campus	436
University of Minnesota, Twin Cities	83	Ohio State University	206	University of Washington	426
Ohio State University	80	University of Texas at Austin	204	University of Minnesota, Twin Cities	417
University of Michigan	80	Iowa State University	201	Georgia Institute of Technology	407
University of Iowa	79	Michigan State University	198	Ohio State University	398
University of Pennsylvania	71	University of California, Los Angeles	187	University of California, Los Angeles	377
Princeton University	65	Harvard University	186	University of California, Davis	376
Massachusetts Institute of Technology	56	University of Washington	156	Texas A&M University, main campus	373
Stanford University	41	Columbia University in the City of New York	155	University of California, San Diego	344

University		University		University	
George Washington University	36	Case Western Reserve University	151	Cornell University	339
Clark University	33	U. of Maryland, College Park	146	University of Texas at Austin	328
New York University	29	Pennsylvania State University	143	University of Maryland, College Park	325
University of Pittsburgh	26	Johns Hopkins University	140	Johns Hopkins University	317
Iowa State University	23	Northwestern University	138	University of North Carolina at Chapel Hill	302
Washington University in St. Louis	21	University of Pennsylvania	136	North Carolina State University at Raleigh	300
Indiana University Bloomington	20	Texas A&M University	135	Columbia University in the City of New York	297
Rutgers University	20	New York University	131	Virginia Polytechnic Institute and State University	288
Total top 10	1855		3047		4814
Percent	68.1		27.1		17.12
Total top 25	2535		5454		10002
Percent	93.1		48.63		35.67
Total PhDs in S&E awarded	2724		11215		28042
Total number of institutions awarding a PhD in S&E	55		194		326
HHI	426		139		83

Source: 1920–1924 data, Lori Thurgood, correspondence of unpublished tabulations; 1968 and 2011 NSF Webcaspar.

Table 10.3   Registered time to PhD degree, selected years

Year	Physical sciences	Engineering	Life sciences
1958–1960	4.9	5.0	5.0
1963	5.1	5.1	5.3
1968	5.1	5.1	5.3
1973	5.7	5.6	5.5
1978	5.9	5.8	5.9
1983	6.1	5.9	6.2
1988	6.3	6.0	6.6
1993	6.7	6.5	7.0
1998	6.7	6.7	7.0
2003	6.8	6.9	6.9
2008	6.7	6.7	6.9

Source: Survey of Earned Doctorates. The NSF/NIH/USED/USDA/NEH/NASA Survey of Earned Doctorates, updated data. Source for 1958–1960 (National Science Board 1969, 24).

and fellowships completed graduate training one to two years earlier than those not supported on these grants (Coggeshall and Brown 1984). It is also consistent (see below) with a view expressed in the Seaborg report.

Increased capacity meant greater demand for research grants as newly minted PhDs came of professional age and joined their elders in submitting grants. By way of example, the number of proposals received by BMS at the NSF grew from approximately 300 in 1952 to 2,462 by 1968 (Appel 2000). The number of competing research project applications at the NIH went from 2,750 in 1956 to 7,975 in 1960 (Divison of Research Grants 1996). Not surprisingly, success rates began to decline (figure 10.2). The increase in submissions continued to grow. In 1987, for example, the Division of Research Grants at NIH received 33,804 proposals (Strickland 1989). The number of institutions supported by the NIH grew as well, going from 120 in 1948 to 330 in 1971 (Figure 10.4). At the NSF, the figure went from 75 in 1953 to 314 in 1971. Grants became less concentrated as measured by the percent of funds that the top ten institutions received, going from 42 percent in 1953 to the low 30s in the early 1970s at the NSF (figure 10A.1). At the NIH it went from 36.3 percent in 1948 to the mid-20s (figure 10A.2). The DOD funds, which were highly concentrated among just three institutions in the late 1940s, were more evenly spread by 1971 (figure 10A.3). The top ten institutions received 41.7 percent of the research funds; overall, 244 institutions received DOD contracts or grants (figure 10.4).

By the early 1960s universities, nurtured by the federal government in the 1950s, had begun to depend upon federal support and to press for more. The 1960 report of the President's Scientific Advisory Committee (PSAC), *Scientific Progress, the Universities, and the Federal Government,* often referred to as the Seaborg report after its chairman, Glenn T. Seaborg, made the case for increased federal support on a variety of fronts (The President's Scientific

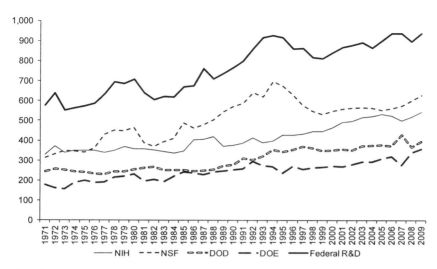

Fig. 10.4 Number of universities reporting research funds, various federal agencies (1971–2010)

Source: NSF Webcaspar.

Advisory Committee 1960).[15] Included were federal support for salaries of new hires (allowing universities to make long-term commitments), increased indirect rates on grants, and additional funds for university research so that the nation could double its fifteen to twenty "centers of excellence" to thirty or forty in fifteen years. The report also pressed for more fellowships for graduate study in science, recommending fellowships over research assistantships or teaching assistantships, which it saw as legitimate part-time work but cautioned that "these instruments are not without hazard: it is possible to do much harm to a young scientist, either by subordinating his need for a lively research experience to the requirements of a large organization or by exploiting his first enthusiasm for teaching by assignment exclusively to routine pedagogical tasks" (The President's Scientific Advisory Committee 1960, 17). It also expressed the concern that increased time to degree reflected the practice of taking part-time positions while training.

The request for across-the-board salary support went nowhere. The Seaborg report, however, met with some success when it came to indirect rates and funds for centers of excellence. In 1966, for example, the NSF announced a policy of negotiating the overhead rate university by university (Appel 2000). In 1964 the NSF created the Science Development Program with the goal of creating additional "centers of excellence." Later in the decade, it

15. Available at http://babel.hathitrust.org/cgi/pt?id=purl.32754081232229;page=root;view =image;size=100;seq=1. Last accessed June 25, 2013.

changed its goal to providing support to programs that already showed some existing strengths (Appel 2000). Other agencies also supported "upgrading" initiatives. The Department of Defense, for example, had project THEMIS, NASA created the Sustaining University Program, and NIH created Health Science Advancement Awards (Appel 2000).

Although government agencies resisted providing long-term funds to universities in support of salaries, federal agencies became increasingly sympathetic to the request that grants cover academic-year salary for the time faculty spent on funded research.[16] As early as 1960, the NSF yielded to the demand of college administrators to cover salaries, allowing faculty to charge off academic-year faculty salaries as a direct cost on grants (Appel 2000). By the end of the 1960s, the NIH regularly paid salaries of tenured faculty (Appel 2000). Indeed, in 1968–69 almost half the medical school faculty in the country received some salary support from the federal government. The salary argument was given ballast by the fact that mission agencies, such as the Army and the Air Force, were willing to pay up to 100 percent of faculty salaries (Appel 2000).

Faculty were not uniformly supportive of the push to put salaries on grants. The PSAC report noted the concern, stating that "We recognize that many university scientists are strongly opposed to the use of federal funds for senior faculty salaries. Obviously we do not share their belief, but we do agree with them on one important point—the need for avoiding situations in which a professor becomes partly or wholly responsible for raising his own salary."[17] It went on to say, "If a university makes permanent professorial appointments in reliance upon particular federal project support, and rejects any residual responsibility for financing the appointment if federal funds should fail, a most unsatisfactory sort of 'second-class citizenry' is created, and we are firmly against this sort of thing" (The President's Scientific Advisory Committee 1960, 24).[18]

The Seaborg report also met with some success with regard to increased federal support for fellowships, especially from the NSF and NIH. The NIH increase has been noted above. But the NSF also provided more fellowships: between the mid-1960s and the late-1960s, the number of fellowships it awarded rose by approximately two-thirds (Freeman, Chang, and Chiang 2005).

One cannot leave a discussion of university science in the 1960s without noting that the 1960s is arguably a period in science in which, to use Steven

---

16. The press for salary coverage was made not only by the PSAC, but also by an earlier report of the Committee on Sponsored Research of the American Council on Education.

17. Alan T. Waterman, the first director of the NSF, shared this concern, recognizing "that salary support led to such undesirable consequences as university pressure on faculty to cover their salaries through grants" (Appel 2000, 161).

18. The PSAC report also expressed the concern that paying for salary on grants could lead to the redistribution of income. Some university and federal administrators also expressed the concern that federal support for faculty salaries and research was leading faculty to become more loyal to Washington, DC than to their home institution.

Weinberg's terminology, "the logic of discovery" changed, especially in the physical sciences, forcing several disciplines to become big. In physics, the Berkeley Bevatron, which had become operational in 1954, rapidly became obsolete: "to make sense of what was being discovered, a new generation of higher-energy accelerators would be needed" (Weinberg 2012). The new accelerators would be too large for one laboratory and increasingly the new facilities that were required were too big for one institution, or one country. National and international laboratories such as Fermilab and CERN became important. The same logic was leading astronomers to request larger and larger instruments.

The logic of discovery was to transform the biomedical sciences, as well—but several decades later—with the invention of "designer" mice (Murray 2010) and the ability to automate the sequencing of genomes (Stephan 2012). Much of the equipment associated with these shifts in logic, although expensive, was still affordable at the lab or institutional level. Some, however, such as nuclear magnetic resonance (NMR),carried sufficiently large price tags to encourage, if not demand, collaboration across institutions.

10.5 The 1970s–2012

10.5.1 The 1970s

University administrators associated with the Seaborg report acknowledged that universities would be in a difficult position if the federal government were to back away from its support for research, but they dismissed the possibility (Beadle 1960). Yet only eight years after the report had been issued, universities were to find themselves in a precarious position when the brakes were put on and federal funding for research remained virtually flat in real terms for almost a decade (figure 10.3). Indeed, between 1968 and 1972, real federal expenditures for university R&D declined by 6 percent. Over the longer period, between 1968 and 1978, they increased by only 5 percent, in stark contrast to the fivefold increase between 1958 and 1968. The "golden age" of university science had ended.[19]

University research was sustained during this period in large part because funding from other sectors grew. A major source of growth came from institutions themselves, whose self-contributions to research increased by 55 percent, and by contributions from all other sources ("other"), which includes philanthropic organizations, that grew by 68 percent. Industry's expenditures on academic research increased by almost 70 percent; that from state and local governments grew as well, but by a modest 30 percent.

---

19. The war in Viet Nam was a factor in the federal government putting on the brakes for university research, as was the Mansfield Amendment of 1969. Declining tensions with the Soviet Union also led the DOD to award less funding to universities for research.

The cut in federal programs was reflected in federal support for fellow-ships. The number awarded for graduate study by the NSF was halved (Free-man, Chang, and Chiang 2005); the number of training positions that the NIH supported at the predoctoral and postdoctoral level fell by almost one-quarter.[20] Not surprisingly, PhD enrollments declined,[21] and by 1972 the number of PhDs awarded had begun to decline; PhD production was not to catch up with the 1971 high of almost 14,000 until 1987 (figure 10.1). Particularly hard hit were the fields of physics (60 percent decline), mathe-matics (33 percent decline), and chemistry (30 percent decline). The fields of engineering and biology experienced modest declines at most. Time to degree increased by 0.6 to 0.8 years depending upon broad field, reflecting, perhaps, the shift from training grants and fellowships to graduate research assistantships (table 10.3). Despite the decrease in PhD production, the number of institutions awarding the PhD in science and engineering con-tinued to increase, growing in most fields by 25 percent between 1965–1969 and 1970–1974 (table 10.1). The increase in PhD programs was fueled in part by newly emerging universities which, in a buyer's market, were able to hire well-trained PhDs, who, in turn lobbied for and often got new PhD programs—another indication of the positive feedbacks in the system that Bush had not foreseen.

Competition for contracts and grants intensified. Success rates at the NIH, which had plummeted during the 1960s, increased in the mid-1970s only to fall again by the end of the decade (figure 10.2). The concentration of NIH grants remained virtually unchanged. The Herfindahl-Hirschman Index (HHI) measure of concentration for the period, for example, varied by at most 5 percent (figure 10.5).[22] The share of funds received by the top ten institutions remained constant, with but one exception, throughout the decade at around 27 percent (figure 10A.2). The number of universities and medical schools receiving funding stayed almost constant as well, just shy of 350 (figure 10.4).

Things played out somewhat differently at the NSF, where the number of institutions receiving grants grew considerably, especially during the late 1970s (figure 10.4). Resources became less concentrated, as well (figure

20. The NIH supported 16,000 training grants in 1969. In the early 1970s, when the Nixon administration tried to eliminate the award, Congress responded with the National Research Service Award (NRSA) Act of 1974, providing funds for training in areas where "there is a need for personnel." In 1976, 11,500 trainees received support (National Research Council 1994, 93).
21. The decline in PhD enrollments also reflected poor market conditions for scientists and engineers in the late 1960s and early 1970s and the abrupt halt to draft deferments for graduate study (Levin and Stephan 1992).
22. The Herfindahl-Hirschman Index (HHI), a commonly accepted measure of concentra-tion, is calculated by squaring the share of each university and then summing the resulting numbers. The Department of Justice considers a share between 1,500 and 2,500 points to be moderately concentrated, and considers markets in which the HHI is in excess of 2,500 points to be highly concentrated. See http://www.justice.gov/atr/public/guidelines/hhi.html.

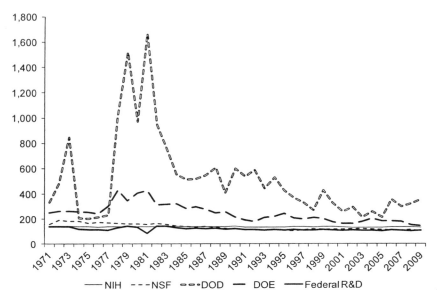

Fig. 10.5 The HHI index of concentration, NIH, NSF, DOE, DOD, all federal funds (1971–2009)
Source: NSF Webcaspar.

10A.1). The HHI index, which initially increased, fell by more than 10 percent; the top ten institutions saw their share decrease from 35 percent in 1972 to almost 30 percent in 1980. At the DOD, funds were considerably more concentrated, and patterns were considerably more sporadic, reflecting both "lumpy" contracts and stop-and-go funding (figure 10A.3). Even in the most equal of times, funds at DOD, as measured by the HHI index, were considerably more concentrated than at the other federal agencies (figure 10.5). The share that the top ten institutions received stayed above 35 percent throughout the period and at times exceeded 60 percent. The Department of Energy (DOE) was yet a different story. Although the number of universities receiving funds increased during the latter 1970s, the DOE funded fewer universities than did the other three agencies (figure 10.4). Moreover, funds were slightly more concentrated than at the NSF or NIH and during the end of the 1970s the degree of concentration increased, as measured by the HHI index (figure 10.5). See also figure 10A.4.

10.5.2 The 1980s–1998

The relative importance of federal funding for university research continued to decline during the 1980s and most of the 1990s (figure 10.3). This time it was not because the federal government's expenditures for university research were flat, however, but rather that they were increasing at a slower rate than the contributions of other sectors—especially those of business and industry, whose expenditures for university research grew by a factor of 3.7 during the period, and of universities themselves, whose contributions to their own research grew by a factor of 3.9 during the period. During the same period, funds from state and local government for research, funds from the federal government, and funds from other sources increased by a factor of 2.2.

The number of universities and colleges receiving research contracts and grants from federal agencies rose during the 1980s, especially the number receiving NSF funds (figure 10.4). The number receiving NIH funds, which had remained remarkably constant for many years, finally began to increase. The concentration of resources, as measured by the HHI index and the percent received by top ten institutions continued to decline for all agencies, save the NIH, where it stayed constant (figure 10.5 and figures 10A.1–10A.5).

The PhD production, which had initially declined and then been almost flat during the 1970s and early 1980s, began to increase. Growth was particularly notable in engineering, and slightly later in the period in the biological sciences. Growth was also notable at non-Research I institutions. The PhD production became increasingly less the domain of elite institutions.

Registered time to degree continued to increase in all fields. In 1993, for example, it was 6.7 years in the physical sciences, 6.5 in engineering, and 7.0 in the life sciences compared to 6.1, 5.9, and 6.2, respectively, ten years earlier (table 10.3). Increasingly, graduate students were supported as graduate research assistants rather than on fellowships or training grants. At the NIH, the number of training positions for predoctoral support remained almost constant; the number of individuals supported on faculty grants as research assistants more than doubled between 1980 and 1990 (figure 10.6).

The percent of new PhDs in engineering and in the physical sciences with definite plans at the time they received their PhD declined substantially in the early 1990s, only to increase dramatically in the mid-to-late 1990s as the dot-com industry began to hire aggressively (figure 10.7.) By 2001, however, with the demise of the dot-com bubble, the career prospects of newly trained engineers had begun to deteriorate considerably; prospects for those in the physical sciences also had begun to deteriorate, but to a lesser extent. A decreasing proportion had definite commitments at the time they graduated and an increasing percent of these definite commitments began to be for postdoctoral positions (figure 10.8).

Definite commitments for new PhDs in the life sciences also deteriorated during the middle part of the 1990s. The percent taking a postdoctoral position increased and/or remained high. Sufficient concern was expressed regarding their career prospects to cause the National Research Council (NRC) to form a committee to study trends in the early careers of life scientists. The chair was Shirley Tilghman of Princeton University. The committee made several recommendations, including restraint in the growth of

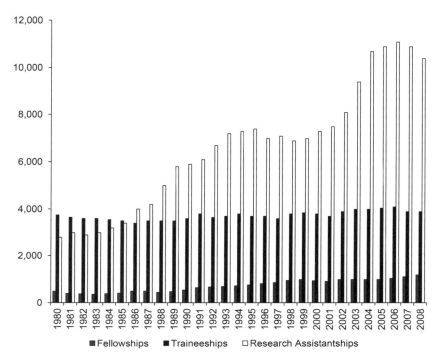

Fig. 10.6 NIH support of graduate students
Source: National Research Council (2011).

the number of graduate students in the life sciences, the dissemination of information on the career outcomes and prospects of young life scientists, the improvement of the educational experiences of graduate students, and enhancement of opportunities for independence of postdoctoral fellows.

The university community did not rush to embrace the committee's recommendations. Graduate programs continued to grow, the ratio of individuals supported on graduate research assistants to training grants and fellowships inched upward, and no effort was made to disseminate job market information. The reason for the failure is clear: the incentives of principal investigators and the university community were incompatible with the recommendations, and the committee had virtually no control over levers that could influence these incentives—such as the requirement that metrics for evaluating a faculty's grant include information on the career outcomes of those trained in his or her lab.

10.5.3 University Contributions to Research and the Cost of Equipment

Before turning to a discussion of the doubling of the NIH budget and the period that followed, two trends of the 1980s and 1990s that continue

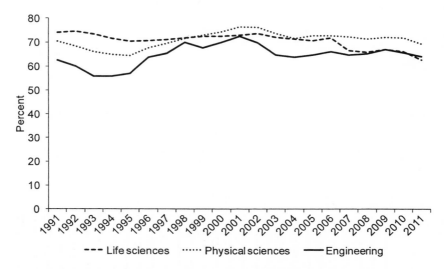

Fig. 10.7 Percent of doctorate recipients from US universities with definite commitments, 1991–2011

Note: Definite commitment refers to a doctorate recipient who is either returning to predoctoral employment or has signed a contract (or otherwise made a definite commitment) for employment or a postdoc position in the coming year (National Science Foundation 2012).

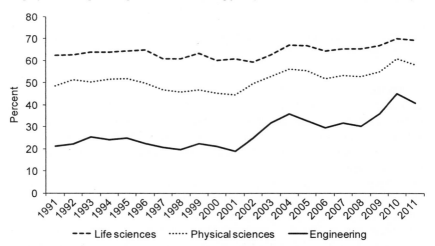

Fig. 10.8 Percent of doctorate recipients from US universities with definite commitments taking a postdoctoral position, 1991–2011

Source: National Science Foundation (2012).

today deserve special comment. One is the increasing share that universities contribute to research and development (figure 10.3), and the second is the increasing expenditures that universities make for research equipment.

At least two factors have contributed to universities picking up a larger and larger share of research funding since the mid-1960s. First, and as noted above, a constant theme of university administrators has been that indirect cost rates fail to cover the institution's costs for research, a problem that became more acute after the Office of Management and Budget (OMB) established limitations on federal indirect costs in 1991 and caps were put on expenses that universities could claim in a number of areas.

A second reason that universities began to pick up a larger and larger share of the cost for research relates to the growing practice of providing start-up packages for newly hired faculty.[23] Such packages not only play an important role in recruiting senior faculty, but they also provide the time and resources that newly minted faculty need to develop the preliminary results to place them in a competitive position for receiving grants, containing funds for graduate research assistants, postdoctoral researchers, supplies, and, in many instances, equipment. At Cornell University, for example, equipment expenditures represent 60 or more percent in one-third of the start-up funds provided new hires recently; in one-half of the start-up packages they represent between 25 and 40 percent.[24]

Start-up packages can be quite large. A 2003 survey found that the average of mean start-up packages offered by institutions for an assistant professor in chemistry was $489,000; in biology, it was $403,071.[25] These are not modest sums. They represent four to five times the starting salary that the institution paid a junior faculty member at the time. At the high end, it was $580,000 in chemistry and $437,000 in biology. For senior faculty, start-up packages averaged $983,929 in chemistry (high end: $1,172,222) and $957,143 in biology (high end: $1,575,000) (Ehrenberg, Rizzo, and Jakubson 2007). More recent data for start-up funds at a private Research I university show packages between $500,000 and $1,178,000 between FY08 and FY10 for assistant professors in biochemistry and biology.[26] Those in chemistry for the same period were between $535,000 and $635,000. Start-up funds for an

---

23. The growing requirement of federal agencies that universities provide matching funds in grant proposals is a third factor that has led to increased contributions of universities toward research (Ehrenberg 2012).

24. Data provided by Robert Buhrman, Cornell University.

25. The survey was administered to three to six science and engineering departments at 222 research and doctoral institutions. The average means reported are drawn from the responses of the 572 department chairs who replied (with a response rate of 55 percent) (Ehrenberg, Rizzo, and Jakubson 2007).

26. The range in the value of the packages is due in part to the practice of the institution to often make an offer with two start-up package numbers: a guaranteed support level and an additional amount that would be made available if the candidate had difficulty getting funding within three years.

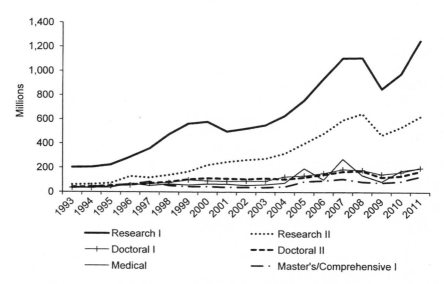

Fig. 10.9 Median of endowment funds in constant 2011 dollars by Carnegie classification, 1993–2011

Source: Data provided by the Common Fund for Nonprofit Organizations.

associate professor of chemistry were $1,178,000. Start-up packages can be considerably higher at medical schools where a full professor reportedly can receive a start-up package of $5 million or more in 2013.

No one has done the accounting regarding where institutional funds for research come from, but research by Ehrenberg and coauthors supports the view that students pick up part of the costs, especially at private institutions, where the student-faculty ratio grows as internal funding for research grows, and where tuition levels increase as internal funding for research grows (Ehrenberg, Rizzo, and Jakubson 2007). The first effect is smaller at public institutions, and the tuition effect is not discernable for public institutions.

The question remains, however, as to where universities get the majority of funds to invest in research, since clearly only a small portion is borne by students in the form of higher tuition and larger class size. One obvious source is endowment income, especially given that endowments have grown significantly over time, as can be seen in figure 10.9. Indeed, despite the beating that endowments took in 2009 and regardless of Carnegie classification, endowments are currently at an all-time high at many institutions in terms of 2011 constant dollars.

Figure 10.10 explores how the growth in internal university R&D expenditures relates to this growth in endowment, plotting the median ratio of institutional expenditures on R&D to the value of the institution's endowment over time by Carnegie classification. We would not, of course, expect to find

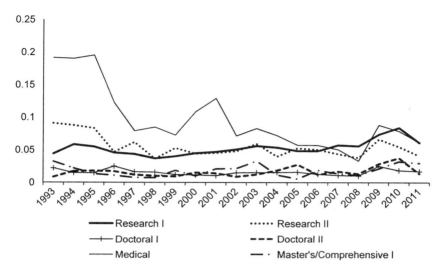

Fig. 10.10 Median ratio of institutional expenditures for university research and development to endowment value, 1992–2011
Source: Webcaspar and the Common Fund for Nonprofit Organizations.
Note: Ratio computed for institutions reporting in that year a positive R&D expenditure value.

a high ratio, given spending rules associated with most endowments. And we find, on the whole, that the ratios are fairly modest—except at medical institutions, where in the early years they approached 0.2. Furthermore, we find that on the whole, at least through the late 1990s, the ratio declined over time. Thereafter the ratio of research expenditures to endowment rose slightly at Research I and Research II institutions. The ratio for Research I continued to increase, matching in certain years that at medical institutions. Research II institutions plateaued or slightly declined, only to increase as a result of the spectacular fall in endowment values in 2009. Reflecting perhaps the desire to move up in the rankings, the ratio of expenditures to endowment increased at master's-level institutions during certain periods, as did that at Doctoral and Doctoral II.

We cannot, of course, conclude from this exercise that endowment is the source of university expenditures on research. But our findings suggest that there has not been a dramatic increase in the research expenditures of universities relative to their endowments. At most institutions, at least up to the middle of the first decade of the twenty-first century, expenditures grew at a slower pace than did the value of the endowment. Our findings are consistent with the importance universities place on fundraising for scientific research (Murray 2012; Mervis 2013).

The second trend that deserves comment relates to the amount that universities spend on equipment for research—either out of their own funds or

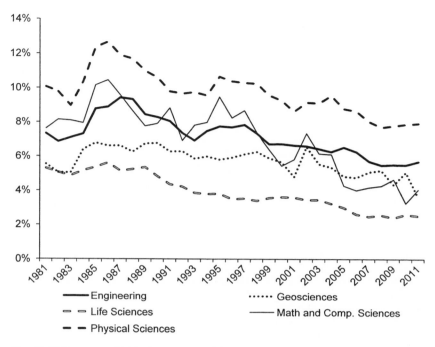

Fig. 10.11 Percent of R&D funds spent on equipment by field, 1981–2011
Source: NSF Higher Education Research and Development survey; Webcaspar.

the funds provided by others—which, in real terms, almost doubled in the six-year period between 1984 and 1990 and almost doubled again in the 1990s. Today it stands at approximately 2.5 billion dollars in 2005 dollars. In terms of level, expenditures for equipment are the greatest in the life sciences, reflecting strong funding, followed by engineering and the physical sciences. Equipment intensity also varies considerably by field (figure 10A.6). Not surprisingly, the physical sciences typically spend the largest portion of their research budgets on equipment—anywhere from 9 to 12.5 percent. The life sciences, which range from 2.5 percent to 5.0 percent, spend the least (figure 10.11).

Faculty and administrators often express the concern that the cost of equipment is rising and that as a result they are forced to spend greater amounts of their research funds on equipment. Not only is the price going up, but new types of equipment, such as sequencers and confocal microscopes, have become necessary, if not for the lab, for core facilities at a university.

While equipment prices have undoubtedly risen over time—one researcher bemoaned how X-ray equipment, which used to cost $250,000, now costs about $1.5 million—the data do not support the idea that the percent of total research and development expenditures spent on equipment has been increasing over time. Indeed, as figure 10.11 shows, with the exception of the mid-1980s, the trend has been definitely downward. There are at least two possible explanations as to why this fact is at odds with the perceptions of deans and faculty. First, the capability and efficiency of the equipment has been rising faster than cost. As a result, universities are able to run core facilities where faculty share a common piece of (expensive) equipment. Second, some of the major costs occur outside the R&D equipment accounting system of universities. For example, a share in SER-CAT, which allows members access to a synchrotron beamline at Argonne National Labs, costs $250,000. Yet the synchrotron beamline built at Argonne cost approximately $7 million to construct. Neither the share price nor the actual cost of construction is likely to show up in the university R&D expenditure accounts for equipment.[27]

10.5.4 NIH Doubling and Years Following the Doubling

It is tempting to assume that more funding is the answer to many of the problems that plague the university research system. One would expect additional funds to translate into higher success rates and be accompanied by improved job prospects, especially for young researchers. But anyone who thinks so should be careful what they wish for. The doubling of the NIH budget in nominal terms between 1998 and 2002 ushered in a number of problems. By the time it was over, success rates were no higher than they had been before the doubling. By 2009, and in part because of the real decrease that the NIH experienced in the intervening years, success rates were considerably lower than they had been before the doubling (figure 10.2). Faculty were spending more time submitting and reviewing grants, in part because an increased proportion of grants were not approved until their last and final round.[28] Moreover, there is little evidence that the increase translated into a substantial improvement in the job prospects of newly minted PhDs, as had been the case in the 1950s and 1960s when government support for research expanded. Yes, the doubling brought more jobs, but the supply of new PhDs grew faster than the demand for new hires. The percent of newly minted PhDs in the life sciences with definite commitments declined from 2002 on (figure 10.7), and the percent taking postdoctoral positions rose (figure 10.8).

A major cause of this seeming paradox was the response of universities to the doubling. Some universities saw the doubling as an opportunity to

27. The Southeast Regional Collaborative Access Team (SER-CAT) was established in 1997. Several universities and research groups purchased more than one share. See Stephan (2012).
28. Early in the twenty-first century, 60 percent of all funded R01 proposals were awarded the first time they were submitted. By the end of the decade, only 30 percent were awarded the first time. More than one-third were not approved until their last and final review (Stephan 2012). This not only took time and delayed careers, but the perception was that these "last chance" proposals were favored over others, creating a system that, according to Elias Zerhouni, awarded "persistence over brilliance sometimes" (Kaiser 2008a).

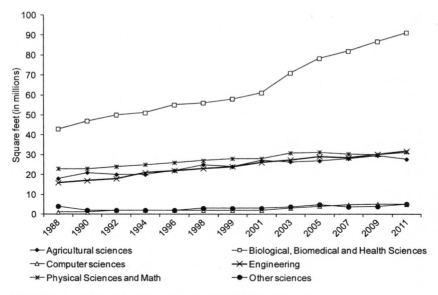

Fig. 10.12 Net assignable square feet for research by field and year
Source: National Science Foundation (2013).

move into a new "league" and establish a program of "excellence." Others saw it as an opportunity to augment the strength they already had. For others still, expansion of their existing programs was simply necessary if they were to remain a player in biomedical research. Regardless, the end result was that the majority of research universities went on an unprecedented building binge. Research space in the biological, biomedical, and health sciences increased by one-third during the six-year period between 2001 and 2006 (figure 10.12).

Not surprisingly, the number of applicants for new and competing research projects grew. Success rates, which were over 30 percent at the beginning of the doubling, fell to 20 percent by 2006 (figure 10.2). One reason for the decline in success rates was the substantial growth in budgets accompanying the proposed research: in 1998, the average annual budget of the typical grant was $247,000; by 2009, it had grown to $388,000 (Stephan 2012). One reason for the increase was that more faculty were on soft-money positions and thus writing off a larger proportion of their salary.

Some of the new grants during the doubling went to researchers who had heretofore not received funds. But the vast majority of new grants went to established researchers: the percentage of investigators who had more than one R01 grant grew by one-third during the doubling, going from 22 percent to 29 percent. The number of first-time investigators grew by less than 10 percent. Young researchers were at a disadvantage competing against more seasoned researchers who had better preliminary data and more grantsmanship expertise. The increased number of grants for experienced investigators and minimal growth in grants for first-time investigators resulted in a dramatic change in the age distribution of PIs. In 1998, less than one-third of awardees were over fifty years old, and almost 25 percent were under forty. By 2010, almost 46 percent were over fifty, and less than 18 percent were under forty. More than 28 percent were over fifty-five years old (Stephan 2012). Faculty staffed these labs with postdocs and graduate students. The number of postdocs in the life sciences grew by almost 33 percent between 1997 and 2008. The PhD production grew by almost 38 percent (figure 10.2).

10.6 Taking Stock

From the perspective of the early twenty-first century, it is clear that *The Endless Frontier* contributed to building the university research enterprise. It also set in motion forces that would transform it. Universities in the early twenty-first century are a far cry from those of the 1940s, having been transformed from a focus on educating students and taking care of patients to placing a high—if not the highest—value on research. The incentives that have evolved over time have encouraged this transformation. Universities are routinely ranked on the amount of federal funds they receive; membership in the prestigious American Association of Universities (AAU) puts considerable emphasis on federal funds, as does the Carnegie classification. The number of doctoral degrees awarded also plays a key role in certain rankings.

Bush would be astonished at the capacity that has been built: The number of research universities has grown from a mere ten to fifteen to more than one hundred, depending upon definition (National Research Council 2012). The number of institutions that are funded has grown considerably, from the 120 universities and medical schools supported by the NIH in 1948 to the 556 supported today. At the NSF, the growth has been even more impressive, going from 60 to 628. Overall, the number of institutions receiving federal funds has grown from slightly fewer than 600 in 1971, the first year for which data are readily available, to over 900 in 2009 (figure 10.4). By any measure, funds are less concentrated. The percent of federal research funds going to the top ten institutions has decreased by almost 50 percent; the HHI index, which has always been relatively low, with minor exceptions for DOD funding, has declined by about 30 percent (figure 10.5). The decrease in concentration has occurred at all agencies, save the NIH, where top universities and medical schools have been remarkably successful at holding on to their share, despite the increased number of universities and medical schools supported by the NIH.

Concomitantly, the number of universities offering doctoral training in science and engineering has grown by more than fivefold. The number of degrees awarded has grown by a factor of 17. The percent of degrees awarded by top ten and top twenty-five institutions has decreased substantially (table 10.2) as has the percent awarded by Research I institutions.

The Endless Frontier also set in motion forces that transformed the relationship between universities and the federal government. No longer need the federal government cajole universities and faculty into submitting grants. Long ago the tables turned. Universities now spend considerable energy and funds convincing federal agencies to provide more resources, hiring lobbyists to work on the university's behalf to direct federal (and state) funds to the university,[29] and joining forces to issue reports that make the case for more support from the federal government. The tradition was established more than fifty years ago with the Seaborg report. It was reaffirmed in 2012 with the NRC report "Research Universities and the Future of America," which pressed, among other things, for moving certain costs covered by indirect to direct costs, federal funding for a "strategic investment program," and a reduction or elimination in regulations that increase administrative costs (National Research Council 2012).

As the tables turned, the way in which graduate students are supported by federal funds dramatically changed as well. The system Bush envisioned was designed to build future research capacity by supporting graduate students and postdocs on fellowships or training grants. While these mechanisms for federal funding remain in place, their importance has paled as increasing numbers of students are supported as research assistants on faculty members' grants and postdocs on stipends paid from grants. The shift means that federal funds are no longer directed at building future research capacity but toward getting the research done today. This shift in mechanisms of support is likely reflected in lengthened time to degree.

Universities increasingly expect faculty to cover part, if not all, of their salary on grants. The practice began sometime in the 1950s and spread fairly rapidly, so that by the late 1960s almost half the medical school faculty in the country received some salary support from federal grants. Today, nonclinical medical school faculty, even those who are tenured, routinely cover close to 100 percent of their salaries on grants. Universities, except for a handful of elite institutions such as Princeton University and Caltech, routinely expect faculty to write off part of their salaries on grants and hire faculty in soft money positions with the expectation that they will cover all of their salary on grants.

In many ways, universities in the United States have come to resemble high-end shopping malls. They are in the business of building state-of-the art facilities and a reputation that attracts good students, good faculty, and resources. They turn around and lease the facilities to faculty in the form

29. See, for example, the work of De Figueiredo and Silverman (De Figueiredo and Silverman 2007).

of indirect costs on grants and the buyout of salary. Some of these faculty are in soft money positions, in essence paying for the opportunity to work at the university, receiving no guarantee of income if they fail to bring in a grant. To help faculty establish their labs—their firm in the mall—universities provide start-up packages for newly hired faculty. After three years, faculty are on their own to get the necessary funding for their lab to remain in business.

The shopping mall model has led universities to spend an increasing amount of their resources in support of research. Some of this is for start-up packages, some is for matching funds required by federal agencies, and some is to defray costs not covered in indirect costs. Not only are universities spending more, but their share of research costs has increased, going from a low of 8.1 in 1963 to a high of 20.4 percent in 2009.

The shopping-mall model also encourages universities to construct new research facilities, increasing their capacity to rent out space to faculty. The expectation is that "the space will be paid from a combination of direct and indirect costs funded by the federal government."[30] In the past ten to fifteen years, this new space has been heavily concentrated in the biomedical sciences. Indeed, two-thirds of the increase in net assignable square feet for research that has occurred in the past ten years was in the biological, biomedical, or health sciences (figure 10.12). Faculty use the space and equipment to create research programs, staffing them with graduate students and postdocs who contribute to the research enterprise through their labor and fresh ideas.

External funding, which was once viewed as a luxury, has become a necessary condition for tenure and promotion. It is even more important for faculty in soft money positions or for those whose tenure does not come with a salary guarantee. Yet external funding has become increasingly more difficult to get as federal funds, excluding the American Recovery and Reinvestment Act (ARRA),have remained almost flat during much of the first decade of the twenty-first century and the number of individuals seeking funding has continued to increase. Reflecting this situation, success rates at the NIH and NSF stood at close to historic lows, hovering around 20 percent (figure 10.2).

10.7 Stresses to the System

The university research system that has grown and evolved since the publication of *The Endless Frontier* almost seventy years ago faces a number of challenges that threaten the health of universities and the research enterprise and have implications for discovery and innovation. Five are discussed in this closing section.

30. Shirley Tilghman as quoted in (Mervis 2013, 1,399).

10.7.1 Risk Aversion

In today's environment, grants are often scored for "doability," selected because they are "almost certain to work" (Alberts 2009). At the time a proposal is submitted, it is routine that two of the three objectives have been completed (Azoulay, Zivin, and Manso 2012). To quote the Nobel laureate Roger Kornberg, "If the work that you propose to do isn't virtually certain of success, then it won't be funded." Yet, as Kornberg continues, "the kind of work that we would most like to see take place, which is groundbreaking and innovative, lies at the other extreme" (Lee 2007, A06). This was not always the case: there is a perception among older scientists that peer review used to be a different game, with reviewers focused on "ideas, not preliminary data" (Kaiser 2008b). It is not only the peer-review system that fosters risk aversion. The Defense Advanced Research Projects Agency (DARPA), which once boasted that "it took on impossible problems and wasn't interested in the merely difficult," has increasingly shifted to funding research that is more near-term and less risky (Ignatius 2007).

The preference to fund research that is "doable" increases when funding is difficult to come by, which has been the case for the last ten years as measured by success rates at the NIH and NSF (figure 10.2). One reason is that agencies feel pressed to report successful research (Petsko 2012). Another is that it is easier to justify funding safe bets when funding is in short supply. The recently released ARISE report (Advancing Research in Science and Engineering) from the American Academy of Arts and Sciences concluded that in tight times "reviewers and program officers have a natural tendency to give highest priority to projects they deem most likely to produce short-term, low-risk, and measurable results" (American Academy of Arts and Sciences 2008, 27).

The preference on the part of agencies to fund doable research need not, of course, translate into faculty taking up less risky lines of research, since the receipt of funding can be viewed as a prize awarded to individuals who have almost completed the research before applying for funding (Azoulay, Zivin, and Manso 2012). But the pressure on faculty to receive funding quickly in their academic career—at the end of their third year at many universities, if not sooner—means that faculty can ill afford to follow a research agenda of an overly risky nature. They need tangible results and they need them quickly. The pressure is even greater for those in soft money positions. Moreover, the fact that grant renewals have a much higher chance of being positively reviewed, be the renewals formal or de facto, discourages faculty from taking up new research agendas once they have established a line of research.

Should this proclivity for risk aversion be of concern to the university community and more importantly to society in general? Yes. First, there is the issue of the composition of the research portfolio. It is pretty clear that if everyone is risk averse when it comes to research there is little chance that transformative research will occur or that the economy will reap significant returns from investments in research and development. Incremental research yields results, but in order to realize substantial gains from research not everyone can be doing incremental research. Second, one of the main reasons that Bush, and those who adopted his proposed course of action, placed research in the university sector was the view that society needed to undertake basic research of an unpredictable nature and that universities were precisely the place to conduct risky research because they "provide the environment which is most conducive to the creation of new scientific knowledge and least under pressure for immediate, tangible results" (Bush 1945, 7). Yet the system that has evolved does precisely the opposite of this, placing pressure on faculty for quick, predictable, results. Finally, and more generally, a fundamental rationale for government support of research is the notion that research is risky. As laid out by Kenneth Arrow, society has a tendency to underinvest in risky research without government support (Arrow 1955).

10.7.2 The Tendency to Produce More PhDs Than the Market for Research Positions Demands

A primary reason of Vannevar Bush for advocating the establishment of the National Science Foundation and ratcheting up funding for the National Institutes of Health was the concern that the United States had exited World War II with a severe lack of research capacity. Thus, a goal of the federal government, operating in cooperation with universities and medical schools, was to build research capacity by training new researchers. It was also to conduct research. However, it was never Bush's vision that training be married to funding for research. Yes, good training required a research environment and good research required assistance, but Bush did not see research grants as the primary way to support graduate students. Nor did he see them as the source of support for postdoctoral study. Rather, he argued that, in order to build capacity, graduate students and postdocs should be supported on fellowships.

It did not take long for the system to change. Faculty quickly learned to include graduate students and postdocs on grant proposals and, by the 1960s, PhD programs had become less about capacity building and more about the need to staff labs and teach classes. The caution of the Seaborg report regarding the harm that research assistantships and teaching assistantships could do to a young scientist went unheeded (The President's Scientific Advisory Committee 1960). The structure of a university lab, with the principal investigator at the top, followed below by postdocs and then by graduate students, began increasingly to resemble a pyramid scheme where, in order to staff their labs, faculty recruit PhD students into their graduate programs, providing them tuition and a research assistantship and the implicit assurance of interesting research careers.

The pyramid scheme works as long as the number of jobs grows sufficiently to absorb the newly trained. Yet by most indications, the system that has evolved, with demand based on the need to staff faculty labs with trainees, is producing more PhDs than the market for future research positions demands, given current and projected levels of funding for research. One indication of this is the rising percent of individuals who do not have a definite commitment at the time they receive their PhD (figure 10.7). Another is the rising percent of those definite commitments that are for postdoctoral positions (figure 10.8). While in certain fields, such as engineering and the physical sciences, a sizable component of this is cyclical, in the field of the biomedical sciences it is chronic and has been so at least since 1976 when an NRC report evaluating training grants concluded that a "slower rate of growth in labor force in these fields was advisable" (National Research Council 1994, 98). The PhD recipients, as a recent NIH workforce study committee documented, increasingly must find jobs that do not utilize their research training (National Institutes of Health 2012).

Such a model for staffing labs is inefficient in the sense that substantial resources have been invested in training these scientists and engineers. The trained have foregone other careers—and the salary that they would have earned—along the way. The public has invested resources in tuition and stipends. If these "investments" then enter careers that require less training, resources have been used inefficiently. There are less expensive ways to train high school science teachers, as a recent NRC report suggested as a career alternative (National Research Council 2011), or a better way to create venture capitalists with a sufficient understanding of science, or a better way to train individuals to represent and service new pharmaceutical products.

Yet questions concerning training outcomes often fall on deaf ears among faculty and university administrators, who, as one report stated, see the current system as "incredibly successful" and resist recommendations such as those put forward by the Tilghman committee in the late 1990s (National Research Council 2011). The alternative, to employ long-term staff scientists in the lab, is resisted. One reason is that a permanent staff would cost more. While this is indisputable in the short run, it fails to account for the cost savings that would be realized if the system were not constantly staffing labs with a new crop of graduate students and postdocs. Adherence to the system also threatens the long-run health of the research system, by discouraging individuals who take career outcomes into their decision-making process from entering careers in science.[31]

31. Smart young people put up with this system for several reasons. First, until recently, there has been a ready supply of funds to support graduate students and research assistants and to hire postdocs. Second, factors other than money play a role in determining who chooses to become a scientist. One factor in particular is a taste for science. Dangle stipends and the prospect of a research career in front of star students who enjoy solving puzzles and it is not surprising that some keep coming, discounting the muted signals that research positions are in short supply. Overconfidence also plays a role: students in science persistently see themselves as better than the average student in their program—something that is statistically impossible. Fourth, when it comes to promoting PhD study, faculty are good salesmen. Finally, PhD programs, despite recommendations of national committees, have been slow to make placement information available.

10.7.3 Overexpansion of Research Facilities

In recent years, universities have gone on a building binge, constructing a substantial amount of new research space that led to a 30 percent increase in net assignable square feet for research between 2001 and 2011. Most of this increase is for facilities in the biological, biomedical, and health sciences—a response of universities to the doubling of the NIH budget (figure 10.12). Some of this space has been paid for by private philanthropy. At MIT, for example, David Koch contributed $50 million to the construction of an institute for cancer research that bears his name (Murray 2012). But in a number of instances, campuses did not have the funds to construct the new buildings but instead did so by floating bonds, assuming that the debt would be recovered through increased grant activity engendered by better facilities housing more research-active faculty. A 2003 survey of medical schools by the Association of American Medical Colleges (AAMC) found that the average annual debt service for buildings in 2003 was $3.5 million; it grew to $6.9 million in 2008 (Heinig et al. 2007).

The brakes were applied to the NIH budget beginning in 2004, and in constant dollars the NIH budget shrank by about 4.4 percent between 2004 and 2009. It has continued to decline since, with the exception of ARRA. Success rates for NIH grants, as we have seen, declined, and universities found that revenues from grants did not live up to their expectations. The situation is not likely to improve in the near future, given sequestration. This means that the only way a university can hope to cover the costs of these buildings is to outdo other academic institutions in bringing in grants. But, as Princeton University's president Shirley Tilghman notes, "this just can't be true for every academic medical center. It does not compute" (Mervis 2013, 1,399). Moreover, given that very top institutions have continued to maintain their share of NIH funding, the pain is most likely to be felt by institutions that historically have not received top funding. Somebody, especially at lower-tiered institutions, is going to have to pay for this substantial expansion and it is unlikely to be the federal government. It is more likely to come through a reallocation of resources within the university.

10.7.4 Mix of Research Funding

In the steady state that Bush envisioned, funding for the natural sciences was to be 2.5 times higher than that for the medical sciences. Yet Bush's vision was never close to being realized. For the period since 1973, for which data are readily available, the share of federal university research and development obligations going to the life sciences has, at a minimum, been above 55 percent and, after the doubling of the NIH, for a short period, approached 70 percent (figure 10.13). It is relatively easy to understand the politics of why this is so. It is far easier for Congress to support research that the public perceives as directly benefiting their well-being. Moreover, a large number of interest groups constantly remind Congress of the importance of medical research for "their" disease.

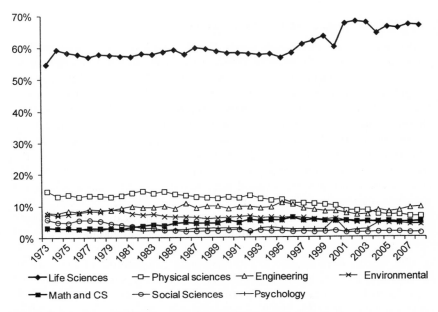

Fig. 10.13 Share of university R&D obligations by field, 1973–2009
Source: Stephan (2012).

One can question whether this mix of funding is efficient. Are the marginal benefits coming from another dollar spent on the biomedical sciences greater than the marginal benefits coming from another dollar spent on the physical sciences? The fourteen-year increase in life expectancy in the past seventy years makes a good case that research in the biomedical sciences has a high marginal product (Stephan 2012). But the slowed rate at which new drugs are being brought to market makes one wonder whether the marginal productivity of resources spent in the biomedical sciences is diminishing (Stephan 2012). Furthermore, one can make a good case that spillovers from the physical sciences have made significant contributions to the economy. Some of these contributions are even in the area of health, such as the laser and magnetic resonance imaging technology.

Although no analysis is sufficiently precise to calculate the degree to which the research portfolio is out of balance, three observations lead one to think that the research enterprise might benefit if the biomedical sciences were to receive a smaller share. First, the heavy focus on the biomedical sciences is propelled by a lobbying behemoth composed of universities and nonprofit health advocacy groups that constantly remind Congress of the importance of funding health-related research. There is no comparably well-established and well-focused lobbying group on the part of other disciplines. Second, portfolio theory leads one to think that the current allocation might be out of balance. A basic tenet of investing is to rebalance one's portfolio if a change in market valuations results in a change in the composition of the portfolio that one is holding. Yet efforts to fund research in engineering and the physical sciences, through the America COMPETES Acts, have met with limited success (Furman 2012). Third, and particularly relevant for the discussion here, the heavy focus on the biomedical sciences affects the life of universities in a number of ways. For example, the heavy push to construct new research facilities for the biomedical sciences has consequences for facilities in other disciplines, which got pushed to the back of the queue. It also has consequences for hiring. Moreover, and as noted above, there are long-run consequences, because some of the funding for these buildings was raised from the sale of bonds, and many universities are not reaping grants or the indirect cost they had expected. It is likely that other disciplines will end up footing part of the bill.

10.7.5 Heavy Reliance on Federal Funds

For many years universities have been heavily reliant on federal funds for research. Yet the future for a steady increase of federal funds looks dim. Congress has been slow to fund the America COMPETES Acts (Furman 2012) and sequestration means that expenditures for research may well decline in real terms. Public institutions face the added challenge that funds from state and local governments for education, and research in particular, have been flat or have declined in recent years (figure 10.3) and are likely to remain low in the future.

This places universities in the position of looking for alternative sources of funding for research. One source is industry, whose contribution is likely to grow as the economy picks up. But given past experience, industry is unlikely to substantially increase its share of university R&D. This leaves only two (related) sources: universities and philanthropic organizations/gifts. The first has been discussed above, the second only briefly alluded to. With regard to the latter, the percent of university research funds coming from philanthropic organizations has been growing and now exceeds that coming from industry (Feldman, Roach, and Bercovitz 2012). Murray provides an overview of the important role that philanthropic gifts are making to university research, arguing that they account for $4 billion of the research funds of the top fifty universities in the United States (Murray 2012).

While the increasing role of philanthropy may address a sizable portion of the resource gap, several factors lead one to wonder if it may place new stresses on the research enterprise. First, as outlined by Murray, the majority of these philanthropic gifts are for research in the biomedical sciences. Far fewer gifts are for research in other fields, although certain foundations, such as the Gordon and Betty Moore Foundation and the W. M. Keck Foundation, routinely support research in the physical sciences and engineering. To the extent the research portfolio is out of balance, philanthropy will only add to the imbalance. Second, and related, much of philanthropic support is directed at applied medical research, with short-term research goals

(Murray 2012). Third, gifts generally supplement federal funding, rather than fill gaps in funding. Philanthropists like the idea that their gifts can be leveraged into federal funding and they share many of the health concerns of the public. Fourth, the push for gifts raises the concern that universities will focus their skills and their research on the rich and their diseases. The medical school at Johns Hopkins has sixty-five full-time fundraisers. Their "caseloads" range from twelve to thirty doctors, with whom they discuss patients who might be potential donors, or to help staff identify a donor with a "qualifying" interest and connect it to their "capacity" to make a donation. The goal: "to turn 'grateful patients' into support for new research, faculty chairs, academic scholarships, bricks and mortar, or simply defraying the cost of running a multibillion-dollar medical center" (Mervis 2013, 1,397). Finally, the philanthropy "answer" is less readily available to publicly funded and nonelite institutions, whose endowments have grown at a considerably slower pace than those at elite private and top-tier research institutions.

10.8 Concluding Thoughts

A widely held belief among university faculty and administrators is that the contract between federal funders and universities has changed dramatically in the past sixty-five years. Initially federal agencies fostered research by providing funds for equipment, supplies, and facilities, and investing in future researchers through the provision of fellowships and training grants. Summer salaries were allowed as a legitimate research expense, but support for academic-year salaries was not common and was resisted. But very early on, in the 1950s, the system began to change. Faculty academic-year salaries began to be written off grants; graduate students and postdocs increasingly were supported on assistantships on faculty grants and less on fellowships and training grants. In the process, graduate programs became less about training future researchers and more about getting the research done now.

Yes, the contract changed. But a careful reading of the record suggests that the change was orchestrated more by universities than by the federal government. Bush established a funding system that faculty and university administrators were adroit at adapting to their ends. The modern university research system evolved. Many of the stresses that the system now faces are a result of these adaptations. We are reaping not so much what Bush sowed but what universities and faculty pressed to put in place in the 1950s and 1960s in response to *The Endless Frontier* and the opportunities it offered. Some of Bush's key insights regarding research and the research process got lost in the process of adaptation. To name but three: the importance of funding and conducting risky research at universities, the focus on fellowships as a method of supporting graduate students, and, implicitly, the need to strike a balance between support of the medical sciences and other fields of science and engineering.

Many of the stresses on the university research system result from a fixation on the part of universities with increased funding for research. Yet, as the doubling of the NIH budget so aptly shows, increased funding does not address problems that are structural and that are reinforced by positive feedbacks. As we move forward, the time may have come, as Princeton's Shirley Tilghman says, "to have a conversation between the government and the research universities on how to live at steady state" (Mervis 2013, 1,935). Such a conversation is unlikely to take place, however. The steady state that Bush had envisioned has long been eclipsed by an addiction on the part of universities to growth for growth's sake. This may be the biggest threat to the health of the university research system.

Appendix

Table 10A.1   **Public health service research grants (NIH) in aid, 1948**

Institution	Amount (1,000s current dollars)	Number of projects	Percent of total
Columbia University	428,000	37	5.26
Johns Hopkins University	402,000	36	4.96
New York University	320,000	26	3.95
Harvard University	315,000	21	3.89
University of Minnesota	310,000	29	3.82
University of California	297,000	29	3.66
University of Chicago	284,000	20	3.50
University of Michigan	209,000	20	2.58
Washington University	191,000	20	2.35
Memorial Hospital, NYC	189,000	12	2.33

Source: http://history.nih.gov/research/downloads/PHSResearchGrantsinAID-June30th1948.pdf.

Table 10A.2   **NSF awards FY1953**

Institution	Amount (1,000 current dollars)	Number of awards	Percent of award dollars
Harvard University	108.2	6	6.5
Yale University	105.1	8	6.3
University of California, Berkeley	95.1	8	5.7
University of Minnesota	77.7	4	4.6
University of Chicago	73.8	6	4.4
University of Illinois	54.3	7	3.2
University of Pennsylvania	51.3	6	3.1
University of Iowa	49.3	4	2.9
University of Indiana	58.5	3	2.9
Northwestern University	40.0	4	2.4

Source: The NSF provided data.

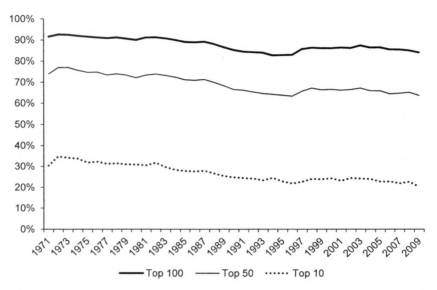

Fig. 10A.1 Share of NSF funding

Source: NSF Webcaspar.

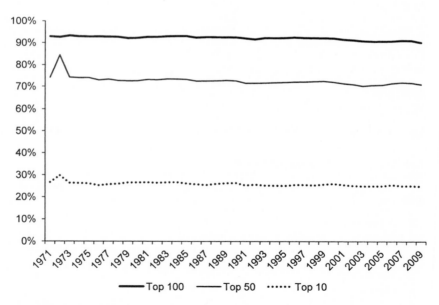

Fig. 10A.2 Share of NIH funding

Source: NSF Webcaspar.

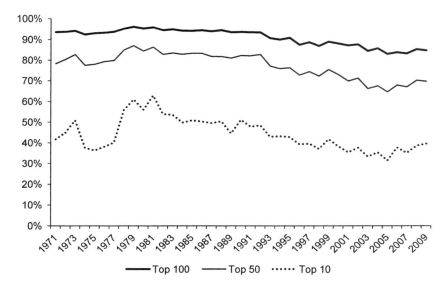

Fig. 10A.3 Share of DOD funding
Source: NSF Webcaspar.

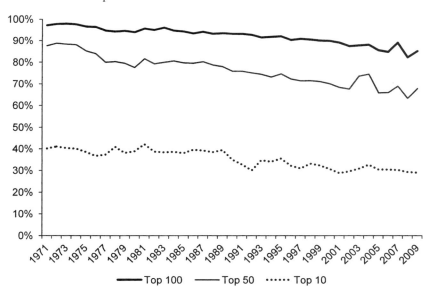

Fig. 10A.4 Share of DOE funding
Source: NSF Webcaspar.

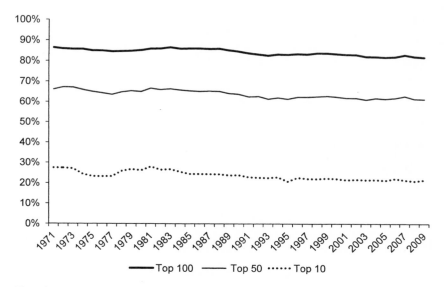

Fig. 10A.5 Share of federal R&D expenditures
Source: NSF Webcaspar.

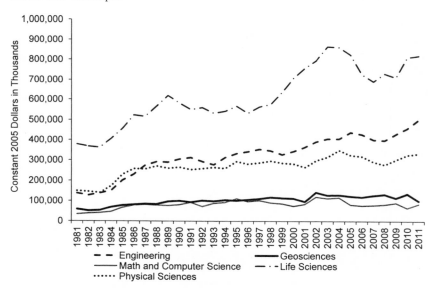

Fig. 10A.6 Equipment expenditures by field
Source: NSF Webcaspar.

References

Alberts, B. 2009. "On Incentives for Innovation." *Science* 326:1163.

American Academy of Arts and Sciences. 2008. *ARISE: Advancing Research in Science and Engineering: Investing in Early-Career Scientists and High-Risk, High-Reward Research.* Cambridge, MA: American Academy of Arts and Sciences.

Appel, T. A. 2000. *Shaping Biology: The National Science Foundation and American Biological Research, 1945–1975.* Baltimore: Johns Hopkins University Press.

Arrow, K. 1955. *Economic Aspects of Military Research and Development.* Santa Monica, CA: Rand Corporation.

Assmus, A. 1993. "The Creation of Postdoctoral Fellowships and the Siting of American Scientific Research." *Minerva* 31 (2): 151–83.

Azoulay, P., J. S. Zivin, and G. Manso. 2012. "National Institutes of Health Peer Review: Challenges and Avenues for Reform." In *Innovation Policy and the Economy*, vol. 13, edited by J. Lerner and S. Stern, 1–22. Chicago: University of Chicago Press.

Beadle, G. 1960. "Scientific Progress, the Universities and the Federal Government." *Engineering and Science* XXIV (3): 11–15.

Bush, V. 1945. *Science: The Endless Frontier.* Washington, DC: US Government Printing Office.

Chubin, D., and E. Hackett. 1990. *Peerless Science: Peer Review and US Science Policy.* Albany, NY: SUNY Press.

Coggeshall, P. T., and P. W. Brown. 1984. *The Career Achievements of NIH Predoctorate Trainees and Fellows.* Washington, DC: National Academy Press.

De Figueiredo, J. M., and B. Silverman. 2007. "How Does the Government (Want to) Fund Science? Politics, Lobbying, and Academic Earmarks." In *Science and the University*, edited by P. Stephan and R. G. Ehrenberg, 36–54. Madison: University of Wisconsin Press.

Division of Research Grants. 1996. *A Half Century of Peer Review 1946–1996.* Bethesda, MD: National Institutes of Health.

Ehrenberg, R. G. 2012. "American Higher Education in Transition." *Journal of Economic Perspectives* 26:193–216.

Ehrenberg, R. G., M. J. Rizzo, and G. H. Jakubson. 2007. "Who Bears the Growing Cost of Science at Universities?" In *Science and the University*, edited by P. Stephan and R. G. Ehrenberg, 19–35. Madison: University of Wisconsin Press.

Feldman, M., M. Roach, and J. Bercovitz. 2012. "The Evolving Research Enterprise: The Role of Foundations and Strategic Funding." Presentation made to the National Academies, September 21.

Freeman, R., T. Chang, and H. Chiang. 2005. "Supporting 'The Best and Brightest' in Science and Engineering: NSF Graduate Research Fellowships." NBER Working Paper no. 11623, Cambridge, MA.

Furman, J. L. 2012. "The American COMPETES Acts: The Future of US Physical Science and Engineering Research?" In *Innovation Policy and the Economy*, vol. 13, edited by J. Lerner and S. Stern, 101–45. Chicago: University of Chicago Press.

Heinig, S. J., J. Y. Krakower, H. B. Dickler, and D. Korn. 2007. "Sustaining the Engine of US Biomedical Discovery." *New England Journal of Medicine* 357:1042–47.

Ignatius, D. 2007. "The Ideas Engine Needs a Tuneup." *Washington Post*, June 3, p. B07.

Kaiser, J. 2008a. "The Graying of NIH Research." *Science* 322: 949–849.

———. 2008b. "NIH Urged to Focus on New Ideas, New Applicants." *Science* 319: 1169.

Lee, C. 2007. "Slump in NIH Funding is Taking Toll on Research." *Washington Post*, May 28, p. A06.

Leslie, S. 1993. *The Cold War and American Science: The Military Industrial Academic Complex at MIT and Stanford.* New York: Columbia University Press.

Levin, S., and P. Stephan. 1992. *Striking the Mother Lode in Science: The Importance of Age, Place, and Time.* New York: Oxford University Press.

Mervis, J. 2013. "How Long Can the US Stay on Top?" *Science* 340:1394–99.

Munger, M. G. 1960. *Growth of Extramural Programs of the NIH.* National Institutes of Health, document no. 021960. http://beta.worldcat.org/archivegrid/collection/data/50003150.

Murray, F. 2010. "The Oncomouse That Roared: Hybrid Exchange Strategies as a Source of Productive Tension at the Boundary of Overlapping Institutions." *American Journal of Sociology* 116:341–88.

———. 2012. "Evaluating the Role of Science Philanthropy in American Research Universities." In *Innovation Policy and the Economy*, vol. 13, edited by J. Lerner and S. Stern, 23–60. Chicago: University of Chicago Press.

National Academy of Sciences. 1978. *A Century of Doctorates: Data Analyses of Growth and Change.* Washington, DC: National Academies Press.

National Institutes of Health. 2012. *Biomedical Research Workforce Working Group.* Bethesda, MD: NIH.

National Institutes of Health. 1959. *History of Extramural Research and Training Programs at NIH.* Bethesda, MD: NIH.

National Research Council. 1994. *Meeting the Nation's Needs for Biomedical and Behavioral Scientists.* Washington, DC: National Academies Press.

———. 2011. *Research Training in the Biomedical, Behavioral, and Clinical Research Sciences.* Washington, DC: National Academies Press.

———. 2012. *Research Universities and the Future of America: Ten Breakthrough Actions Vital to Our Nation's Prosperity and Security.* Washington, DC: National Academies Press.

National Science Board. 1969. *Graduate Education Parameters for Public Policy.* Washington, DC: NSB.

National Science Foundation. 1994. *Science and Engineering Degrees: 1966–91.* Arlington, VA: NSF.

———. 2012. *Doctorate Recipients from US Universities: 2011.* Arlington, VA: NSF.

———. 2013. *Science and Engineering Research Facilities.* Arlington: NSF.

November, J. 2012. *Biomedical Computing: Digitizing Life in the United States.* Baltimore: Johns Hopkins University Press.

Petsko, G. A. 2012. "Goodbye, Columbus." *Genome Biology* 13:155.

Roosevelt, F. D. 1945. Letter, November 17. Washington, DC: The White House.

Samuelson, P. 2009. "Three Moles." *Bulletin of the American Academy* (Winter): 83–84.

Stephan, P. 2012. *How Economics Shapes Science.* Boston: Harvard University Press.

Stephan, P., and R. G. Ehrenberg. 2007. *Science and the University.* Madison: University of Wisconsin Press.

Strickland, S. P. 1989. *The Story of the NIH Grants Programs.* New York: University Press of America.

Teitelbaum, M. 2014. *Falling Behind? Boom, Bust, and the Global Race for Scientific Talent.* Princeton, NJ: Princeton University Press.

The President's Scientific Advisory Committee. 1960. *Scientific Progress, the Universities, and the Federal Government.* Washington, DC: US Government Printing Office.

Weinberg, S. 2012. "The Crisis of Big Science." *New York Review of Books*, May 10. http://www.nybooks.com/articles/archives/2012/may/10/crisis-big-science/.

Comment   Bruce A. Weinberg

Paula Stephan presents a fascinating and compelling story of the evolution of federal support for science in the United States. Her thesis is that in his *Endless Frontier* report, Vannevar Bush emphasized building scientific capacity, while universities were initially standoffish. Today, of course, universities are tremendously dependent on that federal support at a time when federal support is looking increasingly precarious.

How did we get from "there" to "here"? Although Stephan's story is far richer, it has real elements of basic supply and demand. Federal agencies initially sought to build capacity by supplying resources. With supply high, the "generosity" of grants (in terms of support for indirect costs and salaries) increased and the quantity of grants "demanded" naturally increased too. When the supply of funds began to flatten in the 1970s, success rates fell, leading to considerable rent-seeking effort, such as lobbying for support, universities competing for researchers with the ability to bring in large grants, and multiple rounds of revisions. In this environment, an increase in funding, perhaps most notably the doubling of (nominal) funding at the NIH from 1998 to 2002 lead to a large, "speculative" boom in building and hiring.

One of Bush's primary goals was to support the training of graduate students and postdoctoral fellows. As Stephan describes, there has been a striking shift of emphasis from training *per se* to training as part of the production (perhaps even a byproduct) of the production of current research itself. At the National Institutes of Health, as recently as 1980 as many graduate students and postdocs were being supported on training grants and fellowships (combined) as research grants (National Institutes of Health 2012). In the years since, support on training grants and fellowships has held constant or increased somewhat (for graduate students). By contrast, the numbers supported on research grants has tripled (in the case of graduate students) and quadrupled (in the case of postdocs). As a consequence, the number of people being trained (many from abroad) is determined as much by the current research needs of labs as by the long-term market demand for researchers after their training is completed. There is also the perception that the quality of the training on research grants may not be as strong as the training on fellowships, potentially accentuating the extent to which training is deemphasized relative to the production of current research (National Institutes of Health 2012).

Bruce A. Weinberg is professor of economics at Ohio State University, a research fellow of the Institute for the Study of Labor (IZA), and a research associate of the National Bureau of Economic Research.

For acknowledgments, sources of research support, and disclosure of the author's material financial relationships, if any, please see http://www.nber.org/chapters/c13035.ack.

The expansion of the research enterprise that Stephan documents is striking, with the size of awards, the length of awards, the amount of indirect costs covered, and the number of awards all increasing substantially. However, one of the most striking facts that Stephan presents is a decline in the real starting salaries of postdoctoral fellowships, which were (in 2013 dollars) $48,000, compared to under $40,000 in 2012. (Although, she notes that historically a large share of the postdocs were physicians.)

Stephan analogizes the current research university to a high-end shopping mall, which builds buildings speculatively and then competes for researchers to lease the space. There are a number of aspects to this behavior. Although the data are spotty, by all accounts, there has been a shift from supporting researchers on "hard money" to a model where researchers are paid on "soft money," expected to raise their salaries through grants. As Bruce Alberts (2010) has noted, the willingness of federal agencies to cover indirect costs actually compounds the incentive to shift researchers to soft money. Intuitively, an institution that pays people on soft money not only does not have to pay salaries, but also receives the indirect costs paid on the salary. It is striking to see that even at the time when these practices were being put in place, some researchers anticipated the availability of this level of support could be two-edged, with the increased availability of support likely to lead to expectations that researchers would generate more support. The competition for researchers can be seen in the large start-up packages that universities provide to researchers, especially in the natural and biomedical sciences. It can also be seen in the willingness of universities to support newly trained assistant professors on institutional funds for the ever-lengthening time that it takes for researchers (in biomedicine, at least) to obtain independent grant support. Turning to the speculative construction of space, Michael Teitelbaum has vividly pointed out that the calculation of indirect costs favors borrowing to finance construction relative to paying for construction directly, providing a further incentive for Stephan's speculative mallification of academia (National Institutes of Health 2012).

Stephan is to be commended for assembling a tremendous wealth of information and weaving it into a compelling story. Here the evidence from changes over time are complemented by discussions of differences across fields. Although it clearly goes beyond the scope of the current chapter, a formal analysis might compare fields based on the mix of support coming from each of the federal agencies, which differ dramatically in terms of their support parameters— private funding and institutional support.

Stephan rightly notes the tremendous increase in the cost of equipment (as well as the increase in its power). I think that the dissenting voices would focus on this increase in equipment costs as an alternative explanation for some of the trends Stephan highlights. Perhaps the increased reliance on federal support is not driven by its availability but rather by the increased cost of research. That said, Stephan presents some evidence that the share of R&D budgets spent on equipment have trended down over time.

Whatever one makes of it, the transformation of US science is truly remarkable. Bush wrote in an environment where US science was still very much on the trajectory to scientific leadership, with a small number of universities playing large roles in the scientific enterprise (Weinberg 2013). Today the United States has a leading role in research and far more institutions are actively involved in research. No assessment of science in the United States can overlook the tremendous successes of US science. But, as successful as US science has been, Stephan's piece provides valuable insights into how that success was achieved and how the system could be improved.

References

Alberts, Bruce. 2010. "Overbuilding Research Capacity." *Science* 329 (10 September): 1257.

National Institutes of Health. 2012. "Biomedical Research Workforce Working Group Report." http://acd.od.nih.gov/Biomedical_research_wgreport.pdf.

Weinberg, Bruce A. 2013. "Scientific Leadership." Working Paper, Ohio State University.

Algorithms and the Changing Frontier

Hezekiah Agwara, Philip Auerswald, and
Brian Higginbotham

> Everywhere, economic activity is turning outward by embrac-
> ing shared business and technology standards that let busi-
> nesses plug into truly global systems of production.
> —Sam Palmisano, former CEO of IBM (2006, 130)

11.1 Introduction

"What is the frontier?" Frederick Jackson Turner asked in his seminal
work, *The Frontier in American History* (1893). "In the census reports it is
treated as the margin of that settlement which has a density of two or more
to the square mile"(3). When Turner wrote of the closing of the American
frontier, he was referring to the end of an interval that lasted over three
hundred years, as the first European settlements in the North American
continent grew and expanded westward. The frontier was viewed as a place,
bounded on one side by the easternmost fields cleared for agriculture and
on the other by the westernmost wilderness. In between, Turner argued,
was a marginal space in which necessity was, even more than elsewhere, the
mother of invention.

By the time Franklin Delano Roosevelt wrote to Vannevar Bush in
November 1944 to request the report celebrated in this research volume,
the frontier itself had changed. "New frontiers of the mind are before us,"
Roosevelt wrote, "and if they are pioneered with the same vision, boldness,

Hezekiah Agwara is affiliated faculty at the School of Policy, Government, and International
Affairs of George Mason University. Philip Auerswald is associate professor at the School of
Policy, Government, and International Affairs of George Mason University. Brian Higginbo-
tham is a PhD candidate at the School of Policy, Government, and International Affairs of
George Mason University.

We thank Adam Jaffe, Ben Jones, Tim Simcoe, and participants in the NBER "The Changing
Frontier: Rethinking Science and Innovation Policy" preconference and conference for their
comments. We also thank Lewis Branscomb, Stuart Kauffman, José Lobo, and Karl Shell for
their contribution to developing ideas central to this chapter via jointly authored work, and
W. Brian Arthur for helpful conversations, insights, and inspiration. For acknowledgments,
sources of research support, and disclosure of the authors' material financial relationships, if
any, please see http://www.nber.org/chapters/c13030.ack.

and drive with which we have waged this war we can create a fuller and more fruitful employment and a fuller and more fruitful life" (Bush 1945b). Much as Thomas Jefferson had charged Meriwether Lewis and William Clark to survey the previously unexplored domains of the West in 1803, so Roosevelt tasked Bush to survey previously unexplored domains of human inquiry. The desired endpoint of the undertaking was the same in both cases: to improve lives and increase prosperity.

The title of the report Bush produced, *Science: The Endless Frontier,* expressed succinctly how societal progress was defined by the middle of the twentieth century. Released in July 1945, a month after the Allied victory in Europe and a year before George Doriot created the world's first publicly owned venture capital firm, Bush's report was about how best to maintain in peacetime a rate of scientific progress that had been unprecedented when driven by the necessities of war.

The frontier of scientific knowledge has advanced at least as dramatically in the nearly seventy years since 1945 as the frontier of the American West advanced in the seventy years after 1803. In both cases, the advancement was part of "the changing frontier" that has been a central feature of American economic history, which in turn is the title of this volume. The real change related to the evolution of the frontier itself.

It is significant that America's first World's Fair opened in Philadelphia exactly seventy years after Lewis and Clark returned to St. Louis at the end of their two-year expedition. The International Exhibition of Arts, Manufacturers, and Products of the Soil and Mine, as it was officially called, was a sort of museum in reverse in which inventions that signaled the creation of major new industries were first exhibited to the general public. These included Alexander Graham Bell's telephone (communications technology), the Remington typewriter (office services), the Wallace-Farmer Electric Dynamo (electric power), and Heinz ketchup (food processing). Indeed, there is considerable poetic significance in the fact that Frederick Jackson Turner first presented his renowned paper on "The Significance of the Frontier in American History" before the American Historical Society at a subsequent World's Fair—the 1893 World's Columbian Exposition in Chicago.

What of today? Where is the changing frontier of societal advance situated in 2013, both in the United States and globally? Is that frontier expanding or closing? These are the questions we seek to answer in this chapter.

To be clear, we recognize from the outset the parallel relevance of multiple notions of the "frontier." From the standpoint of a single firm or of a nation, we can think of the frontier as the boundary of economic production given existing technology and techniques—the "production possibilities frontier" (PPF) whose origins trace back to the early nineteenth century and the work of David Ricardo. We can also define the frontier in a global sense in terms of the boundaries of scientific advance, much in the way that Vannevar Bush employed the term. Or, as Cesar Hidalgo, Ricardo Hausmann, and coauthors have recently explored,[1] we can define the economic frontier for any region or country in terms of the country's existing (and constantly evolving) production capabilities.

In this chapter we are arguing that these three concepts of the frontier—the frontier of industry, the frontier of science, and the frontier of what we will call algorithms—actually define an advancing frontier of their own. That advancing "frontier of frontiers" is global.

The way we think about the frontier obviously affects how we seek to measure it. Total factor productivity serves to measure the advance of aggregate production possibilities. Patents can (with well-known limitations) measure the advance of the scientific frontier. Measures of the advance of the algorithmic frontier are less well developed. We propose some potential proxies in this chapter.

We develop our argument in three stages. In section 11.2 we introduce the idea of "the algorithmic frontier" through a summary of the progression of the idea of the frontier in American history. In particular we summarize different dominant interpretations of the frontier over the past four hundred years of US history: agricultural (1610s-1880s), industrial (1890s-1930s), scientific (1940s-1980s), and algorithmic (1990s-present). We then go back and motivate the idea of the algorithmic frontier a second time, from the standpoint of the evolution of economic theory. We argue (as suggested above) that the progression of historical frontiers (or, more precisely, of the frontier of frontiers) finds its direct parallel in the progression of economic theory.

In section 11.3 of the chapter we set the stage for proposing potential measures of the advance of the algorithmic frontier by discussing the relationship of "production recipes" (a term with historical resonance in economics that we use interchangeably with "production algorithms"), standards, and interoperability (both within and between firms). This section of the chapter provides a bridge to sections 11.4 and 11.5, in which we argue that the last thirty years of the centuries-old process referred to as "globalization" has been, more than ever before, defined by the adoption of standards and associated improvement in the interoperability of production algorithms. This is where our argument connects directly with Hidalgo and Hausmann (2009) and Hausmann et al. (2011), as well as with considerable prior literature that emphasizes the algorithmic substructure of the global exchange economy.

Finally, in section 11.6, we tie the chapter back to the core theme of this volume by discussing how the advance of the algorithmic frontier has affected the process of scientific discovery and technological innovation. Section 11.7 concludes.

---

1. Hidalgo and Hausmann (2009) and Haussman et al. (2011).

11.2 Changing Frontiers in the United States

11.2.1 Historical Context

The first American frontier requires little description. The map in figure 11.1 illustrates the movement of the frontier westward from 1803 through the nineteenth century. The social complexity of the process of westward movement—a subject of active scholarly inquiry in the century since Turner presented his paper—yields to remarkable simplicity when looked at from a cartographic perspective. Inexorably, the frontier moved westward until European settlements covered a continent. The economy of the United States during this lengthy interval was defined by two industries: agriculture and the extraction of natural resources. Accordingly, we refer to this first, most famous frontier in American history as the agricultural frontier.

The second American frontier was not the scientific one that formed the subject of the Bush report, but its industrial precursor. The World's Fair was to the era of the industrial frontier what the earliest precursors of the rodeo were to the era of the agricultural frontier: places where successful experimentation could be recognized and rewarded. The inventive wave that had been building since the 1870s continued to gain force. In the first decade of the 1900s alone the Wright brothers flew the first plane at Kitty Hawk, Henry Ford sold his first Model A, Samuel Insull merged Commonwealth Electric with Chicago Edison to create Commonwealth Edison—the world's first large-scale electric utility—and major breakthroughs were made in the development of the radio.

The frontier for the United States in the first third of its history was thus about realizing economies of scale afforded by the combination of new technologies and new modes of social organization. The era from the 1890s to the 1930s (in particular from roughly 1910 to the start of World War II) was the one in which the basic infrastructure of the modern United States was developed. The high-level industrial classifications that experienced the greatest growth during this interval include utilities, electric equipment and supplies, rubber and plastic products, petroleum and coal products, and printing and publishing.[2] The inventions listed above were among the sparks that ignited the industrial engine of the early twentieth century.

Bibliometric analysis provides a particularly vivid lens through which to view the changing industrial frontier. Figure 11.2 presents data on word frequencies created using Google Ngram, which is based on a digital database of more than 5.2 million books published worldwide between 1500 and 2008 comprising more than 500 billion words. Around that database Google created an interface they call the Ngram Viewer, which enables users to plot the frequency with which words and phrases appear in this data set over time.[3] The Ngram tool can

---

2. Data from the Historical Statistics of the United States and the Census of Manufacturers.
3. See http://books.google.com/ngrams.

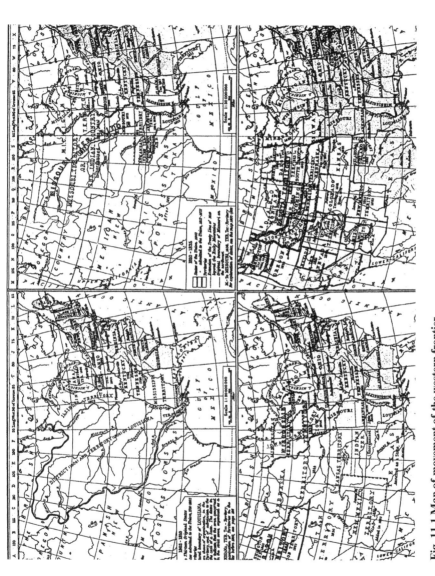

Fig. 11.1 Map of movement of the western frontier

Source: Shepherd (1923).

Note: I (top left), 1803–1810; II (top right), 1810–1835; III (bottom left), 1835–1855; and IV (bottom right), since 1855.

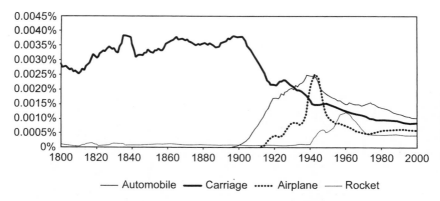

Fig. 11.2 Ngram, "carriage, automobile, airplane, and rocket"

be used to get a sense of the intensity of interest in particular technologies over time—put simply, the relative frequency with which particular words appear in published works of any type, for every year since 1500. Figure 11.2 presents a sample plot using the words "carriage, automobile, airplane, and rocket." The pattern shown for each of these words is consistent with the "hype cycle" hypothesized by Gartner Consulting, illustrated in figure 11.3. In the Gartner model, societal interest (which we conjecture is correlated with word frequencies in the Ngram database) in a technology grows rapidly after its first introduction. Interest soon reaches a peak, after which an era of disillusionment sets in. Interest falls off, usually just as the foundation for widespread societal adoption is setting in. By the time a technology is ubiquitous, its everyday usage is roughly constant; economic stability is reflected by this bibliometric stability.[4]

Although the plot in figure 11.2 is a simple representation of word frequency over time, it has some interesting characteristics. First, from 1900 to 1940, use of the word "carriage" decreases at just about the same rate that use of the word "automobile" increases; this is consistent with our intuition about the introduction of a more powerful substitute technology. Second, consistent with the Gartner hypothesis, the peak of relative intensity of usage comes well before technological maturity and market ubiquity. Finally, for these two words at least, the "hype cycle" seems to become increasingly compressed over time. This is consistent with considerable data that documents the increasing rates of adoption of new technologies and shorter product life cycles over time.

The inventions that defined the industrial frontier from 1890 through the 1930s represented major advances not just for the United States, but for humanity on a global scale. However, these inventions were the out-

4. For more on the Gartner hype cycle see http://www.gartner.com/technology/research/methodologies/hype-cycle.jsp.

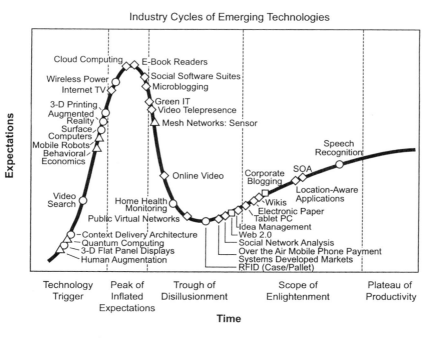

Industry Cycles of Emerging Technologies

Fig. 11.3 Gartner "hype cycle"

Note: For more on hype cycles, see http://www.gartner.com/technology/research/methodologies/hype-cycle.jsp.

come not of scientific research but of systematic tinkering. In the middle of the twentieth century the nature of invention began to change; invention became more scientific, with scientific research playing an increasing role in motivating major advances.[5]

5. Joseph Schumpeter wrote of the innovation as early as 1928 that within the emerging "trustified" capitalism "innovation is no longer . . . embodied typically in new firms, but goes on, within the big units now existing. . . . Progress becomes 'automatised,' increasingly impersonal and decreasingly a matter of leadership and individual initiative." (384–85). By 1958, John Jewkes, David Sawers, and Richard Stillerman wrote:
"In the twentieth century . . . the individual inventor is becoming rare; men with the power of originating are largely absorbed into research institutions of one kind or another, where they must have expensive equipment for their work. Useful invention is to an ever-increasing degree issuing from the research laboratories of large firms which alone can afford to operate on an appropriate scale. . . . Invention has become more automatic, less the result of intuition or genius and more a matter of deliberate design" (31).
This world of systematic innovation—if not based on science, per se, then on research more generally understood—represents the frontier that Vannevar Bush described, and sought to advance, in *Science: The Endless Frontier.*

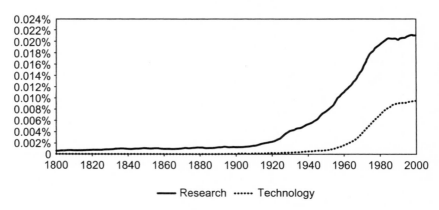

Fig. 11.4 Ngram, "research, technology"

Again, simple word-frequency plots help illustrate. Figure 11.4, also created with Google's Ngram tool, illustrates the intensity of usage of the words "research" and "technology" from 1800 to 2000. Rarely used before 1900, "research" begins to gain currency only in the 1920s, rises steadily, and then levels off starting in the 1980s. Technology follows a similar trend, but the period of rapid rise begins in the 1960s.

Figure 11.5 illustrates the system of science-based innovation that came into being following World War II. Advances in fundamental knowledge—the basic science column on the left-hand side—undergird the system of science-based innovation in a modern economy. Of course, advances in basic science will have no impact on economic growth or human well-being if they do not translate first into technologies, and ultimately into goods and services. The core technologies represented in the second column are the direct translation of science into a capability for innovation. Core technologies may be developed within the university, in an entrepreneurial start-up, or, most commonly, in the existing corporation. Core technologies typically are combined to create new goods and services. Industry production networks organized around existing goods and services are represented in the third column from the left. Industry production networks, or industry "clusters" when localized, are defined in terms of goods and services, not in terms of technologies. On the far right-hand side are the product markets themselves, where consumers and workers are situated. Innovations that renew or recreate existing industries not infrequently originate with workers near a production process, or with consumers of a product or service, rather than in a lab or university.

As Bush foresaw, this system has yielded significant dividends for American society. Tremendous scientific advances were made at Bell Telephone Laboratories, DuPont, General Electric, RCA Laboratories, the IBM T. J. Watson Research Center, the Xerox Palo Alto Research Center, together defining a

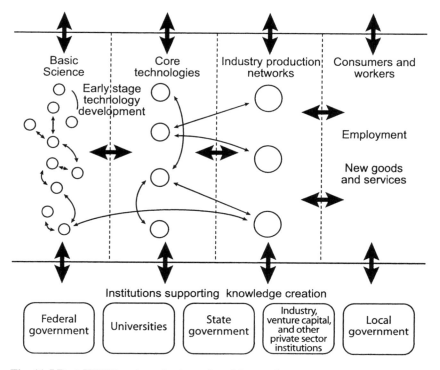

Fig. 11.5 Post-WWII system of science-based innovation

golden age of corporate research and development (R&D) in the United States (Auerswald and Branscomb 2005). It was the work that took place in these laboratories—arguably at least as much as that in universities, which was more distant from market applications—that defined the scientific frontier that drove the advance of the US economy in the post–World War II era (Trajtenberg, Henderson, and Jaffe 1992).

Coming fifty years after the publication of Turner's classic work on the closing of the western frontier, the title of the Bush report was significantly expansive: *Science: The Endless Frontier* pointed to a dimension of human attainment that would not be subject to limitation, as the prior era had been. In the case of the westward expansion, an insurmountable obstacle ultimately was reached: the Pacific Ocean. To Bush and those of his generation, no such obstacle was foreseeable when it came to the scientific frontier. An end to science-based innovation was essentially inconceivable. Yet by the 1970s, fewer private firms—regardless of their size—found it to be in their interest to invest in the sort of basic research that the Bush report had championed. One by one, the great corporate laboratories either closed or sharply narrowed their focus.

Macroeconomic data also suggests that a significant structural shift took place in the economic frontier in the 1970s. Using a methodology developed by the Bureau of Economic Analysis (BEA), Lee and Schmidt (2010) calculate the changes to the gross domestic product (GDP) that result from treating R&D as an investment rather than as an expense (table 11.1). They find that recategorizing R&D in this manner adds 0.13 percent to GDP growth rates from 1959 to 1973. However, from 1973 to 1994, the impact vanishes. This coincides with the much-discussed productivity slowdown, as well as the "conglomerate" discount experienced by the largest and most diversified US corporations starting in the late 1960s.

In the BEA analysis, the recategorization of R&D as investment once again begins to change the calculation of GDP growth rates appreciably from 1995 to 2007. As Jorgenson and Stiroh (2000) have documented, the primary vehicle by which R&D was contributing to GDP growth in the late 1990s and early in the twenty-first century was via innovations in information and communications technology (ICT). The new frontier was, and is, algorithmic rather than research-based.

This is not to say that the system of science-based innovation described in figure 11.5 has vanished. Far from it: it is larger and more robust than ever. As in prior eras, the infrastructure developed during the advance of one frontier remains fundamental to society as the next develops. Agricultural output increased for decades after the agricultural frontier was overtaken by the industrial frontier. Manufacturing output increased for decades after the industrial frontier was overtaken by the scientific frontier. Similarly, the output of science-based innovation has continued to increase even as that frontier has been overtaken by the algorithmic frontier.

11.2.2 Theoretical Context

We can readily describe the difference between agricultural, industrial, scientific, and algorithmic frontiers using standard production theory. Each of the first three frontiers has an associated domain within economics. The economics of the agricultural frontier were Malthusian, the economics of the industrial frontier were those of neoclassical growth theory, and the economics of science-based innovation are those of the particular variant of growth theory advanced in Romer (1986, 1990). In this section, we argue that the algorithmic frontier may require different economics as anticipated long ago by Simon (1967), Winter (1968), and Arrow (1974), and partially articulated more recently by Aghion and Howitt (1992), Kremer (1993), Romer (1996), Weitzman (1998), Hidalgo and Hausmann (2009), Hausmann et al. (2011), Bloom and Van Reenen (2010), and Arthur (2009, 2011), among others.

In a Malthusian world, land is the fundamental fixed factor, whereas populations are variable. Accordingly, rents accrue to land, and an interval of growth (though ephemeral) can be realized only through geographical expansion. Long-term growth is infeasible.

In an industrial model, capital replaces land. Investment can increase the capital-to-labor ratio. This increases the marginal product of labor, and thus the wage rate. Both the rate of population growth and the rate of technical change are exogenous. Growth is a matter of reaching the steady state level of per-capita consumption, which in turn is limited by the rate of technological advance. This, writ large, is the familiar world of the neoclassical growth model.

In the science-based model, technical change is the result of active investment. Knowledge is nonrival and nonexcludable, so the outcomes of R&D investments spill over to the economy as a whole.[6] Achieving economic equilibrium in the presence of aggregate increasing returns to knowledge is feasible so long as the research technology exhibits locally decreasing returns. Long-term growth rates can be increased by subsidizing research or the accumulation of human capital. This, writ large, is the familiar growth model set forth in Romer (1986, 1990).

How does the algorithmic model differ from the science-based model? Where the science-based model (like the Bush report) is built on the assumption that the transmission of economic knowledge is (or at least can be) costless and error free, the algorithmic model takes the costly process by which ideas are created, stored, shared, combined, and, of course, connected to economic exchange as the central problem of economic life.

The algorithmic model recognizes the possibility of a global best practice, or optimum, but does not assume that all firms, or all countries, will operate

---

6. This notion is associated with the work of Romer (1986, 1990), though it is present many other places. "Nonrival" means that one person's use of an idea does not keep another person from using the idea, "nonexcludable" means that it is impossible to keep a person from using an idea once it is "out in the open," and "knowledge spillovers" refers to the costless transmission of ideas that are nonrival and nonexcludable. Romer (1996) also employs the term "recipes" to refer to production algorithms, following both Simon (1967) and Winter (1968).

at this point. Even within the United States not all firms within an industry operate at the production possibilities frontier. While some firms maintain an advantaged position through nonmarket competition, there is a deeper trend at work in this new frontier.

The algorithmic model recognizes that research and development is an important element inside the black box of productivity, but that research and new knowledge creation are necessary but insufficient conditions to allow firms to operate at the boundary of the production possibilities frontier. This reflects the fact that there is diversity in the productivity levels of knowledge creation, which forms the basis of comparative advantage. This diversity in knowledge—itself the outcome of path-dependent processes—leads to diverse levels of productivity in product innovation.[7]

New knowledge creation through research and development plays a crucial role in economic growth, but the diversity in productivity levels within or between countries is also a function of the diversity of productivity levels in process innovation. The algorithmic model internalizes the insight that management quality varies and that management heterogeneity is central to understanding observed differentials between regions and nations.[8] When firm-level managers oversee the evolution of a production algorithm, they may emphasize different components of a management strategy. As an example, Bloom and Van Reenen (2010) find that American firms are better at providing incentives but are worse at monitoring than are managers in Sweden. Process innovation has always mattered but because of increasing organizational complexity from dispersed production networks, production algorithms and the organization of information are now defining the frontier of economic progress as never before.

In this light, consider the notion, central to the science-based model, that ideas are both "nonrival" and "nonexcludable," and economically relevant innovations are characteristically subject to "knowledge spillovers." In the algorithmic model, the ideas that actually propel growth and development are overwhelmingly uncodified, context dependent, and transferable only at significant cost—which is to say that tacit knowledge dominates, information asymmetries are the norm, and transactions costs are significant.[9]

While knowledge spillovers of the type that are central to the science-based model clearly exist, they are unlikely to be of significant relevance in the practical work of creating the new business entities that drive twenty-first-century

---

7. Hidalgo and Haussman (2009), and Hidalgo et al. (2011).

8. Goldfarb and Yang (2009).

9. Important early work by Mansfield (1961, 1963) on the subject of technological change related to imitation by one firm of the production methods of another. This work advanced the studies by Griliches (1957) on technological adoption. Where Griliches had used published data to study the adoption of essentially modular agricultural technologies, Mansfield (1961) used questionnaires and interviews to study the adoption of new production techniques by large firms in four industries.

global value chains.[10] The reason for this is that most productive knowledge is firm specific and producers far from dominant production clusters must learn to produce through a process of trial and error. Market-driven innovation involves the search for ideas that are rivalrous and excludable (at least temporarily), out of which ventures with proprietary value can be created. The impediments to innovation that matter most are not a lack of appropriability of returns but the everyday battles involved in communicating ideas, building trust, and making deals across geographically disparate regions and diverse economic units.[11] To the extent that the public benefits not captured by the investing firm (resulting from knowledge spillovers or other mechanisms) are temporally far off or uncertain, it is unlikely that they will be of greater importance to innovation-related decision making than will be the immediate, first-order challenges of organizing and financing the firm's operations.[12]

In the next section we develop this idea further. We first define "production recipes," a term that has historical resonance in economics that we will use interchangeably with "production algorithms." We then consider how the adoption of standards enables the interoperability of firm-scale production recipes. We argue that improvements in interoperability have been essential to the functioning of complex supply chains, and in this manner have been central to the story of unprecedented growth experienced globally in the past decade.

11.3 Production Recipes, Standards, and Interoperability

11.3.1 Production Recipes

Schumpeter (1912) famously wrote, "Technologically as well as economically considered, production 'creates' nothing in the physical sense. In both cases it can only influence or control things and processes, or 'forces.'. . . [T]o produce means to combine the things and forces within our reach. Every method of production signifies some definite combination. . . . The carrying out of new combinations we call 'enterprise'; the individuals whose function it is to carry them out we call 'entrepreneurs'" (14). The "new combinations" that entrepreneurs create are combinations of interdependent activities that jointly constitute the organization. These are "routines" in the language of Nelson and Winter (1982), "organizational capabilities" in the language of Chandler (1990, 1992), and "production recipes" in the language of Winter (1968) and Auerswald et al. (2000).

---

10. We emphasize that the focus here is not on web pages and pirated music videos. These digitized products—even including patents—are not the same thing as production algorithms or recipes.

11. Auerswald (2008).

12. Bloom and Van Reenen (2010).

In this chapter, we employ "production recipes" as our term of choice to describe combinations of interdependent activities that jointly constitute an organization. In this language, we can readily think of mangers as "cooks" who oversee the execution of existing, well-known recipes while entrepreneurs are more like "chefs," improvising new recipes—which is to say, creating new combinations.[13]

The use of the term "recipe" serves as a reminder that the sort of routinized processes of production we seek to describe with this term are as old as human society itself. Among the oldest Sumerian tablets are ones that describe actual recipes (e.g., for the production of beer) as well as numerical algorithms.[14] Indeed, we will employ the term "recipe" interchangeably with "production algorithm."

The idea of a recipe was clearly expressed by Winter (1968, 9): "'Knowing how to 'bake a cake' is clearly not the same thing as 'knowing how to bring together all of the ingredients for a cake.' Knowing how to bake a cake is knowing how to execute the sequence of operations that are specified, more or less closely, in a cake recipe." In the algorithmic model, this distinction takes on first-order importance: knowing how to bake a cake is different from knowing how to bring together all of the ingredients for a cake.

Figure 11.6, drawn from Auerswald et al. (2000), illustrates this point.[15] A neoclassical production plan is a particular input-output relationship. In its simplest rendition, it is a point (x, y) where $x \geq 0$ is the quantity of the input and where $y \geq 0$ is the quantity of the output. Figure 11.6 shows the production possibilities of the firm, the shaded area T, and three specific possible production plans labeled A, B, and C. The production function in this figure exhibits constant returns to scale, such that the best a firm can do is

$$y = \theta x,$$

where θ is a positive scalar that can be thought of as the organizational capital of the most productively efficient firm. The production function is comprised of the set of input-output pairs that lie on the boundary of the production possibilities set.

All of this is just a restatement of standard theory. Now, however, assume further that the approach utilized by the firm to convert inputs to outputs is encoded as a program. This program runs inside the "black box" of the standard production function to convert inputs to outputs.

For the sake of illustration, let us say that a given production process is comprised of three operations, each of which can be conducted in one

13. We thank Irwin Feller for sharing this analogy. See also Auerswald (2012).

14. Knuth (1972) and Auerswald (2012).

15. The description of figure 11.6 that follows in the next three pages is drawn from Auerswald (2010) and Auerswald and Branscomb (2005).

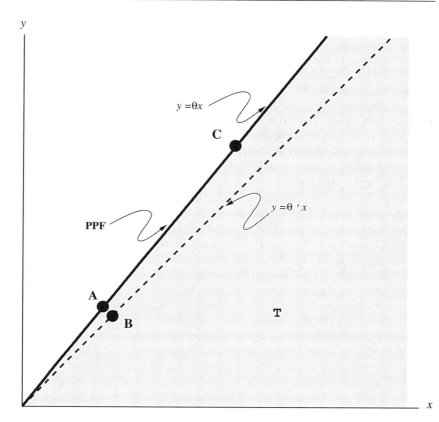

Fig. 11.6 Production possibilities frontier
Source: Auerswald et al. (2000).

of just two ways. We can exhaustively enumerate all possible production recipes as the set of eight binary strings {000, 001, 010, 100, 011, 101, 110, 111}. Each of these recipes will be associated with its own scalar measure of effectiveness. Let us refer to the level of effectiveness as the "organizational capital" associated with the recipe. For example, recipe 010 might be associated with organizational capital θ, and recipe 101 might be associated with organizational capital θ'. Let us arbitrarily say that recipe 010 is the best of the bunch, so its associated level of organizational capital is greater than the organizational capital associated with any of the other recipes.

Referring back to figure 11.6, input-output pair A, which lies on the boundary line as defined by $y = \theta x$, clearly "dominates" input-output pair B; the firm using recipe 010 produces more output with less input than the firm using recipe 101. For all firms to operate on the production possibilities frontier requires (a) that all firms have knowledge of the elements of the set of potentially usable recipes, and (b) that all firms are aware of the effectiveness of each recipe in actual production. Under such conditions, all firms in this example would use production recipe 010.

Figure 11.6 also allows us to clearly see the difference between economic distance and technological distance. From an economic standpoint, input-output pair A is close to input-output pair B but distant from input-output pair C. However, from the standpoint of technology, pairs A and C are the same, as they are produced with the same recipe (010); input-output pair B is maximally different from both A and C, in that the recipe used to produce B differs in every operation from that used to produce pairs A and C. Taking one operation at a time, $0 \neq 1$; $1 \neq 0$; $0 \neq 1$. Since there are three operations in all and the two recipes differ in every operation, the technological distance between the two recipes is 3.

The complexity of a production recipe can be represented either in terms of both the number of "operations" or distinct units involved in the production process or (critically) the extent of the interdependence between those units.[16] The greater the complexity of technology as defined in terms of interdependence, the lower the correlation between the effectiveness of the original production recipes (i.e., the leader's method) and that of the same recipe altered slightly (i.e., an imperfect imitation).[17]

In the Romer (1986, 1990) science-based model, technological distance does not exist; newly discovered recipes add to aggregate knowledge as soon as they are put into practice. In the algorithmic model, the search for better recipes is constrained both by technological distance and by the complexity of the production process. Newly discovered recipes that are not easily imitated are the essence of economic differentiation and the basis for above-normal profits.

11.3.2 The Historical Importance of Standards

According to the International Organization for Standards (ISO 2004, 4), standards are a "document, established by consensus and approved by a recognized body, that provides, for common and repeated use, rules, guidelines or characteristics for activities or their results, aimed at the achievement of the optimum degree of order in a given context." The term "optimum order" is significant. The standardization of interfaces allows for a system-wide optimization of the balance between order and flexibility in the supply

16. Coase (1937) argued that, in the presence of technological interdependencies, firms will expand to realize economies of scope. When firms do expand in such a manner to internalize the externalities, they create what Auerswald et al. (2000) term "intrafirm externalities." Indeed, if one particular unit of a firm is not linked to any other via such intrafirm externalities, then we reasonably wonder why that unit is part of the firm to begin with (rather than, for example, acting as an outside contractor). In this sense, a transactions cost theory of the firm predicts that, in industries where technological interdependencies abound, managers will typically be charged with solving complex coordination problems.

17. See Auerswald (2010) and related prior work, and also Rivkin (2000).

chain.[18] Standards are one form of codified knowledge that have been critical in sharing technical knowledge and expanding the reach of the market.[19]

The first emperor of China, Qin Shi Huangdi (221 to 206 BC), standardized both writing and weights and measures, with the aim of increasing trade within the newly unified country. From the invention of bills of exchange in the Middle Ages, to the development of universal time in the late nineteenth century and container shipping in the mid-twentieth century, innovations in standards have lowered the cost and enhanced the value of exchanges across distances. In the process, they have created new capabilities and opportunities on a global scale.

In our most recent period of globalization, the role of standards grew in importance as trade resumed in the post–World War II era. High transactions costs initially impeded trade, despite the emphasis on open markets and the resumption of (mostly) free trade through the Bretton Woods institutions. Some of the impediments stemmed from difficulties at the transition points of the global economy rather than tariff levels per se. One of the dominant forces driving the algorithmic frontier has been an unceasing quest to harmonize standards globally.

Thus, the growth of supply chains and the role of standardization in facilitating efficient chains have been critical to the functioning of global markets. Standardization of containerized shipping[20] and pallets,[21] two seemingly innocuous and generally unheralded developments, combined to transform global trading patterns. Entrepreneurs Malcolm McLean in shipping and Norman Cahners in pallets, performing an essential operations research task, were responsible for these two transformations.[22] As a result of these standards, global shipping costs fell from over $5.86 per ton in the 1950s to about $0.16 today.[23]

The European adoption of the global system for mobile communications (GSM) standards is another example of the benefit of standards harmonization. While Europe achieved rapid advances in mobile technology, in the United States the Federal Communications Commission (FCC) decided not to adopt an official cellular standard but to allow competition to select the optimal technology.[24] As a result, the market became segmented in the United States, with different companies each lobbying for their proprietary

---

18. Auerswald (2012).
19. It is evident, of course, that standards can also restrict trade. For a thorough discussion see, for example, World Trade Organization, *2005 World Trade Report: Exploring the Links between Trade, Standards and the WTO.*
20. Levinson (2008).
21. Vanderbilt (2012), Raballand and Aldaz-Carroll (2007).
22. Levinson (2008), Vanderbilt (2012).
23. Murphy and Yates (2009, 50). One result of the decline in transportation costs has been the rise of just-in-time manufacturing practices that have been the drivers of modern growth for firms like Honda, Toyota, and Walmart (Levinson 2008).
24. Guasch et al. (2007).

standards. Adoption of cellular technology was slower in the United States as a result.[25]

More generally, the diffusion of mobile phones based on the two dominant standards (GSM and code division multiple access [CDMA]) is one of the most astounding cases of the expansion of the algorithmic frontier. In 2000 there were fewer than 740 million mobile phone subscriptions, or roughly 16 per 100 inhabitants;[26] by 2012 there were more than 6.3 billion subscriptions (101 per 100 inhabitants). The rapid diffusion was facilitated by the adoption of technical standards that enable communications to occur over a common network. Once the technology was standardized, it was comparatively easy for firms like Vodafone to move into untested markets.

Instead of building extensive landline networks, developing countries built mobile towers and "leap-frogged" the older technology. The fastest growing mobile markets between 2002 and 2008 were in Africa, India, and China.[27] This is even true when we look at conflict-ridden environments such as Afghanistan and Pakistan.[28] In Afghanistan, the number of subscriptions rose from under 30,000 in 2000 to more than 18 million in 2012. The case of Pakistan is even more remarkable: in 2000 the number of mobile subscriptions was 360,000, and rose to more than 120 million by 2012.

The standards underlying GSM and CDMA encompass one layer in what engineers refer to as the internet protocol stack. The five layers of the TCP/IP protocol stack are: applications, transport, Internet, link (or routing), and physical.[29] The modular design of the protocol stack allows engineers to design standards for one stack independent of the others. Thus, at the application level, for instance, the World Wide Web Consortium (W3C) can focus on web design and applications standards like HyperText Markup Language (HTML) and Cascading Style Sheets (CSS). This allows for an efficient division of labor in standard creation and allows firms within standard-setting organizations to develop expertise at a given layer.[30]

In many areas of government regulation, but particularly in environmental regulation where standards are binding, firms operating in multiple jurisdictions may face different regulations. In order to comply, firms will often choose to adhere to the most stringent, an effect colloquially referred

25. Over time this has been important, but because mobile standards are updated almost every ten years, the lock-in effect from settling on a potentially inferior standard is reduced; the United States appears to have become slightly more innovative recently (Dodd 2012).

26. Data from ITU (2013). Retrieved from http://www.itu.int/en/ITU-D/Statistics/Documents/statistics/2013/Mobile_cellular_2000–2012.xls.

27. Kalba (2008), Sauter and Watson (2008, 20).

28. Auerswald (2012).

29. The Internet Engineering Task Force (IETF) defines the standard in RFC 1122, *Host Requirements*, and defines four layers. Authors frequently refer to the Link and Physical layers separately, although in RFC 1122 they are considered one layer.

30. Simcoe (forthcoming).

to as "so goes California." The phenomenon applies to many areas beyond environmental standards and regulations.

For the past thirty years, as global ties have deepened, firms have found that dealing with the financial and accounting rules in different jurisdictions can be a regulatory hurdle that advantages large firms with extensive financial and accounting departments. Two global efforts have gradually pushed the international financial system toward a harmonized set of standards.

In the United States, FASB (Financial Accounting Standards Board), at the direction of the US Securities and Exchange Commission (SEC), has led the transition from US generally accepted accounting principles (GAAP) accounting standards to the International Financial Reporting Standards (IFRS). The IFRS began as a system to harmonize financial accounting standards within the European Union, but as with California's environmental regulations, the value of global harmonization was quickly appreciated. Progress has been slow, but the efforts are ongoing, and the United States is gradually transitioning to the globally recognized standards.

In addition to accounting standards, the Bank for International Settlements has coordinated efforts to harmonize capital standards in the banking industry. There are currently three sets of accords, Basel I, Basel II, and Basel III. The United States and other industrialized countries are in the implementation phase of Basel III, while developing countries are typically at earlier stages. Most low-income countries are making progress at implementing Basel II, especially when large global financial institutions are present.[31]

Of comparable significance to these global standards have been within-firm standards that define and hold together global supply chains. From the sourcing of raw materials to the marketing of final goods, procurement contracts between buyers and suppliers depend on clear communication of expectations and specifications, all facilitated by standards.

11.3.3 Types of Standards

Our theoretical understanding of standards is limited and the categories that are most commonly identified are somewhat arbitrary because standards can blur formal boundaries. Kindleberger (1983) provided one early attempt at understanding the economic role of standards. He identified two primary purposes: "to reduce transactions costs and to achieve economies of scale through product interchangeability" (395). David (1987) observed that standards could perform both functions and that it might be preferable to classify standards based on the economic problem they solve (e.g., compatibility standards). One common characteristic intrinsic to all standards is that they codify technological knowledge.[32]

---

31. Gottschalk and Griffith-Jones (2010).
32. Blind and Jungmittag (2008).

Standards are classified based on their function and also on whether they are formal or informal. Informal or *de facto* standards are norms or requirements that may be voluntarily adopted and that frequently arise as a result of path dependence. Formal, or *de jure*, standards have the force of law behind them, either as laws, regulations, or contracts.[33] These are flexible categories and there may be some movement between types over time.

Whether standards have the force of law or simply gain network effects and de facto status, they are typically classified in one of four categories: reference, compatibility, interchangeability, and quality standards, with some room for overlap[34] (Blind 2004; David and Greenstein 1990; Guasch et al. 2007).

Reference standards (information or measurement standards) tie the value of one object to a reference base (NIST/SEMATECH 2012.) A weight measurement serves as a metaphor or simile. For example, a standardized pound in a scale is used to measure a comparable weight of another object, such as a bag of oranges (Busch 2011). Standardized weights and measures are a typical example of these standards (NIST/SEMATECH 2012.) Reference standards can also serve as coordination mechanisms. Landes (1983) describes the historical importance of establishing the measurement of time. Today dates and time have been codified by ISO 8601—data elements and interchange formats.

Compatibility (interface) standards enable different components of a system to work together because they are based on common characteristics. The extensive network of railroads is a clear example because commercial and passenger rail both work on the same tracks.[35]

Compatibility standards are among the most ubiquitous. According to Farrell and Simcoe (2013), compatibility standards account for more than 40 percent of the total stock of American National Standards. Because the infrastructure encompassing ICT is inherently modular, compatibility standards dominate in this field. Biddle, White, and Woods (2010) estimate that a modern laptop utilizes at least 251, and perhaps more than 500, unique compatibility standards.

The next two standards, interchangeability and quality, can be thought of as subsets of compatibility standards. Interchangeability (variety reducing) standards refer to parts that are interchangeable and for the most part identi-

33. Rycroft and Kash (1999).
34. There are other classification systems. For example, ISO/IEC (2014) defines three categories. "Standards can be broadly subdivided into three categories, namely product, process and management system standards. The first refers to characteristics related to quality and safety for example. Process standards refer to the conditions under which products and services are to be produced, packaged or refined. Management system standards assist organizations to manage their operations. They are often used to help create a framework that then allows the organization to consistently achieve the requirements that are set out in product and process standards."
35. Blind (2004), David and Greenstein (1990), and Guasch et al. (2007).

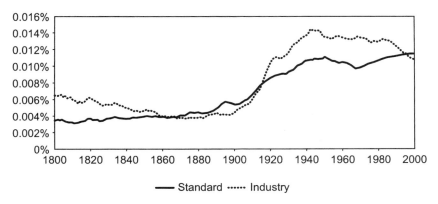

Fig. 11.7 Ngram, "industry, standard"

cal. The industrial revolution and the rise of assembly line processes at the beginning of the twentieth century required that tools in engineering be standardized.[36] As a result, standardized components from one manufacturer can be expected to work just as well as another. This is not limited to engineering but extends to many practical products, such as paper products.[37]

Quality (and safety) standards certify to consumers that a product or service was produced in a specific manner with a consistent minimum allowable quality. The best known standards are the ISO quality management standards, which are discussed in section 11.5. Health and safety standards for toys, food, drugs, and electrical appliances fall into this category as well.[38]

11.3.4 The Creation and Maintenance of Standards

The creation and maintenance of standards is a social activity at least as much as it is technical. While basic research, invention, and technological innovation can create ever-expanding menus of technological options, the process of standardization involves inducing a potentially large set of stakeholders to agree on a single choice. Solving such coordination problems is difficult and costly. Furthermore, those most knowledgeable about the relative strengths and weaknesses of various standards options are also often those with the most at stake in deciding which standards are selected.

The role of standards grew in importance following the American Civil War and the rise of science-based industry during the Second Industrial Revolution. The first standards institutions arose during this time. Figure 11.7, also created with Google Ngram, shows the concomitant rise in the

36. Brady (1961).
37. Guasch et al. (2007) and Blind (2004).
38. Guasch et al. (2007).

use of the words "industry" and "standard."[39] The gradual development of standards institutions followed three stages in which: (a) standards were localized within the firm or other administrative unit, (b) standards were agreed at the national level by one industry, and (c) standardization was carried out at an international scale.

In practice, because of the benefits of sponsoring a standard, international standard-setting is an almost unavoidably contentious and difficult process. It usually occurs in one (or a combination) of four ways.[40]

First, standards may arise through the decentralized choice of market participants. Second, firms may find it easier to negotiate, typically through a formal standard setting organization (SSO) or independent industry consortia. Third, a dominant market leader may set a standard, which smaller firms may follow. Finally, a standard may be set ex post through converters or multihoming.[41]

It is important to note that the convergence of international standards and global supply chain footprints has offered opportunities to both small and large players. Within these opportunities lie leverage points for systemic change. Increased efficiencies in the operation of global supply chains have enabled Walmart to grow over the past fifty years to become the world's largest private employer.[42] Yet the same shared global infrastructure famously has allowed the flat-screen TV manufacturer Vizio to develop the best logistics partnerships in its industry—from manufacturers in China to distribution through Costco—despite employing a minimal core staff.[43]

11.3.5 Standards and Interoperability

Global enterprise requires interoperability. This interoperability is being developed in a cumulative manner, piece by piece, and new standards are required at every stage.[44]

Standards have become increasingly important in the era of the algorithmic frontier because they enable the interoperability of firm-level recipes. The existence of standards turns a firm-level recipe into a subroutine of a larger program comprising many different recipes. That larger program enumerates the full instructions for the operation of a supply chain. As Paul Agnew, an early proponent of international standards, pointed out, compatibility standards resolve the difficulties that arise "at the transition points—points at which the product passes from department to department

39. One important trend to notice is that standards have continued to grow in importance even though industry trails off starting in the 1980s.
40. This section draws on Farrell and Simcoe (2013).
41. Converters allow different platforms to work together. Mutlihoming refers to the process of making a product, such as a video game, available on two different platforms, like the Sony PlayStation or Microsoft Xbox.
42. If Walmart was a country it would be China's eighth-biggest trading partner, ahead of all but the largest economies. See http://www.edf.org/page.cfm?tagID=2101.
43. See http://hbswk.hbs.edu/pdf/item/6424.pdf. The apparel manufacturer Li & Fung is among companies that have pursued a similar strategy (Hagel and Brown 2005).
44. Baldwin and Clark (2000).

within a company, or is sold by one company to another or to an individual" (Agnew, quoted in Murphy and Yates [2009], 7). Without standards, the interdependencies between the firm-level recipes comprising a supply chain would grow at a greater rate than the length of the chain, and the operation of the supply chain would be unmanageable.

Note that standardization to enable the interoperability of recipes as subroutines is different from the standardization of firm-level recipes themselves. In the food service industry, standardized recipes and production processes have famously been the success of franchise-based firms like McDonald's and KFC. Such firms have contributed to advancing the algorithmic frontier, but the particularly strict form of encoding that the franchise model represents is not the focus of this chapter.

Along similar lines, we can argue that the domain of analysis relating to general purpose technologies (GPTs) is really about the nesting of production recipes. Following from Schumpeter's observation that "[t]o produce means to combine the things and forces within our reach," we can readily observe that technologies themselves are frequently combined into other technologies, as subcomponents. Such technologies-acting-as-subcomponents themselves encode production recipes.[45] The recombination of technologies is thus, implicitly, the recombination of subroutines within a recipe. The GPTs are technologies that operate at lower—more fundamental—levels of such an algorithmic hierarchy (aka supply chain).[46]

In the next section we continue this line of argument by proposing that the recent phenomenon referred to as "globalization" is actually better understood as the progression of the algorithmic frontier, enabled by standards that in turn facilitate interoperability—both among and within supply chains.

11.4 Globalization Is Really Standardization

The recent wave of global integration termed "globalization" tends to be described in terms of the international integration of commodity, capital, and labor markets (Bordo, Taylor, and Williamson 2003). If this is what we mean by the term, however, then it is clear that our current period is not

---

45. The automated teller machine (ATM) or the rice cooker are two examples of physical technologies (hardware) encoding specific subroutines of a production recipe—in the first case, as executed by a teller sitting at a window in a physical bank, and in the second case as executed by a cook in a kitchen. In the case of ATMs, the process has iterated forward again in the past three to four years as significant elements of the functioning of the ATMs (software/hardware) have been encoded as "apps" on mobile phones, making them subroutines of the functioning of another technology (the phone).

46. With regard to general purpose technologies, the seminal reference is Bresnahan and Trajtenberg (1995). Doyne Farmer and James McNerney have sought recently to formalize the idea that a supply chain constitutes an instance of a production ecology, and further that the hierarchies of nested subroutines that, in the aggregate, comprise the production algorithm for the supply chain are equivalent to "trophic levels" in such an ecology. The lower the trophic level of a given technology (which is to say, of the subroutine the technology encodes) the more "general purpose" the technology.

the first example of globalization. There have been two major periods of globalization (Baldwin and Martin 1999) since the mid-nineteenth century. The first began in the mid-nineteenth century and ended with the onset of World War I. After an interlude between the wars that included the Great Depression, the second era of globalization began during the reconstruction after World War II. Growth in trade accelerated following the end of World War II, but in the past century we have seen trade flows of comparable magnitude to our current experience.[47] This type of global integration has actually been occurring for at least one thousand years, although the flow of information has not always been from the West to the East (Sen 2002).[48]

What makes the current era unique and different from prior eras of globalization? Alternatively, what has been the driver of the shift from the scientific frontier to the algorithmic? The primary difference between the algorithmic frontier and the earlier era of the scientific frontier is the rise of distributed networks of production and innovation (Auerswald and Branscomb 2008). In this view, globalization is really a process of interdependence and interconnectedness (Acs and Preston 1997). The real driver expanding the algorithmic frontier is the increasing reach of collaborative networks of all kinds—particularly production, but also research.[49] As Branstetter, Li, and Veloso write in this volume (chapter 5), "The important role of multinational corporations in the international invention explosions in China and India may help to explain why they are occurring at an early stage of economic development." The production networks themselves are the direct result of standardization. For that reason, we argue that globalization is really standardization.

Shared standards and business practices have been a precondition to this process of economic integration. In contrast with the traditional multinational assembly of subsidiaries, the global enterprise is a flexible assembly of firms around the world, with skills and capacity that can be drawn upon for the most efficient combination of business processes. The rapid globalization and economic integration witnessed in recent years has, in this manner, created the need for standardization of management systems, which are essentially the interface layer between production subroutines. As then-CEO of IBM Palmisano wrote in 2006:

> [S]tarting in the early 1970s, the revolution in information technology (IT) improved the quality and cut the cost of global communications and business operations by several orders of magnitude. Most important, it

47. Foreign direct investment as a share of GDP, starting with the third wave of democracy in 1974, accelerated from 5.2 percent between 1950 and 1973 to 25.3 percent from 1974 to 2007 (WTO 2008).
48. Sen (2002) cites as examples the transfer of knowledge of mathematics (decimal system) from India to the West, and of paper, gunpowder, and the printing press from China to Western Europe, among other technologies.
49. The resulting diversity of production levels is thus a result of the degree of incumbency and competition in an industry (Auerswald 2012).

standardized technologies and business operations all over the world, interlinking and facilitating work both within and among companies. This combination of shared technologies and shared business standards, all built on top of a global IT and communications infrastructure, changed the sorts of globalization that companies found possible.[50]

With diverse productivity levels among firms, companies in "ascending markets" within the developing world have faced significant signaling challenges in the global marketplace. In addition, they must manage information and compliance costs and adopt a common language of exchange. The result has been a remarkable increase in certain standards, or norms, issued by international organizations. We discuss this trend in the next section.

11.5 Using Quality Management Standards to
Map the Movement of the Algorithmic Frontier

Despite the importance of standards, empirical research on standardization has made only limited progress since the late 1990s.[51] The adoption of technical standards has proved difficult to measure. Data limitations and the potential for endogeneity have plagued the empirical study of the effects of adopting standards.[52]

Standards are well known to be associated with both costs and benefits from a social welfare standpoint. On the side of social costs, standards may serve as impediments to technical advancement or function as a nontariff barriers to trade.[53] On the social benefit side, the literature suggests at least three potential categories of positive impact. First, the existence of internationally recognized process standards may lower barriers of entry into production and distribution networks on a global scale, thereby enabling trade and making it more inclusive. Second, achieving functional compatibility and interoperability according to global norms may facilitate the adoption of platform technolo-

---

50. Palmisano (2006, 130). The transformation of IBM, which embodied the large-scale research-based firm of the scientific era, epitomizes the structural evolution that has taken place on a global scale. Once the epitome of the industrial giant with an international reach driven by science-based innovation—at its peak in the 1960s–1970s, IBM was investing half of its net income on developing new products and spent more money on computing research than the federal government (Acs 2013, 72)—IBM is today best "understood as global rather than multinational" (Palmisano 2006, 127). The change involves sourcing production from a variety of firms in different countries, and marketing the resulting products globally as well. Palmisano describes the integration of China and India into the global economy as the "most visible signs of this change." Between 2002 and 2003, he writes, foreign firms built sixty thousand manufacturing plants in China, many of them targeting global markets. Similar ties with firms in India are expanding the base from which global products and services are created.
    51. Among the few firm-level studies of the decision to seek certification from global standards bodies are Chen, Wilson, and Otsuki (2008) and Guasch et al. (2007).
    52. Clougherty and Grajek (2012).
    53. World Trade Organization (2005).

gies. Third, the process of obtaining and maintaining standards certification may serve as an important learning tool and lead to increased productivity through standardized routines; such learning expands the set of capabilities present in the economy, expanding the set of pathways for growth in the manner described by Hidalgo and Hausmann (2009) and Hausmann et al. (2011).

One source of data to help map the movement of the algorithmic frontier on process certifications is the International Organization for Standardization. Formed by the United States along with the other leading powers in 1947—just two years after Vannevar Bush published *Science: The Endless Frontier*—the International Organization for Standardization (abbreviated as "ISO" for reasons explained in footnote)[54] was designed to complement the functions of existing national standards bodies[55] by providing a wider forum for the agreement, adoption, and dissemination of standards.

Today the two most common ISO management standards are the ISO 9000 and 14000 series, which have been supplemented in recent years by standards for information security management, food security, and, most recently, social responsibility (ISO 26000), among others.[56]

The ISO 9000 addresses "quality management," which covers what an organization does to fulfill quality and regulatory requirements, enhance customer satisfaction, and achieve continual performance improvement.[57] The ISO 9000 consists of internationally accepted principles and requirements for managing an enterprise so as to earn the confidence of customers and markets.[58] Among the ISO standards the ISO 9000 series of quality management standards are the most general standards, and thus particularly interesting from our standpoint as they most plausibly relate to the management of integration with other subroutines within a global supply chain.

The adoption of ISO 9000 series of standards has occurred on a massive, global scale. The ISO 9000 series quality-management standards are diffused across more than 170 countries, but certification remains concentrated. Table 11.2 presents the top ten countries by certified firms, which account

54. According to the ISO, "because 'International Organization for Standardization' would have different acronyms in different languages ('IOS' in English, 'OIN' in French for *Organisation internationale de normalisation*), its founders decided to give it an all-purpose shortened name. They chose 'ISO,' derived from the Greek *isos,* meaning 'equal.' Whatever the country, whatever the language, the short form of the organization's name is always ISO." From http://www.iso.org/iso/about/discover-iso_isos-name.htm.
55. Notably, the American National Standards Institute in the United States and the British Standards Institute in the United Kingdom.
56. The appendix provides a summary description of the ISO quality-management standards.
57. The immediate predecessor to ISO 9000 was BS 5750, a quality-management standard in Great Britain. Since its inception, ISO 9000 quality-management standards have transitioned beyond manufacturing and have become widespread across industries, including the service sector, as can be seen in table 11.3. Despite the widespread adoption of these standards across industries, the existing literature has been concerned with the trade effects from the adoption of these standards in agriculture (Swann 2009).
58. Furusten (2002).

Table 11.2   Top ten countries for ISO 9001 certificates, 2011

Rank	Country	Number of certificates
1	China	328,213
2	Italy	171,947
3	Japan	56,912
4	Spain	53,057
5	Germany	49,450
6	United Kingdom	43,564
7	India	29,574
8	France	29,215
9	Brazil	28,325
10	Korea, Republic of	27,284
	Sum	817,631 (73.5%)
	All Others	294,067 (26.5%)
	Total	1,111,698

Source: ISO survey (2013).

Table 11.3   Top five industrial sectors for ISO 9001 certificates 2011

	Sector	Number of certificates
1	Services	203,970
2	Basic metal and fabricated metal products	101,848
3	Construction	83,864
4	Electrical and optical equipment	79,237
5	Machinery and equipment	58,427

Source: ISO survey (2013).

for more than two-thirds of the total certifications in 2011. There are two notable trends in these data. First, the new frontier in quality processes, or process design algorithms, has expanded globally through distributed networks of production. The fast-growing BRIC countries (Brazil, Russia, India, and China) constitute more than a third of total certifications, and three countries are in the top ten: Brazil (no. 9), China (no.1), and India (no. 7). Russia (no. 14) and South Africa (no. 39), which are increasingly identified with these emerging markets, also rank quite highly. South Korea, the most rapidly growing country in the world during the past half century,[59] rounds out the top ten.

The change in the composition of the top ten countries between 1993, the first year for which data is available, and 2011 is striking. In 1993 the top ten countries were, in order, the United Kingdom, Australia, the United States,

59. We omit Equatorial Guinea, which experienced the greatest rate of growth in per capita income of any country in the world in the fifty years after 1960, but without any appreciable advance in economic development measured along other dimensions.

France, Germany, the Netherlands, South Africa, Ireland, Italy, and Denmark; South Africa was the only nondeveloped country included and one of only a few outside Western Europe. The wide acceptance and adoption of the ISO series has expanded the algorithmic frontier and enhanced the capabilities and opportunities for firms in the developing world.

The distributed nature of production networks is clear from these data. More than one-quarter of firms with foreign ownership (the majority-owned foreign affiliates of parent companies) are ISO certified.[60] However, the ISO story is not limited to the case of the parent companies of multinationals in developed countries imposing quality standards on their foreign subsidiaries. Interestingly, the top-certified developing countries, or ascending markets, did not dominate firms by country in 2011. Instead, industrialized firms in Japan, Western Europe, and the United States (no. 11) also found benefits from adopting management standards, such as the ISO 9000 series. Firms seeking ISO certification in the developed world include those at the technological frontier, such as General Electric (in energy, health care, and related services) and Netgear (ICT), which proudly proclaim their ISO certifications on their websites. Even in cases where product quality is undisputed, managers find the process of codifying the production process to be a useful activity.[61] This survey evidence may imply that management standards are an important link in global supply chains but that they are not simply a procurement standard.

Figure 11.8 graphs the adoption rates of a broader range of ISO quality-management standards. The data follow a similar pattern, with initial adoption in Western Europe followed by gradual adoption outside the region. This process appears to have accelerated following the successful implementation of the ISO 9000 set of standards.

While the potential benefits of adopting ISO standards have been studied extensively, the literature has produced few clear results. Pathways of benefits vary among studies as well. The ISO adoption is alternately conjectured to function as a signal of competitiveness, to be associated with productivity enhancements and firm learning, or to lead to enhanced compatibility via a common-language effect.

If there is a consensus in the literature on ISO adoption, it is that the benefit of certification to the certified firm is at least as much in the widening of market opportunities as it is in the achievement of process-quality

---

60. Authors' calculations from World Bank Enterprise Survey (enterprisesurveys.org); majority-owned foreign affiliates are defined as businesses in which an investor of another country holds at least 10 percent voting ownership (BEA 2013).

61. Corbett and Luca (2002).

Terlaak and King (2006), however, argue that there must be other tangible financial incentives to justify the outlay of substantial organizational resources to obtain certification, so that the decoupling of the certification effect and the quality effect is not sustainable in a longer-run equilibrium.

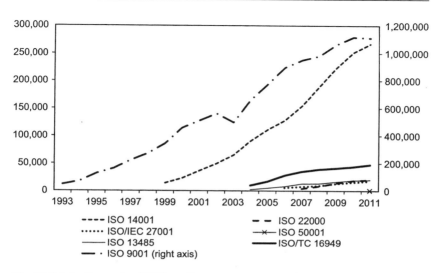

Fig. 11.8 Adoption path of ISO quality management standards

improvements.[62] A model developed by Terlaak and King (2006) suggests that certification with a management-quality standard may reduce information asymmetries in supply chains and bestow a competitive advantage on certified firms. The pursuit of certification may also, on its own, communicate desirable but otherwise unobserved organizational attributes.[63]

While the ISO certification process nominally requires companies to undergo restructuring of their organization and operational processes, a question remains as to whether certification results in actual improvements in firm-level capabilities. Some past studies suggest that achieving ISO certification induces organizations to adopt practices that improve operational performance.[64] Corbett and Luca (2002) and others have found the process of obtaining ISO certification increases functional compatibility and interoperability according to global norms, thus easing adoption of platform technologies. To be sure, if ISO certification conveys no information at all about quality or capabilities (in expectation), then buyers in the supply chain should tend to ignore it and suppliers should, as a consequence, ultimately cease to seek it. In this sense, we would conjecture that a full decoupling of

62. Breka (1994), Litsikas (1997), and Rao, Ragu, and Solis (1997).
63. Terlaak and King (2006) find that tangible financial incentives justify the outlay of substantial organizational resources to obtain ISO certification The authors examined the effects of ISO 9000 certification on the competitive advantage of US manufacturing firms. Using a panel of firms, they found that firms grew faster after ISO certification, ceteris paribus. Importantly, firms' growth effect was greater in situations where buyers faced greater difficulties in acquiring information about suppliers.
64. Litsikas (1997) and Rao, Ragu, and Solis (1997).

the certification effect and the quality effect (including revealed quality) is not sustainable in a longer-run equilibrium.

Whether financial benefits clearly accompany certification is another unresolved empirical question. In addition to prior empirical and theoretical work, some evidence is available from survey data. Corbett and Luca (2002) surveyed business executives in fifteen countries in the winter of 2001, receiving just under 3,000 responses. The responses indicated that ISO 9000 certification was considered "important" to the firms' continued success, the most favorable survey choice.[65] In a study of the impact of ISO quality management standards in Asia, survey data from the United Nations Industrial Development Organization (UNIDO) revealed that 36.1 percent of respondents cited "internal improvement" as their primary reason for choosing to implement a quality-management system (UNIDO 2012). This was the most common reason cited, followed by "customer pressure" (26.1 percent) and "corporate or top management objective" (18.5 percent). With regard to the link to certification and performance, the purchasers surveyed in this report responded that ISO 9000-certified firms performed "better" or "much better" than noncertified suppliers. The World Bank Enterprise Surveys data show a strong positive correlation between ISO certification and exporting activity.[66] Eighty-eight percent of the time (significant at .01 level), regions with higher proportions of firms with standards certification also recorded greater exporting activity.[67] Further, as shown by Prakash and Potoski (2007), the certification intensity of trading partners has a strong influence on adoption patterns.

Regardless of the direction of the specific causality running between trade and ISO certification, there is ample reason to believe that process standards such as ISO and technical standards such as GSM and TCP/IP (discussed above) have, together, had a dramatic impact on the global economy during its past thirty years of unprecedented growth.

The promise of the next standards to drive change on a global scale lies outside the bounds of reasonable conjecture. But while the nature of standards in the next—and most substantial—phase of the centuries-old process of global economic inclusion is uncertain, the continued centrality of standards is not. A core challenge of global development in the twenty-first century is thus the creation and maintenance of standards that accelerate, rather than impede, innovation at the same time they expand, rather than contract, economic inclusion on a global scale.

65. The authors report that "the average score for ISO 9000 certification across all industries was 3.95 (where 3 corresponds to 'somewhat important' and 4 to 'important')." (Corbett and Luca 2002, 10) The responses for ISO 14000 certification were not as robust and varied between "somewhat important" to "important" for the firms continued success. Demand for ISO 14000 certification typically originated in a firms marketing department rather than a quality control office or from top management, as was the case with ISO 9000.

66. Prakash and Potoski (2007).

67. Authors' pairwise correlation analysis of data from the World Bank Enterprise Surveys, available at http://www.enterprisesurveys.org.

In the next section we consider a dimension of the advance of the algorithmic frontier of particular significance not only to the past development of the global economy, but also to its future: the impact of the advance of the algorithmic frontier on the discovery of new ideas.

11.6 Algorithms and the Process of Discovery

So far in this chapter we have:

- considered the development of the economy of the United States as the frontier has shifted from agricultural, to industrial, to scientific, and finally to algorithmic;
- proposed a theoretical unit of analysis in the study of the algorithmic economy, which we term the "production recipe"; and
- noted the importance of standards in enabling the interoperability of production recipes, with particular attention to how such interoperability has facilitated international trade and accelerated global growth.

What we have not yet done is consider sources of novelty in the algorithmic economy, and how they may be different from those that have driven economic advance in the past. That is the goal of this last section in the chapter.

Following directly from the previous section we certainly can observe that, just as improvements in information and communications technologies associated with the advance of the algorithmic frontier have enabled the decentralization of processes to produce things, so have they led to a decentralization of processes to produce ideas. From the 1940s to the 1980s, when economic advance was bounded by the scientific frontier, a handful of corporate research labs—Bell Telephone Laboratories ("Bell Labs"), RCA Laboratories, the IBM T. J. Watson Research Center, and the Xerox Palo Alto Research Center (PARC), being the leaders—produced an astoundingly disproportionate share of the world's scientific discoveries and technological advances.[68]

The "golden age" of the large-scale corporate research laboratory supported by monopoly rents came to an end in the 1980s, due to a combination of factors including deregulation, the emergence of the conglomerate discount, and changes in the technology of discovery. In its place has emerged a less concentrated system of "divided technical leadership" (Ozcan and Greenstein 2013) involving a global network of smaller-scale corporate laboratories, government laboratories, academic institutions, and start-up ventures. The primary role of large corporations in this system is not to

---

68. For example, William Shockley and colleagues at Bell Labs developed the silicon transistor, and then went on to understand the underlying science of their innovation, a process that earned them the Nobel Prize in Physics in 1956. For more on Bell Labs, see Gertner (2012); regarding the "golden age" of corporate research labs, see Auerswald and Branscomb (2005).

generate new innovations themselves, but rather to produce and market such innovations at mass scale.[69]

The decentralization of the production of ideas enabled by the advance of the algorithmic frontier is one dimension of change in the process of discovery. A second—and more fundamental—one relates to methodologies of discovery themselves.

Among biologists the notion that life is fundamentally *algorithmic* earned a central place in theory decades ago. Most of the great discoveries in biological sciences in the twentieth century—the discovery of the double helix, most notably among them—were inspired by the work of pioneers of information theory notably including Norbert Wiener and Claude Shannon. As Sydney Brenner, the 2002 Nobel Laureate in medicine, said in 1971: "I feel that this new molecular biology has to go in this direction—to explore the high-level logical computers, the programs, the algorithms of development" (Gleick 2011, 300). The result of biologists' shift to focusing on "the algorithms of development" has been a revolution in the life sciences that continues to unfold today, with dramatic effects.

Advances in understanding about the nature of biological systems have translated directly into advances in actual technologies of discovery in the life sciences. Techniques such as combinatorial chemistry and high-throughput screening are today encoded in "lab-on-a-chip" technologies,[70] turning Schumpeter's vision of the search for economically valuable "new combinations" into algorithmically driven pieces of hardware capable of doing the work of dozens, if not hundreds, of postdocs using the methods of two decades ago.

In contrast, the advance of discovery in economics was, from the 1950s to the 1980s, firmly rooted in the logical positivist project of deriving theories from axiomatic foundations, then subjecting them to empirical test using econometric methods. The development and increasing influence within economics of lab experiments, field experiments, and even applications of neuroscience in the 1990s to the present has greatly increased the methodological diversity of the discipline.

Until recently, however, the social sciences (economics in particular) have remained separated from the life sciences and the physical sciences by orders-of-magnitude differences in both the availability and the reliability of data. The relatively poor quality of data available to social scientists and the perceived impracticability of conducting controlled economic experiments at large scale has necessitated the development by economists of methods of causal inference far more elaborate than those employed in other sciences.

---

69. Auerswald and Branscomb (2005).
70. A pioneer in this work is Caliper Technologies, founded by Larry Bock and Michael Knapp in 1995.

However—as is well known and widely discussed as the advent of "big data"—data on people's behavior in social contexts is increasing in quantity and improving in quality at an astounding rate. As a consequence, there is ample reason to believe that continued advances in the algorithmic frontier over the next two decades will transform the process of discovery in economics and other social sciences just as they have been doing for the past two decades in the life sciences.

11.7 Conclusion

Vannevar Bush is best known as the author of *Science: The Endless Frontier*, which provides the inspiration for this volume, for his work during World War II as director of the US Office of Scientific Research and Development, and for his part in the development of analogue computers. However, one of Bush's most powerful and enduring contributions may have been that which he made via a July 1945 magazine article in *The Atlantic* titled, "As We May Think," which was published just weeks before V-J Day. As he looked ahead to the frontier of societal advance in the postwar era, his emphasis was not on the products of publicly funded science but on the capacities of privately produced tools: "The world has arrived at an age of cheap complex devices of great reliability; and something is bound to come of it." In that essay, he envisions how existing low-cost technologies might be further advanced and networked into a system for the storage and retrieval of ideas, which he called the "memex": "Wholly new forms of encyclopedias will appear, ready made with a mesh of associative trails running through them, ready to be dropped into the memex and there amplified." The existence of this tool would allow for a continuation of the forward progress of human inquiry: "[Man] has built a civilization so complex that he needs to mechanize his records more fully if he is to push his experiment to its logical conclusion and not merely become bogged down part way there by overtaxing his limited memory."

Among those who read this essay in *The Atlantic* magazine was a twenty-five-year-old aerospace engineer named Douglas Engelbart. Engelbart was so taken by the vision set forth in "As We May Think" that he redirected his career to making that vision a reality. In 1968, at the fall Joint Computer Conference (a semiannual meeting of the then-major computing societies held in San Francisco), Engelbart delivered to over a thousand participants a presentation that set forth for the first time the core elements of the user architecture that would define the information revolution in decades to come: the computer mouse, text editing, hypertext, windowing, and video conferencing.

This story of serendipitous inspiration and invention illustrates the fundamental link between the science-based frontier and the algorithmic frontier, as well as the differences between the two. Although Vannevar Bush conceived and led the most ambitious and large-scale R&D programs ever undertaken at that time (notably including the Manhattan Project), he was able to look ahead to an era where the greatest progress in science-based discovery would be enabled by lowering the cost of storing and sharing ideas horizontally among scientists. Douglas Engelbart further democratized that vision, prototyping an architecture of interaction through standardized interfaces that we have come to know simply as information and communications technology.

This anecdote also suggests that the legacy of Vannevar Bush is arguably not about the importance of science-based research per se, any more than it is about the creation of the National Science Foundation and the decades of discovery it has enabled. Rather, it is about the importance of understanding that, at any point in time, the frontier of social attainment is changing. When Bush led the committee that produced *Science: The Endless Frontier* in 1945, the changing frontier consisted of the transition of an economy based on industrial growth through economies of scale to one based on improved goods and services through science-based innovation. Today, as we have sought to describe above, the frontier is changing again.

Just as the advent of science-based innovation motivated an earlier generation of economists to create new theoretical frameworks and analytic techniques to understand the rate and direction of technical change, so the advance of the algorithmic frontier is challenging the current generation of economists to respond in a like manner. The existence of this volume and the work it contains provides some evidence of the will that exists to meet that challenge.

Appendix
ISO Management Standards (ISO 2012)

ISO 9001:2008

ISO 9001:2008 gives the requirements for quality-management systems. Certification to the standard is used in global supply chains to provide assurance about suppliers' ability to satisfy quality requirements and to enhance customer satisfaction in supplier-customer relationships.

Up to the end of December 2011, at least 1,111,698 certificates had been issued in 180 countries and economies, two more than in the previous year. The 2011 total represents a decrease of 1 percent (−6,812) over 2010.

The top three countries for the total number of certificates issued were China, Italy, and Japan, while the top three for growth in the number of certificates in 2011 were Italy, China, and Romania.

ISO 14001:2004

ISO 14001:2004, which gives the requirements for environmental management systems, retains its global relevance for organizations wishing to operate in an environmentally sustainable manner.

Up to the end of December 2011, at least 267,457 ISO 14001:2004 certificates had been issued, a growth of 6 percent (+15,909), in 158 countries, two more than in the previous year.

The top three countries for the total number of certificates were China, Japan, and Italy, while the top three for growth in the number of certificates in 2011 were China, Italy, and France.

ISO/TS 16949:2009

ISO/TS 16949:2009 gives the requirements for the application of ISO 9001:2008 by suppliers in the automotive sector. Up to the end of December 2011, at least 47,512 ISO/TS 16949:2009 certificates, a growth of 8 percent (+3,566), had been issued in eighty-six countries and economies, two more than in the previous year.

The top three countries for the total number of certificates were China, the Republic of Korea, and the United States, while the top three for growth in the number of certificates in 2011 were China, India, and the Republic of Korea.

ISO 13485:2003

ISO 13485:2003 gives quality-management requirements for the medical device sector for regulatory purposes. Up to the end of December 2011, at least 20,034 ISO 13485:2003 certificates, a growth of 6 percent (+1,200), had been issued in ninety-five countries and economies, two more than in the previous year.

The top three countries for the total number of certificates were the United States, Germany, and the United Kingdom, while the top three for growth in the number of certificates in 2011 were the United States, Israel, and Japan.

ISO/IEC 27001:2005

ISO/IEC 27001:2005 gives the requirements for information security-management systems. At the end of December 2011, at least 17,509 ISO/IEC 27001:2005 certificates, a growth of 12 percent (1,883), had been issued in one hundred countries and economies, eight less than in the previous year.

The top three countries for the total number of certificates were Japan, India, and the United Kingdom, while the top three for growth in the number of certificates in 2011 were Japan, Romania, and China.

ISO 22000:2005

ISO 22000:2005 gives the requirements for food-safety management systems. Up to the end of December 2011, at least 19,980 ISO 22000:2005 certificates, a growth of 8 percent (1,400), had been issued in 140 countries and economies, two more than in the previous year.

The top three countries for the total number of certificates were China, Greece, and Romania, while the top three for growth in the number of certificates in 2011 were China, Italy, and Romania.

ISO 50001:2011

ISO 50001:2011 gives the requirements for energy-management systems. It was published in mid-June 2011. Up to the end of December 2011, at least 461 ISO 50001:2011 certificates had been issued in thirty-two countries and economies. The top three countries for the total number of certificates were Spain, Romania, and Sweden.

References

Acs, Z. J. 2013. *Why Philanthropy Matters: How the Wealthy Give, and What It Means for Our Economic Well-Being*. Princeton, NJ: Princeton University Press.

Acs, Z. J., and L. Preston. 1997. "Small and Medium-Sized Enterprises, Technology, and Globalization: Introduction to a Special Issue on Small and Medium-Sized Enterprises in the Global Economy." *Small Business Economics* 9 (1): 1–6.

Aghion, P., and P. Howitt. 1992. "A Model of Growth through Creative Destruction." *Econometrica* 60 (2): 323–51.

Arrow, K. J. 1962. "The Economic Implications of Learning by Doing." *Review of Economic Studies* 29 (3): 155–73.

———. 1974. *The Limits of Organization*. New York: Norton.

Arthur, W. B. 2009. *The Nature of Technology: What It Is and How It Evolves*. New York: Free Press.

Arthur, W. B. 2011. "The Second Economy." *McKinsey Quarterly* October (4). http://www.mckinsey.com/insights/strategy/the_second_economy.

Auerswald, P. 2008. "Entrepreneurship in the Theory of the Firm." *Small Business Economics* 30:111–26.

———. 2010. "Entry and Schumpeterian Profits: How Technological Complexity Affects Industry Evolution." *Journal of Evolutionary Economics* 20:553–82.

———. 2012. *The Coming Prosperity: How Entrepreneurs are Transforming the Global Economy*. New York: Oxford University Press.

Auerswald, P., and L. M. Branscomb. 2005. "Edwin Mansfield, Technological Complexity, and The 'Golden Age' of US Corporate R&D." *Journal of Technology Transfer* 30 (1/2): 139–57.

———. 2008. "Research and Innovation in a Networked World." *Technology in Society* 30 (3–4): 339–47.

Auerswald, P., S. Kauffman, J. Lobo, and K. Shell. 2000. "The Production Recipes Approach to Modeling Technological Innovation: An Application to Learning by Doing." *Journal of Economic Dynamics and Control* 24:389–450.

Baldwin C., and K. Clark. 2000. *Design Rules: Volume 1*. Cambridge, MA: MIT Press.

Baldwin, R. E., and P. Martin. 1999. "Two Waves of Globalisation: Superficial Similarities, Fundamental Differences." NBER Working Paper no. 6904, Cambridge,
MA. http://www.nber.org/papers/w6904.

Biddle, B., A. White, and S. Woods. 2010. "How Many Standards in a Laptop? (And
Other Empirical Questions)." SSRN Scholarly Paper no. ID 1619440, Social
Science Research Network. http://papers.ssrn.com/abstract=1619440.

Blind, K. 2004. *The Economics of Standards: Theory, Evidence, Policy.* Northamp-
ton, MA: Edward Elgar Publishing.

Blind, K., and A. Jungmittag. 2008. "The Impact of Patents and Standards on
Macroeconomic Growth: A Panel Approach Covering Four Countries and 12
Sectors." *Journal of Productivity Analysis* 29 (1): 51–60.

Bloom, N., and J. Van Reenen. 2010. "Why Do Management Practices Differ across
Firms and Countries?" Occasional Paper no. 26, Centre for Economic Performance. Retrieved July 14, 2013 from http://cep.lse.ac.uk/.

Bordo, M. D., A. M. Taylor, and J. G. Williamson. 2003. "Introduction." In *Globalization in Historical Perspective*, edited by M. D. Bordo, A. M. Taylor, and
J. G. Williamson, 1–10. Chicago: University of Chicago Press.

Brady, Robert A. 1961. *Organization, Automation, and Society: The Scientific Revolution In Industry.* Berkeley and Los Angeles: University of California Press.

Breka, J. 1994. "Study Finds Gains with ISO 9000 Registration Increase over Time."
Quality Progress May:18–20.

Bresnahan, T. F., and M. Trajtenberg. 1995. "General Purpose Technologies: 'Engines
of Growth'?" *Journal of Econometrics* 65 (1): 83–108.

Bureau of Economic Analysis. 2013. "Summary Estimates for Multinational Companies: Employment, Sales, and Capital Expenditures for 2011." April 18. http://
www.bea.gov/newsreleases/international/mnc/2013/_pdf/mnc2011.pdf.

Busch, Lawrence. 2011. *Standards: Recipes for Reality.* Cambridge, MA: MIT Press.

Bush, V. 1945a. "As We May Think." *The Atlantic*, July http://www.theatlantic.com/
magazine/archive/1945/07/as-we-may-think/303881/.

———. 1945b. *Science: The Endless Frontier.* Washington, DC: US Government
Printing Office.

Chandler, A. D. 1990. *Scale and Scope: The Dynamics of Industrial Capitalism.* Cambridge, MA: Belknap/Harvard University Press.

———. 1992. "Organizational Capabilities and the Economic History of the Industrial Enterprise." *Journal of Economic Perspectives* 6 (3): 79–100.

Chen, M. X., J. Wilson, and T. Otsuki. 2008. "Standards and Export Decisions:
Firm-Level Evidence from Developing Countries." *Journal of International Trade
& Economic Development* 17:501–23.

Clougherty, J. A., and M. Grajek. 2012. "International Standards and International
Trade: Empirical Evidence from ISO 9000 Diffusion." NBER Working Paper no.
18132, Cambridge, MA.

Coase, R. H. 1937. "The Nature of the Firm." *Economica* 4:386–405.

Corbett, C. J., and A. Luca. 2002. "Global Survey on ISO 9000 and ISO 14000:
Summary of Findings." Working Paper. http://personal.anderson.ucla.edu/charles
.corbett/papers/iso_survey_report_us.pdf.

David, P. A. 1987. "Some New Standards for the Economics of Standardization in
the Information Age." In *Economic Policy and Technological Performance*, edited
by P. S. Dasgupta and P. Stoneman, 206–39, Centre for Economic Policy Research.
Cambridge: Cambridge University Press.

David, P. A., and S. Greenstein. 1990. "The Economics of Compatibility Standards:
An Introduction to Recent Research 1." *Economics of Innovation and New Technology* 1 (1–2): 3–41.

Dodd, A. Z. 2012. *The Essential Guide to Telecommunications*. Upper Saddle River, NJ: Prentice Hall.

Farrell, J., and T. Simcoe. 2013. "Four Paths to Compatibility." In *The Oxford Handbook of the Digital Economy*, edited by M. Peitz, and J. Waldfogel. New York: Oxford University Press.

Furusten, S. 2002. "The Knowledge Base of Standards." In *A World of Standards*, edited by N. Brunsson and B. Jacobsson. New York: Oxford University Press.

Gertner, J. 2012. *The Idea Factory: Bell Labs and the Great Age of American Innovation*. New York: Penguin Books.

Gleick, James. 2011. *The Information: A History, a Theory, a Flood*. New York: Random House.

Goldfarb, A., and B. Yang. 2009. "Are All Managers Created Equal?" *Journal of Marketing Research* 46 (5): 612–22.

Gottschalk, R., and S. Griffith-Jones. 2010. "Basel II Implementation in Low-income Countries: Challenges and Effects on SME Development." In *The Basel Capital Accords in Developing Countries: Challenges for Development Finance*, edited by Ricardo Gottschalk. Basingstoke, Hampshire, UK: Palgrave Macmillan.

Griliches, Z. 1957. "Hybrid Corn: An Exploration in the Economics of Technological Change." *Econometrica* 25:501–22.

Guasch, J. L., J.-L. Racine, I. Sanchez, and M. Diop. 2007. *Quality Systems and Standards for a Competitive Edge*. Washington, DC: World Bank Publications.

Hagel III, J., and J. S. Brown. 2005. *The Only Sustainable Edge: Why Business Strategy Depends on Productive Friction and Dynamic Specialization*. Boston: Harvard Business School Press.

Hausmann, R., C. A. Hidalgo, S. Bustos, M. Coscia, S. Chung, J. Jimenez, A. Simoes, and M. A. Yildirim. 2011. *The Atlas of Economic Complexity: Mapping Paths to Prosperity*. Cambridge, MA: MIT University Press. http://atlas.media.mit.edu/book/.

Hidalgo, Céssar A., and Ricardo Hausmann. 2009. "The Building Blocks of Economic complexity," *Proceedings of the National Academy of Sciences*, 106: 10570-75.

International Organization for Standardization. 2004. "Standardization and Related Activities: General Vocabulary." http://www.iso.org/iso/iso_iec_guide_2_2004.pdf.

———. 2012. "The ISO Survey of Management System Standard Certifications, 2011." http://www.iso.org/iso/iso-survey.

International Organization for Standardization and International Electrotechnical Commission (2014). "About Standardization and Conformity Assessment." http://www.standardsinfo.net/info/aboutstd.html.

Jewkes, J., D. Sawers, and R. Stillerman. 1958. *The Sources of Invention*. London: McMillan & Co.

Jorgenson, D. W., and K. J. Stiroh. 2000. "Raising the Speed Limit: US Economic Growth in the Information Age." *Brookings Papers on Economic Activity* 31 (1): 125–236.

Kalba, K. 2008. "The Adoption of Mobile Phones in Emerging Markets: Global Diffusion and the Rural Challenge." *International Journal of Communication* 2:631–61.

Kindleberger, C. P. 1983. "Standards as Public, Collective and Private Goods." *Kyklos* 36 (3): 377.

Knuth, D. 1972. "Ancient Babylonian Algorithms." *Communications of the ACM* 15 (7): 671–77.

Kremer, M. 1993. "Population Growth and Technological Change: One Million B.C. to 1990." *Quarterly Journal of Economics* 108 (3): 681–716. doi:10.2307/2118405.

Landes, D. 1983. *Revolution in Time Clocks and the Making of the Modern World*. Cambridge, MA: Belknap Press.

Lee, J., and A. G. Schmidt. 2010. "Research and Development Satellite Account Update, Estimates for 1959–2007." *Survey of Current Business* December:16–56.

Levinson, M. 2008. *The Box: How the Shipping Container Made the World Smaller and the World Economy Bigger*. Princeton, NJ: Princeton University Press.

Litsikas, M. 1997. "Companies Choose ISO Certification for Internal Benefits." *Quality* 36:20–26.

Mansfield, E. 1961. "Technical Change and the Rate of Imitation." *Econometrica* 29: 741–66.

———. "The Speed of Response of Firms to New Techniques." *Quarterly Journal of Economics* 77 (2): 290–311.

Murphy, C., and J. Yates. 2009. *The International Organization for Standardization (ISO): Global Governance through Voluntary Consensus*. London: Routledge.

Nelson, R. R., and S. G. Winter. 1982. *An Evolutionary Theory of Economic Change*. Cambridge, MA: Harvard University Press.

NIST/SEMATECH. 2012. "e-Handbook of Statistical Methods." Accessed August 24, 2012. http://www.itl.nist.gov/div898/handbook/index.htm.

Ozcan, Y., and S. Greenstein. 2013. "The (de)Concentration of Sources of Inventive Ideas: Evidence from ICT Equipment." Unpublished manuscript.

Palmisano, S. J. 2006. "The Globally Integrated Enterprise." *Foreign Affairs* May/June. http://www.foreignaffairs.com/articles/61713/samuel-j-palmisano/the-globally-integrated-enterprise.

Prakash, A., and M. Potoski. 2007. "Investing Up: FDI and the Cross-Country Diffusion of ISO 14001 Management systems1." *International Studies Quarterly* 51:723–44.

Raballand, G., and E. Aldaz-Carroll. 2007. "How Do Differing Standards Increase Trade Costs? The Case of Pallets." *World Economy* 30:685–702.

Rao, S., T. S. Ragu, and L. E. Solis. 1997. "Does ISO 9000 Have an Effect on Quality Management Practices? An International Empirical Study." *Total Quality Management* 8:335–46.

Rivkin J. 2000. "Imitation of Complex Strategies." *Management Science* 46:824–44.

Romer, P. M. 1986. "Increasing Returns and Long-Run Growth." *Journal of Political Economy* 94:1002–37.

———. 1990. "Endogenous Technological Change." *Journal of Political Economy* 98 (5): S71–102.

———. 1996. "Why, Indeed, in America? Theory, History, and the Origins of Modern Economic Growth." *American Economic Review* 86 (2): 202–06.

Rycroft, R. W., and D. E. Kash. 1999. *The Complexity Challenge: Technological Innovation for the 21st Century*. New York: Thomson Learning.

Sauter, R., and J. Watson. 2008. *Technology Leapfrogging: A Review of the Evidence*. Report for UK Department of International Development, SPRU, University of Sussex. http://sro.sussex.ac.uk/29299/.

Schumpeter, Joseph A. 1912. *Theorie der witschaftlichen Entwicklung*. Leipzig: Duncker & Humblot. Translated by Redvers Opie as *The theory of economic development*. (Oxford: Oxford University Press, 1934). Cited edition is New Brunswick, NJ: Transaction Publishers, 1982.

———. 1928. "The Instability of Capitalism." *Economic Journal* 38:361–86.

Sen, A. 2002. "How to Judge Globalism." *American Prospect* 13 (1): 2–7.

Shepherd, W. R. 1923. *Historical Atlas*. New York: Henry Holt.

Simcoe, T. S. Forthcoming. "Modularity and the Evolution of the Internet." In *Economics of Digitization*, edited by A. Goldfarb, S. Greenstein, and C. Tucker. Chicago: University of Chicago Press.

Simon, H. A. 1967. "Programs as Factors of Production." *California Management Review* 10 (2): 15–22.

Swann, G. M. P. 2009. "International Standards and Trade: A Review of the Empirical Literature." Paper presented at the conference on Barriers to Trade: Promoting Good Practice in Support of Open Markets. http://www.oecd.org/trade/non-tariffmeasures/43685142.pdf.

Terlaak, A., and A. A. King. 2006. "The Effect of Certification with the ISO 9000 Quality Management Standard: A Signaling Approach." *Journal of Economic Behavior and Organization* 60:579–602.

Trajtenberg, M., R. Henderson, and A. Jaffe. 1992. "Ivory Tower versus Corporate Lab: An Empirical Study of Basic Research and Appropriability." NBER Working Paper no. 4146, Cambridge, MA. http://www.nber.org/papers/w4146.

Turner, F. J. 1894. "The Significance of the Frontier in American History." State Historical Society of Wisconsin.

United Nations Industrial Development Organization (UNIDO). 2012. "Study into the Value of ISO 9001." Accessed September 25, 2012. http://www.iaf.nu/articles/Study_into_the_value_of_ISO_9001_/279.

Vanderbilt, T. 2012. "The Single Most Important Object in the Global Economy." *Slate*, August 14. http://www.slate.com/articles/business/transport/2012/08/pallets_the_single_most_important_object_in_the_global_economy_.2.html.

Weitzman, M. L. 1998. "Recombinant Growth." *Quarterly Journal of Economics* 113 (2): 331–60.

Winter, S. G. 1968. *Toward a Neo-Schumpeterian Theory of the Firm.* Santa Monica, CA: Rand Corporation.

World Trade Organization. 2005. *World Trade Report 2005: Trade Standards and the WTO.* Geneva, Switzerland. http://www.wto.org/english/res_e/publications_e/wtr05_e.htm.

———. 2008. *World Trade Report 2008: Trade in a Globalizing World.* Geneva, Switzerland. http://www.wto.org/english/res_e/reser_e/wtr08_e.htm.

Comment   Timothy Simcoe

This chapter by Agwara, Auerswald, and Higginbotham (AAH) is an ambitious and thought-provoking attempt to describe how innovation at the "algorithmic frontier" links process innovation to globalization and economic growth. They begin with a historical discussion that emphasizes how ideas about the nature of the frontier have changed over time, gradually shifting from geographic expansion, to industrialization, to the scientific frontier described by Vannevar Bush and commemorated in this volume. The chapter's main thesis is that the scientific frontier has been replaced by an "algorithmic" frontier characterized by IT-enabled business process innovation and increasingly fragmented global supply chains. After describing this new frontier, the authors consider its implications for science and innovation policy.

Timothy Simcoe is associate professor of strategy and innovation at Boston University and a faculty research fellow of the National Bureau of Economic Research.

For acknowledgments, sources of research support, and disclosure of the author's or authors' material financial relationships, if any, please see http://www.nber.org/chapters/c13031.ack.

One major goal of the chapter is to draw economists' attention to several issues that deserve more scrutiny. These issues include business process innovation, the important role of standardization in the economy, and the gradual replacement of industrial R&D by decentralized innovation. Overall, I am sympathetic with this goal and the views expressed by AAH. Their chapter reminds us that process innovations, while difficult to measure, may be just as important as the patents and papers that are more frequently the object of statistical inquiry.

The chapter's second, more ambitious goal is to articulate a theory of "algorithmic production" and to explore its implications for trade, growth, and innovation. While AAH make some interesting progress on this front, it is not clear to me whether the kernel of a theory provided in this chapter can be developed into a full-fledged alternative to existing models of production or innovation.

This short response to AAH is organized into three parts. I begin by noting that the idea of algorithmic production is closely related to the management literature on firm-level routines and capabilities, and shares many of that literature's strengths and weaknesses. My second set of comments considers the hypothesized link between standardization, the algorithmic frontier, and global trade. I conclude by highlighting some potential implications of this chapter's thesis for science and technology policy.

Algorithmic Production

One of AAH's recurring themes is that economics neglects the important role of variation in production processes. Their starting point for this argument seems to be neoclassical production theory, in which perfectly competitive markets push atomistic firms relentlessly toward the most efficient technologies available. AAH argue that in reality, firms rely on different "recipes" or algorithms to produce similar goods and services, and that these differences in the methods of production are closely linked to differences in the rate and direction of technological change.

This is an important idea, though AAH are not the first to suggest it. In their path-breaking work, "An Evolutionary Theory of Economic Change," Nelson and Winter (1982) argue that heterogeneity in firm performance is driven by variation in the underlying methods of production, and that technological change occurs through a process of trial-and-error learning under selection pressure, as opposed to invention followed by rapid adoption of a single profit-maximizing technology. These ideas launched a multidecade research agenda within the field of business strategy to measure firms' routines and capabilities, and to link those constructs to variation in performance. Many of AAH's ideas are closely related to a more recent branch of that literature that models business process innovation as search on a complex landscape (e.g., Rivkin 2000).

AAH's use of the algorithmic metaphor in this chapter is an innovation relative to the business strategy literature, which typically describes the different production processes of seemingly similar firms in terms of "routines" or "capabilities." The algorithmic metaphor has both strengths and weaknesses. One strength of AAH's metaphor is that it highlights how key assumptions of the strategy literature diverge from classical economic theory. Just as many different algorithms can produce a similar computational result (albeit at differing levels of flexibility or efficiency), real firms operating in identical markets do seem to use very different processes to transform a particular mix of capital, labor, and knowledge into final goods and services. Moreover, there is an intriguing parallel between organizational and algorithmic design—in both settings complex problems are often broken into discrete steps that can be addressed independently in order to compartmentalize certain tasks and isolate interdependencies. In general, these design questions have received less attention within organizational economics than more familiar incentive and informational problems that are amenable to traditional modes of theorizing, and perhaps less context dependent.

The algorithmic metaphor also shares some key weaknesses of the management literature on firm capabilities. First, it neglects the idea that business processes are designed and managed by people, as opposed to machines. While individuals may lack either the information or incentives required to move quickly to an idealized production possibilities frontier, they do adapt, learn, and respond to local incentives. These latter ideas are not always easily accommodated within the algorithmic framework.

Second, an algorithmic theory of production typically takes a very reduced-form approach to the problem of demand discovery, often assuming that it can be represented by myopic search on some exogenously shifting landscape. For AAH, this approach to demand discovery strikes me as somewhat ironic, since their chapter suggests that economic frontiers have moved beyond the perfection of mass production techniques that exploit classical supply-side economies of scale. An alternative view of the contemporary frontier is that it rewards firms like Apple or Google that have developed the ability to anticipate consumer needs or rapidly solve difficult demand-side matching problems.

Finally, the strategy literature has struggled for years with the problem of measuring routines or capabilities in a manner that does not require making inferences based on past performance. Recent efforts to systematically survey management practices (e.g., Bloom and Van Reenen 2010) may herald some progress on this front. However, rather than attack this problem directly, AAH propose an alternative measurement strategy based on linking algorithmic innovation to the diffusion of management standards, notably the ISO 9000 series of quality standards.

Standards and Globalization

AAH's idea that the diffusion of business process standards can be used to measure the advancing algorithmic frontier is novel and creative. However, it is not clear that this measure does precisely what the authors would like, and in my view they push the underlying analogy too hard when arguing that process innovation has opened new frontiers in global trade.

A key piece of AAH's argument is the idea that standards "enable the interoperability of firm-level recipes . . . [by turning] a firm-level recipe into a subroutine of a larger program containing many different recipes." In support of this claim, AAH briefly describe several standards, such as uniform shipping containers and pallet sizes, that arguably played an important role in promoting global trade. Their thesis would be strengthened by unpacking these examples in more detail, and by describing some other important business process standards, such as Universal Product Codes (Basker 2012) or Electronic Data Interchange. Focusing on a wider variety of standards would also reduce the chapter's emphasis on ISO 9000. While ISO 9000 is widely adopted, and easily measured because of its certification program, it is not clear whether the specification promotes coordination among firm-level recipes in the sense emphasized by AAH, as opposed to providing a simple method of signaling that adopters have acquired some baseline level of managerial competence.

The chapter should also be careful about claims that increased globalization is "better understood" as the advance of an algorithmic frontier. The implicit baseline for this comparison is a vast literature on trade and development, a large part of which is concerned with firm-level relationships between productivity and trade (see Bernard et al. 2012). Standards are clearly important to trade. However, the idea of the advancing algorithmic frontier needs to be made more explicit if it is to be distinguished from the view that today's disintegrated design and production processes are a natural consequence of increasing returns to specialization, declining transport costs, and falling tariffs.

Concluding Thoughts

Given the focus of this volume, I will conclude with three short observations about this chapter's implications for science and innovation policy.

First, AAH draw our attention to the importance of business process innovation. Much of the economic literature on innovation focuses on easily measured inputs (R&D spending) or outputs (paper and patents). While it has become de rigeur to note that this is an example of "looking for our keys under the lamppost," AAH actually take a position on what we are missing. It is not clear to me that this observation corresponds to a change in the frontiers of innovation. For example, Paul David (1990) shows organizational innovations were an important complement to technical innovation during the late industrial revolution. However, AAH's emphasis on standards may highlight a genuine shift in the direction of inventive activity to the extent that today's IT-enabled frontiers require greater levels of interfirm coordination.

Second, AAH's idea of an advancing algorithmic frontier highlights the role of actual algorithms in contemporary innovation. Digitization is continuing to exert a major influence in the way that science is organized and practiced, both directly—through advances in measurement, computation, and instrumentation—and indirectly, through lowering the costs of collaboration and facilitating new practices such as open-access publishing or real-time remote access to shared facilities. These topics provide grist for the remainder of this volume.

Finally, AAH's chapter poses the interesting question of whether the decline of industrial R&D corresponds to the closing of the scientific frontier as envisioned by Vannevar Bush. At one level, the answer is "surely not." While it is intriguing to ponder the decline of Bell Labs, the data show that large firms still conduct the overwhelming majority of R&D. Nevertheless, there are clear indications that innovation has become more decentralized (e.g., Greenstein and Ozcan 2013). Perhaps this simply reflects a swing of the Schumpeterian pendulum back toward smaller firms, or the maturing of key segments within the IT-producing sector. On the other hand, it could reflect structural changes in the organization of innovative activity that present new challenges and opportunities for policymakers. This is an important question, and a nice contribution to a volume that emphasizes the changing innovation policy landscape fifty years after the idea of the scientific frontier was first put forward.

References

Basker, E. 2012. "Raising the Barcode Scanner: Technology and Productivity in the Retail Sector." *American Economic Journal: Applied Economics* 4 (3): 1–27.
Bernard, A., J. Bradford Jensen, Stephen J. Redding, and Peter K. Schott. 2012. "The Empirics of Firm Heterogeneity and International Trade." *Annual Review of Economics* 4 (1): 283–313.
Bloom, N., and John Van Reenen. 2010. "Why Do Management Practices Differ across Firms and Countries?" *Journal of Economic Perspectives* 24 (1): 203–24.
David, Paul A. 1990. "The Dynamo and the Computer: An Historical Perspective on the Modern Productivity Paradox." *American Economic Review* 80 (2): 355–61.
Greenstein, S., and Yasin Ozcan. 2013. "The (de)Concentration of Sources of Inventive Ideas: Evidence from ICT Equipment." Working paper presented to NBER Summer Institute 2013.
Nelson, R. R., and Sidney G. Winter. 1982. *An Evolutionary Theory of Economic Change*. Cambridge, MA: Harvard University Press.
Rivkin, J. 2000. "Imitation of Complex Strategies." *Management Science* 46:824–44.

Contributors

Ajay Agrawal
Rotman School of Management
University of Toronto
105 St. George Street
Toronto, ON M5S 3E6 Canada

Hezekiah Agwara
School of Public Policy
George Mason University
3351 Fairfax Drive, MS 3B1
Arlington, VA 22201

Philip Auerswald
School of Policy, Government, and
    International Affairs
George Mason University
3351 Fairfax Drive, MS 3B1
Arlington, VA 22201

Lee Branstetter
Heinz College
School of Public Policy and
    Management
Department of Social and Decision
    Sciences
Carnegie Mellon University
Pittsburgh, PA 15213

Timothy F. Bresnahan
Stanford Institute for Economic Policy
    Research (SIEPR)
Landau Economics Building, Room 325
579 Serra Mall
Stanford, CA 94305–6072

Annamaria Conti
Scheller College of Business
Georgia Institute of Technology
800 West Peachtree Street, NW
Atlanta, GA 30308-0520

Jason P. Davis
Department of Entrepreneurship and
    Family Enterprise
INSEAD
1 Ayer Rajah Avenue
Singapore 138676

Maryann Feldman
Department of Public Policy
University of North Carolina at
    Chapel Hill
209 Abernethy Hall, CB 3435
Chapel Hill, NC 27599–3435

Lee Fleming
Department of Industrial Engineering
330B Blum Center
University of California, Berkeley
Berkeley, CA 94720

Chris Forman
Scheller College of Business
Georgia Institute of Technology
800 West Peachtree St., NW
Atlanta, GA 30308

Richard B. Freeman
National Bureau of Economic Research
1050 Massachusetts Avenue
Cambridge, MA 02138

Ina Ganguli
Department of Economics
904 Thompson Hall
University of Massachusetts
200 Hicks Way
Amherst, MA 01003

Joshua S. Gans
Rotman School of Management
University of Toronto
105 St. George Street
Toronto ON M5S 3E6 Canada

Avi Goldfarb
Rotman School of Management
University of Toronto
105 St. George Street
Toronto ON M5S 3E6 Canada

Shane Greenstein
Kellogg School of Management
Northwestern University
2001 Sheridan Road
Evanston, IL 60208

Brian Higginbotham
School of Public Policy
George Mason University
3351 Fairfax Drive
Arlington, VA 22201

Adam Jaffe
Motu Economic and Public Policy
    Research
PO Box 24390
Wellington 6142 New Zealand

Benjamin Jones
Department of Management and
    Strategy
Kellogg School of Management
Northwestern University
2001 Sheridan Road
Evanston, IL 60208

Lauren Lanahan
Department of Public Policy
University of North Carolina at
    Chapel Hill
203 A Abernethy Hall, CB 3435
Chapel Hill, NC 27599-3435

Julia Lane
American Institutes for Research
1000 Thomas Jefferson Street, NW
Washington, DC 20007

Guangwei Li
Heinz College
School of Public Policy and
    Management
Carnegie Mellon University
3013 Hamburg Hall
Pittsburgh, PA 15213

Christopher C. Liu
Rotman School of Management
University of Toronto
105 St. George Street
Toronto, Ontario M5S 3E6
Canada

John McHale
School of Business & Economics
National University of Ireland
Cairnes Building
Nui Galway, Ireland

Raviv Murciano-Goroff
Economics Department
Stanford University
579 Serra Mall
Stanford, CA 94305–6072

Fiona Murray
MIT Sloan School of Management
100 Main Street, E62–470
Cambridge, MA 02142

Ramana Nanda
Harvard Business School
Rock Center 317
Soldiers Field
Boston, MA 02163

Alexander Oettl
Scheller College of Business
Georgia Institute of Technology
800 West Peachtree Street, NW
Atlanta, GA 30308

Timothy Simcoe
School of Management
Boston University
595 Commonwealth Avenue
Boston, MA 02215

Paula Stephan
Department of Economics
Andrew Young School of Policy
    Studies
Georgia State University
Box 3992
Atlanta, GA 30302–3992

Francisco Veloso
Department of Engineering and Public
    Policy
Carnegie Mellon University
131E Bake Hall
Pittsburgh, PA 15213

Bruce A. Weinberg
Department of Economics
Ohio State University
410 Arps Hall
1945 North High Street
Columbus, OH 43210

Pai-Ling Yin
SIEPR
366 Galvez Street
Stanford, CA 94305–6015

Ken Younge
Krannert School of Management
Purdue University
West Lafayette, IN 47907–2076

Author Index

Aboelela, S. W., 293
Abowd, J., 103
Acs, Z. J., 394, 395n50
Adams, J. D., 17, 23, 49, 68
Aghion, P., 109, 381
Agrawal, A., 2, 18, 68n7, 76, 77, 80, 97, 98, 99, 107, 113, 170, 171
Akcigit, U., 215
Alberts, B., 115, 116, 368
Alcácer, J., 154
Aldaz-Carroll, E., 387n21
Anderson, P., 215
Appel, T. A., 327, 329, 329n11, 330, 332, 333, 336, 338, 338n17
Armstrong, M., 269n38
Arora, A., 171
Arrow, K., 108, 355, 381
Arthur, W. B., 381
Auerswald, P., 379, 383, 383n11, 384, 384n13, 384n14, 384n15, 386n16, 387n18, 388n28, 394n49, 401n68, 402n69
Azoulay, P., 50, 53, 99, 108, 116, 354

Baldwin, C., 392n44
Baldwin, R. E., 394
Barefoot, K., 143n11
Barro, R. J., 171, 184
Basant, R., 136
Basker, E., 413
Baumgartner, F. R., 296
Beadle, G., 339

Beaver, D. B., 17, 115
Becker, G. S., 68
Bercovitz, J., 359
Bergemann, D., 223
Berglund, D., 289, 290
Bernard, A., 413
Berry, F. S., 295, 296, 297, 297n6
Berry, W. D., 295, 296, 297, 297n6
Biagioli, M., 112
Biddle, B., 390
Bikard, M., 72, 111
Black, G., 105
Blind, K., 389n32, 390, 390n35, 391n37
Bloom, N., 175n8, 226, 381, 382, 383n12, 412
Blossfeld, H.-P., 297, 297n6
Blum, B., 171
Blume-Kohout, M. E., 296
Bordo, M. D., 393
Boudreau, K., 35
Bozeman, B., 292, 293
Brady, R. A., 391n36
Branscomb, L. M., 379, 384n15, 401n68, 402n69
Branstetter, L., 138n4, 139n7, 140n8, 142
Breka, J., 399n62
Bresnahan, T. F., 84, 118, 175n8, 235n1, 236n4, 242n12, 258n31, 272n43, 280, 282n47, 393n46
Broad, W., 115, 117
Brown, J. S., 392n43

Brown, P. W., 336
Brynjolfsson, E., 175n8
Bunton, S. A., 293
Burgess, S., 103
Busch, L., 390
Bush, V., 322, 326, 372

Cairncross, F., 4, 171
Cameron, A. C., 154, 155, 156
Campbell, D. T., 314
Cantwell, J., 143
Card, D., 117
Catalini, C., 18, 171
Cawkell, A. E., 115
Chandler, A. D., 383
Chang, C.-H., 136n1, 138n3, 340
Chang, T., 330, 338
Chen, J., 136n1, 138n3
Chen, M. X., 395n51
Chiang, H., 330, 338, 340
Christensen, C., 258n30
Clark, K., 215, 392n44
Clougherty, J. A., 395n52
Coase, R. H., 386n16
Coburn, C., 289, 290
Coggeshall, P. T., 336
Cohen, W. M., 173
Cole, J. R., 112
Cole, S., 112
Combes, R. S., 289, 292
Conti, A., 50, 53, 105, 220
Cook, T. D., 314
Cooper, R. N., 137, 143, 152
Corbett, C. J., 398n61, 399, 400, 400n65
Cozzens, S. E., 290
Crane, D., 112
Cummings, J. N., 35

Danguy, J., 139n6
Darby, M. R., 144, 290
Dasgupta, P., 53, 72, 107, 112
David, P. A., 53, 72, 107, 108, 112, 127n6,
    287, 296, 389, 390, 390n35, 413
Davis, J., 278
De Figueiredo, J. M., 352n29
Delgado, M., 169, 171, 184
DellaVigna, S., 117
Denas, O., 50, 105
Dewatripont, M., 109
Diamond, A. M., 296
Ding, W. W., 72, 113, 171
Downes, T., 177n11

Eberhardt, M., 152
Ehrenberg, R. G., 345, 345n23, 345n25, 346
Eisinger, P. K., 289
Engers, M., 115

Fann, R.-E., 203
Farrell, J., 247n20, 278n45, 390, 392n40
Feldman, M., 289, 290, 290n2, 292, 294, 359
Feller, I., 287, 289, 290, 292
Fernandez, J.-M., 225
Fleming, L., 72, 210
Fogarty, M. S., 154
Foley, C. F., 142
Fons-Rosen, C., 18
Forman, C., 68n7, 170, 171, 173, 175, 175n8,
    176, 176n9, 176n10, 190
Fosler, R. S., 289
Franzoni, C., 18
Freeman, R. B., 17, 43, 50, 55n2, 57, 70, 136,
    330, 338, 340
Friedman, R. C., 293
Friedman, R. S., 293
Friedman, T. L., 4, 171
Frische, S., 116
Furman, J., 108, 116, 359
Furusten, S., 396n58

Gaeta, T. J., 115
Galison, P., 58, 113
Gans, J., 72, 111, 115, 128, 225
Gaspar, J., 89
Gaulé, P., 18
Gavetti, G., 215
Geisler, C., 72
Gertner, J., 401n68
Ghosh, S., 200, 220n9, 220n10
Giles, L., 104
Gittelman, M., 154
Glaeser, E. L., 11, 89, 169, 171
Gleick, J., 402
Goldfarb, A., 2, 18, 68n7, 76, 99, 107, 113,
    170, 171, 173, 175, 175n8, 176, 176n10,
    382n8
Golsch, K., 297, 297n6
Gompers, P., 200, 223
Gopalakrishnan, S., 287
Gottschalk, R., 389n31
Gourieroux, C., 155
Graff Zivin, J., 99, 108
Grajek, M., 395n52
Gray, V., 296
Green, J., 109, 118, 118n4, 123

Greenstein, S., 84, 170, 171, 173, 175, 175n8, 176, 176n10, 177n11, 192, 235n1, 242n12, 280, 390, 390n35, 401, 414
Griffith, R., 138n4
Griffith-Jones, S., 389n31
Griliches, Z., 155, 173n2, 382n9
Grossman, G. M., 135, 142, 143
Guasch, J. L., 387n24, 390, 390n35, 391n37, 391n38, 395n51
Guerrero Bote, V. P., 26
Guler, I., 223

Hagel, J., III, 392n43
Hägstrom, W. O., 112
Hall, B. H., 147, 154, 155, 173, 174, 192, 287, 296
Hall, R., 222
Haltiwanger, J., 103
Hampton, K., 170
Harhoff, D., 154
Harrison, R., 138n4
Hascic, I., 202n2, 203, 203n4
Hauger, J. S., 299
Hausman, J. A., 155
Hausmann, R., 373, 373n1, 381, 382n7, 396
Häussler, C., 115
Hayek, F. A., 283
Hecker, D. E., 300
Hege, U., 223
Heinig, S. J., 357
Helmers, C., 152
Helpman, E., 135, 142
Henderson, R., 169, 215, 379
Herr, B. W., 103
Hicks, D., 17, 26
Hidalgo, C. A., 373, 373n1, 381, 382n7, 396
Hitt, L., 175n8
Horn, M. B., 258n30
Howitt, P., 381
Hsu, D., 225
Hu, A. G., 138, 150, 152n19
Huang, C., 150
Huang, K., 113
Huang, W., 17, 43
Hummels, D., 137, 143, 152

Ignatius, D., 354
Ishii, J., 137, 143, 152

Jaffe, A. B., 49, 144, 147, 154, 169, 173, 174, 192, 379

Jakubson, G. H., 345, 345n25, 346
Jang, S.-L., 136n1, 138n3
Jefferson, G. H., 138, 150, 152n19
Jensen, K., 116
Jewkes, J., 377n5
Johnson, C. W., 258n30
Jones, B. F., 2, 3, 17, 43, 49, 50, 55n2, 68, 69, 71, 76, 87, 113, 114, 156, 171
Jones, C. I., 75n1
Jorgenson, D. W., 380
Jungmittag, A., 389n32

Kaiser, J., 349n28, 354
Kalba, K., 388n27
Kaminski, D., 72
Karch, A., 288, 297n6
Kash, D. E., 390n33
Katz, J. S., 17, 26
Kerr, W. R., 11, 169, 200, 215, 222
Khabsa, M., 104
Kiesler, S., 35
Kim, E. H., 2, 76, 99
Kindleberger, C. P., 389
King, A. A., 398n61, 399, 399n63
Klemperer, P., 247n20, 278n45
Klepper, S., 235n2
Knoor-Cetina, K., 113
Knuth, D., 384n14
Kogut, B., 143, 144
Kolko, J., 171
Kortum, S., 200
Kremer, M., 381
Krugman, P., 135, 137, 142, 143, 152
Kumar, K. B., 296

Lai, R., 144, 144n12
Lanahan, L., 289, 290, 290n2, 292, 294
Landes, D., 390
Lane, J., 103
Lawani, S. M., 17
Lee, C., 354
Lee, J., 380
Lei, Z., 152
Lendel, I., 289, 290, 290n2, 292, 294
Lerner, J., 200
Leslie, S., 330
Levin, S., 340n21
Levinson, M., 387n20, 387n22, 387n23
Libaers, D., 293
Litsikas, M., 399n62, 399n63
Liu, C. C., 53
Lo, A. W., 225

Lowery, D., 296
Luca, A., 398n61, 399, 400, 400n65

Magrini, S., 171, 184
Malerba, F., 282n47
Mallon, W. T., 293
Mani, S., 136
Mansfield, E., 49, 382n9
Manso, G., 108, 354
Martin, P., 394
McHale, J., 77, 80, 97, 98
Medhi, N., 202n2, 203, 203n4
Melkers, J. E., 290
Merton, R., 107, 112
Mervis, J., 347, 353n30, 357, 360, 361
Mintrom, M., 297n6
Mokyr, J., 75n1, 107, 108, 113
Monfort, A., 155
Morse, A., 12, 76, 99
Moya-Anegón, F. de, 26
Munger, M. G., 327, 328
Murphy, C., 387n23, 393
Murphy, K. M., 68
Murray, F., 72, 108, 111, 112, 113, 115, 116,
    128, 339, 347, 357, 359, 360
Muzyrya, Y., 278

Nanda, R., 200, 220n9, 220n10, 222, 225,
    226
Nelson, R. R., 108, 173, 383, 411
Nerkar, A., 215
Nookala, B. S., 147

Oettl, A., 77, 80, 97, 98
Olmeda-Gómez, C., 26
Olson, G., 3, 35
Olson, J., 3, 35
Otsuki, T., 395n51
Ozcan, Y., 84, 192, 401, 414

Palmisano, S. J., 371, 394, 395n50
Pancaldi, G., 113
Payne, A. A., 287, 296
Peng, L., 223
Petsko, G. A., 354
Plosila, W. H., 294
Ponzetto, G. A. M., 169, 171
Popp, D., 202n2, 203, 203n4
Porter, M., 169, 171, 184
Potoski, M., 400, 400n66
Prakash, A., 400, 400n66
Preston, L., 394

Price, D. J., 115, 127n6
Puga, D., 11, 136, 143, 152

Raballand, G., 387n21
Ragu, T. S., 399n62, 399n63
Rajaraman, A., 218
Rao, S., 399n62, 399n64
Reiss, P. C., 273n43
Rhodes-Kropf, M., 200, 222, 225
Rigby, J., 26
Rivkin, J., 386n17, 411
Rizzo, M. J., 345, 345n25, 346
Roach, M., 24, 359
Rohwer, G., 297, 297n6
Romer, P. M., 49, 75n1, 143, 381, 381n6, 386
Rosenbloom, R. S., 84
Rosenberg, N., 108, 200
Rosenkopf, L., 215
Rossi-Hansberg, E., 143
Rothenberg, J., 226
Roychowdhury, V. P., 26n4, 127n6
Ruegg, R. T., 287
Rycroft, R. W., 390n33
Rysman, M., 236n4, 268n37

Sadun, R., 175n8
Sahlman, W., 222
Sakakibara, M., 138n4
Sala-i-Martin, X., 171, 184
Samila, S., 200
Samuelson, P., 322n1
Santoro, M. D., 287
Sapolsky, H. M., 288, 289, 299
Sauermann, H., 24, 115
Sauter, R., 388n27
Sawers, D., 377n5
Saxenian, A., 169
Scellato, G., 18
Schaffer, S., 113
Schmidt, A. G., 380
Schumpeter, J. A., 377n5, 383
Scotchmer, S., 109, 118, 118n4, 123
Sen, A., 394, 394n48
Shadish, W. R., 314
Shapin, S., 113
Shapiro, P., 292
Sharma, A., 147
Sharma, P., 147
Shine, K., 115
Sichel, D. E., 236n3
Silverman, B., 352n29
Simcoe, T. S., 115, 388n30, 390, 392n40

Simkin, M. V., 26n4, 127n6
Simon, H. A., 381, 381n6
Sinai, T., 171
Singh, J., 72, 144, 210
Solis, L. E., 399, 399n63
Sood, N., 296
Sorensen, A. T., 240n9
Sorensen, J., 215
Sorenson, O., 200
Spencer, W. J., 84
Srinivasan, T. N., 137, 143, 152
Stein, J., 109
Stein, R. M., 225
Stephan, P., 3, 18, 49, 50, 55, 57, 70, 72, 103,
     105, 339, 340n21, 349n27, 350, 351, 358
Stern, S., 108, 169, 171, 184, 200, 225
Stevens, D., 103
Stillerman, R., 377n5
Stiroh, K. J., 380
Strickland, S. P., 321, 326, 329, 336
Stuart, T., 53, 72, 215
Sun, Z., 152
Sutton, J., 242n13, 258n32
Swann, G. M., 396n57

Talley, E. M., 103
Tang, L., 80
Taylor, A. M., 393
Taylor, C. D., 295
Teich, A., 290
Teitelbaum, M., 326n5
Teodoridis, F., 2, 76
Terlaak, A., 398n61, 399, 399n63
Thursby, J., 220
Thursby, M., 220
Tilghman, S., 55n2
Todd, W. J., 289, 292
Toole, A. A., 287, 296
Trajtenberg, M., 138n4, 144, 147, 154, 169,
     173, 174, 192, 236n4, 379, 393n46
Treeratpitu, P., 104
Trefler, D., 136, 143, 152
Tripsas, M., 215
Trivedi, P. K., 154, 155, 156
Trognon, A., 155
Trost, R. P., 290
Turner, F. J., 371
Tushman, M. L., 215

Ullman, J., 218
Uzzi, B., 17, 49, 68, 87, 113, 114, 171

Vanderbilt, T., 387n21, 387n22
Van Reenen, J., 138n4, 175n8, 381, 382,
     383n12, 412
Van Zeebroeck, N., 68n7, 170, 171, 176n9,
     190
Venkatramen, V., 116
Vergari, S., 297n6
Vernon, R., 135, 142
Vilhuber, L., 103
Visentin, F., 50, 105
Vogel, R. C., 290
Volden, C., 297n6

Waguespack, D. M., 115
Waldfogel, J., 171, 251
Waldinger, F., 53, 97
Walsh, J. P., 80, 173
Wang, J., 50, 99
Watson, J., 388n27
Wayne, L., 295
Weinberg, B. A., 55n2, 369
Weinberg, S., 339
Weitzman, M. L., 43n9, 75n1, 144, 381
Wellman, B., 170
White, A., 390
Williamson, J. G., 393
Wilson, J., 395n51
Winter, S. G., 381, 381n6, 383, 384, 411
Woods, S., 390
Woodward, S., 222
Wooldridge, J. M., 155
Wright, B., 152
Wright, J., 269n38
Wuchty, S., 17, 49, 68, 87, 113, 114, 171

Yang, B., 382n8
Yates, J., 387n23, 393
Yi, K.-M., 137, 143, 152
Yin, P.-L., 278
Yorgason, D., 143n11
Youtie, J., 292, 293
Yu, Z., 152

Zander, U., 143, 144
Zhao, M., 142, 143, 152, 155
Zimmerman, A., 33
Zingales, L., 2, 76, 99
Zivin, J. G., 50, 354
Zucker, L. G., 144, 290
Zuckerman, H., 50

Subject Index

Page numbers followed by *f* or *t* refer to figures or tables, respectively.

Academic knowledge: as collective phenomenon, 49–50; empirical setting for study of, 51–55; industrial innovations and, 49. *See also* Knowledge production; Scientific knowledge, advancement of frontier of; Scientific productivity

Advertising, apps and, 260–61; organizational structures for, 260

Agnew, Paul, 392–93

Algorithmic frontier, 373, 380–83; using quality management standards to map movement of, 395–401

Algorithms, process of discovery and, 401–3

Amazon Kindle platform, 279

American frontier, 371, 374–76

Android apps, 238

App Annie, 245

Apps. *See* Mobile applications (apps)

App stores: problems facing collaborative filters of, 241–42; rankings for, 239–41

Atari Democrats, 295

Authors, corresponding, survey of, 23–25

Authorship, conventions of, 114–16

Authorship "law and order," 114–15

Automated teller machines (ATMs), 393n45

Baliles, Gerald, 295

Basel I, 389

Basel II, 389

Basel III, 389

Basic science, 378, 379f

Bayh-Dole Act (1980), 12, 290, 299

Biotechnology, 20

Brenner, Sydney, 402

Brout, Robert, 128

Bush, Vannevar, 1, 8–9, 10–12, 169, 199, 321–22, 371–72, 379, 403–4; university research and, 351

Cahners, Norman, 387

Cascading Style Sheets (CSS), 388

Cellular technology, adoption of, 387–88

Centers of Excellence programs, 8, 290, 291t, 293–95; discussion of results for study of, 311; empirical results for study of, 306–8, 307t; overlap of, 311–12; as part of portfolio, 312–14

CERN. *See* European Organization for Nuclear Research (CERN)

China: data and descriptive features of rise of innovation of, 144–52; development in, 135–36; empirical models and regression results on quality and quantity of patenting in, 152–61; location of inventors in, 147–49; ownership of patents in, 146–47; patenting in, 5; research and development (R&D) in, 136–37; types of invention in, 147; US multinational R&D in, 137–44. *See also* India

Citations, international collaboration and, 26, 27f
Coauthors: contributions to collaboration of, 35–37, 36f; meetings and communication between, 32–35. *See also* Collaboration
Coauthorship, 17–18, 117
Code division multiple access (CDMA), 388
Coinvention: empirical model and results for, 152–61; lessons from interviews of multinational R&D personnel, 161–63. *See also* China; India; Research and development (R&D)
Collaboration: advantages and challenges of, 37–40, 39t; bias and, 125–26; central role star scientists in, 97–101; changes in, 76; costs of, 43; declining costs of, and star scientists, 88–92; distance and, 76; evolving role of, in science, 2–3; in field of evolutionary biology, 87–92; improvements in technology and, 92–97; issue of getting credit in joint production and, 43–44; level of, 76; local growth in patenting and, 188–92; model examining effects of improved technology and, 92–97; over distance, 25–31; productivity advantage of, 42–43; reasons for increasing, 44–45, 76–77; scientific, economics of, 40–45; supporting technologies and, 76–77; trend of increasing, in evolutionary biology, 87–92; trends, 17–18; types of, 20–21, 21f; US, 19; variation in, and fields of study, 21–23, 22–23f. *See also* International collaboration; Scientific collaboration, economics of
Collaborative filters, app store, problems facing, 241–42, 253
Communication costs, invention and, 170
Compatibility standards, 390
comScore, 242–45
Containerized shipping, standardization of, 387
Converters, 392, 392n41
Core technologies, 378, 379f
Corporate apps, 235, 261–62, 266
Corresponding authors, survey of, 23–25
Cox proportional hazard model, 297
Credit, 3–4; formal model of, 118–25; history of, 114–18; implications of formal model of, 125–28; institutions of, 107–8; Matthew Effect and, 112, 126–28; organizational choices and institutions of, 110–14; organizational choices of science and, 108–10; researchers and, 43–44; role of, and shaping of organization of science, 111–13; "salami slicing" and, 116–18, 126
CSS (Cascading Style Sheets), 388

Darwin, Charles, 78
Department of Defense (DOD), 330–31
Discovery: logic of, 338–39; process of, algorithms and, 401–3
Divided technical leadership (DTL), 401; apps and, 280–81
Doriot, George, 372

Eminent Scholars programs, 8, 290–92, 291t; discussion of results for study of, 308–10; empirical results for study of, 304–5, 304t; overlap of, 211–312; as part of portfolio, 312–14
Endless Frontier, The (Bush). See *Science: The Endless Frontier* (Bush)
Endogenous fixed costs, 242n13
Energy supply, 199–204. *See also* Renewable energy
Engelbart, Douglas, 404
Englert, François, 128
Entrepreneurship: market-based innovation and, 6–9; mobile software applications and, 7
European Organization for Nuclear Research (CERN), 18
Evolutionary biology: changes in spatial organization of, 75–78; collaboration and, 87–92; data for study of, 78–81; decline in skew of distribution of output across departments in, 81–85; defining knowledge in, 78–79; increasing importance of star scientists in, 85–86; trend of increasing collaboration in, 87–92

Financial Accounting Standards Board (FASB), 389
Firm type: and local growth in patenting, 188–92; multihoming by, 277–78
Fixed marketing costs, 242n13
Formal (*de jure*) standards, 390
Freemium apps, 259

Frontier in American History, The (Turner), 371

Frontiers, 371–72; concepts of, 373; historical context of, 373, 374–80; measuring, 373; theoretical context of, 380–83

Gartner, hype cycle, 376, 377f
Generally accepted accounting principles (GAAP), 389
General purpose technologies (GPTs), 235–36, 236n4, 393, 393n46
Globalization, 393; invention and, 169–70; as standardization, 393–95
Global supply chains, 392
Global system for mobile communications (GSM), 387
Google Ngram, 374–76, 375f, 376f, 391
Google Play, 240–41, 241n11
Graduate students: collaboration trends for, 67–69; duration of training of, 55–59; publication trends and, 63–67; time to first publication and, 59–63

Heineman, Dave, 295
Herfindahl-Hirschman Index (HHI), 340–41, 340n22
Higgs, Peter, 128
Higgs mechanism, 128
Hype cycle, Gartner, 376, 377f
HyperText Markup Language (HTML), 388

In-app purchasing (IAP), 25n32, 259
India: data and descriptive features of rise of innovation of, 144–52; development in, 135–36; empirical models and regression results on quality and quantity of patenting in, 152–61; location of inventors in, 147; ownership of patents in, 146–47; patenting in, 5; research and development (R&D) in, 136–37; types of invention in, 147; US multinational R&D in, 137–44. *See also* China
Industrial frontier, 373; inventions defining, 376–77
Industry evolution, mobile apps and, 234
Industry production networks, 378, 379f
Informal (*de facto*) standards, 390
Information and communications technology (ICT) platform industries, 235–36
Information technology data, 175–77

Innovation: geography of, 4–6; market-based, entrepreneurship and, 6–9; mobile software applications and, 7; in platform-based industries, 236–39; state policies and, 7–8; venture-backed, 200–201
Interchangeability standards, 390–91
International collaboration, 18, 19, 20; citations and, 26, 27f; growing trend of, 19–23; quality of science and, 26–31; survey evidence, 31–32. *See also* Collaboration; Scientific collaboration, economics of
International Financial Reporting Standards, 389
International Organization for Standardization (ISO), 396, 396n54; adoption rates of quality-management standards, 398–99, 399f; certification by, 397–400; management standards (ISO 2012), 404–6
International Organization for Standardization (ISO) 1400 series, 396–97
International Organization for Standardization (ISO) 2600 series, 396
International Organization for Standardization (ISO) 9000 series, 396–97, 400n65
Internet: business adoption of, and concentration of patenting, 186–88; business adoption of, and local growth in patenting, 188–92; data for adoption of, and inventive activity, 172–77; geographic concentration of invention and, 170–72. *See also* Patenting
Interoperability, standards and, 392–93
Invention: factors affecting agglomeration for, 169–70; forces for or against geographic agglomeration of, 170
iOS apps, 238
ISO. *See* International Organization for Standardization (ISO)
iTunes Store, 239–40

KFC, 393
Killer apps, 238
Kindle platform, 279
Knowledge production: policy implications of results for, 71–72; results for, 69–71. *See also* Academic knowledge; Scientific knowledge, advancement of frontier of
Knowledge production function, 49

Kornberg, Roger, 354
Kosslyn, Stephen, 116

Least publishable units (LPUs), 116–17
Leavitt, Michael, 295
Logic of discovery, 338–39

Marketing costs, 242, 242n13
Matthew Effect, 112, 126–28
McDonald's, 393
McLean, Malcom, 387
Mobile applications (apps), 233–36; advertising and, 260–61; asymmetrics between platforms for, 267t; competition among platforms for, 278–80; concentration of, and success, 247–51; corporate, 235, 261–62, 266; data sources for, 242–47; divided technical leadership (DTL) and, 280–81; economic return of development of new, 257–67; entrepreneurship and, 7; first stage of innovation for, 237; "Freemiums," 259; innovation and, 7; institutional and conceptual bottlenecks of, 282; killer, 238; matching across platforms, 270; matching customers to, 239–42; monetization of, 235, 262–66; network effects and, 237–38; no (current) revenue stream, 261–62; other (currently) zero-revenue, 262–63; paid, 259; relative attractiveness of platforms, 267–69; short-run dynamics of, 251–53; store rankings for, 239–42; "top list" implications for market development of, 253–57; twenty-first century innovation and, 282–85. See also Multihoming
Multihoming, 269–70, 392, 392n41; analysis of, 270; defined, 270; at firm level, 274–77; by firm type, 277–78; weight rates of, 272–74. See also Mobile applications (apps)
Multinational corporations (MNCs): R&D spending in China and India by, 137–44; US patents awarded to, 136

Nanotechnology, 2, 20
Napolitano, Janet, 295
National Defense Education Act (NDEA), 331
National Governors Association (NGA), state science policies and, 295–97
National Institutes of Health (NIH), 322; cut in fellowships by, in 1970s, 340;

doubling in budget of, 1998–2002, 349–51; early years of, 326–29; universities and capacity-building initiatives of, 331–39. See also *Science: The Endless Frontier* (Bush); Universities
National Science Foundation (NSF), 322, 404; cut in fellowships by, in 1970s, 340; early years of, 329–30; universities and capacity-building initiatives of, 331–39. See also *Science: The Endless Frontier* (Bush)
Network effects, mobile apps and, 237–38
Ngram. *See* Google Ngram
NIH. *See* National Institutes of Health (NIH)
Novelty, new measure of, 230
NSF. *See* National Science Foundation (NSF)

On the Origin of Species by Means of Natural Selection (Darwin), 78
Optimum order, 386–87

Paid apps, 259
Pallets, standardization of, 387
Particle physics, 19
Patenting: business adoption of Internet and concentration of, 186–88; business adoption of Internet and growth in, 184–85; characteristics of, by incumbent vs. venture-capital backed firms, 210–19; collaboration, firm type, and local growth in, 188–92; county-level growth in, 171–72; data for, and inventive activity, 172–75; empirical models and regression results on quality and quantity of, 152–61; empirical strategy and results for, 177–92; explosion of, in China and India, 5, 136; increased concentration of, 177–84; Internet adoption and, 188f; Lorenz curve for, by county, 177–80, 180f; rates of, in renewable energy, 204–10; in United States, 6; venture capital-firms and, 7. *See also* Internet
Platform-based industries, innovation in, 236–39
Platform innovation, mobile apps and, 234
Postdocs: collaboration trends for, 67–69; duration of training of, 55–59; publication trends and, 63–67; time to first publication and, 59–63

President's Scientific Advisory Committee (PSAC), 336–38
Production recipes, 383–86, 393
Publication: time to first, and scientific productivity, 59–63; trends, for graduate students and postdocs, 63–67

Quality standards, 390–91

Reference standards, 390
Renewable energy: characteristics of patenting by incumbent vs. venture capital-backed firms and, 210–19; data for study of, 202–4; patenting rates in, 204–10
Renewable energy start-ups, venture capital financing of, 220–26
Research and development (R&D): funding for, 8, 8n1; "golden age of," 401–2; interviews with personnel and multinational, 161–63; multinational, in China and India, 137–44; vertical disintegration of, 137. See also Coinvention
Research systems, Bush's vision of and present day, 9
Reward structure, scientists and, 110–11
Ricardo, David, 372
Rice cookers, 393n45
Roosevelt, Franklin D., 371–72

Salami slicing, 116–18, 126
Science, 321; evolving role of collaboration in, 2–3; organization of, and credit, 108–14; quality of, and international collaboration, 26–31; role of credit in shaping of organization of, 111–13; spatial organization of, 75–76
Science: The Endless Frontier (Bush), 1, 10–12, 321–22, 372, 379, 403–4; R&D and, 8–9; scientific landscape circa 1940 and, 323–26. See also National Institutes of Health (NIH); National Science Foundation (NSF)
Science frontier, 373
Science institutions, historical perspectives on, 9–10
Sciences, changes in spatial organization of, 75–78
Scientific collaboration, economics of, 40–45. See also Collaboration; International collaboration
Scientific credit. See Credit

Scientific knowledge, advancement of frontier of, 372–73. See also Academic knowledge; Knowledge production
Scientific productivity: duration of training and, 55–59; time to first publication and, 59–63
Scientific Progress, the Universities, and the Federal Government (PSAC), 336–38, 338n18
Scientific research, organization of, 2–4
Scientists, reward structure and, 110–11
Seaborg, Glen T., 336
Seaborg report. See Scientific Progress, the Universities, and the Federal Government (PSAC)
Shannon, Claude, 402
Sputnik, 331
Standardization, globalization as, 393–95
Standards: compatibility, 390; creation and maintenance of, 391–92; defined, 386–87; formal (de jure), 390; historical importance of, 387–89; informal (de facto), 390; interchangeability, 390–91; interoperability and, 392–93; quality, 390–91; quality management, for mapping movement of algorithmic frontier, 395–401; reference, 390; types of, 389–91
Standard setting organizations (SSOs), 392
Star scientists, 76; causal impact of, on departmental performance, 77; central role of, in collaboration, 97–101; declining costs of collaboration and, 88–92; effect of improvements in technology on, 92–97; efficient distribution of, 97–101; increasing importance of, in evolution biology, 85–86
Start-ups, renewable energy, venture capital financing of, 220–26
State Intellectual Property Office (SIPO), 5
State science policies, 7–8, 287–89; background on, 289–95; discussion of study results, 308–15; empirical results for study of, 302–8; methodology for study of, 297–302; motivations for, 295–97
Supply chains, 393, 393n46; growth of, 387; vertical disintegration of, invention and, 169–70

TCP/IP protocol stack, 388
Technology, improvements in, and collaboration, 92–97

Term frequency inverse document frequency (TF-IDF), 230

Training, duration of, and scientific productivity, 55–59

Turner, Frederick Jackson, 371, 372

United States: collaborations in, 19; frontiers and, 374–83; patenting in, 6

Universities: capacity-building initiatives of NIH and NSF and, 331–39; challenges threatening research and health of, 353–60; contributions to research and equipment costs by, 343–49; evaluation of research by, 351–53; overexpansion of research facilities by, 357; PhD production and market for research positions demand, 355–56; reliance on federal funding and, 359–60; research by, in 1970s, 339–41; research by, in 1980s–1998, 341–43; research funding mix and, 357–59; risk aversion and research by, 354–55

University research, Bush and, 351

University Research Grants program, 8, 290, 291t, 292–93; discussion of results for study of, 310–11; empirical results for study of, 305–6, 306t; overlap of, 311–12; as part of portfolio, 312–14

US-only collaborations, 20

Value creation, mobile apps and, 234

Venture-backed innovation, 200–201

Venture capital-backed firms (VCs), 372; characteristics of patenting, by incumbent vs., 210–19; patents and, 7

Venture capital financing, of renewable energy start-ups, 200–226

Vertical disintegration of supply chains, invention and, 169–70

Vizio, 392

Walmart, 392, 392n42

Watson, James, 117

Wiener, Norbert, 402

World Wide Web Consortium (W3C), 388